Thin-Film Organic Photonics

Optics and Photonics

Series Editor

Le Nguyen Binh

Monash University, Clayton, Victoria, Australia

Thin-Film Organic Photonics

Molecular Layer Deposition and Applications

Tetsuzo Yoshimura

Tokyo University of Technology, School of Computer Science, Tokyo, Japan

CRC Press
Taylor & Francis Group
Boca Raton   London   New York

CRC Press is an imprint of the
Taylor & Francis Group, an **informa** business

CRC Press
Taylor & Francis Group
6000 Broken Sound Parkway NW, Suite 300
Boca Raton, FL 33487-2742

First issued in paperback 2017

© 2011 by Taylor and Francis Group, LLC
CRC Press is an imprint of Taylor & Francis Group, an Informa business

No claim to original U.S. Government works

ISBN 13: 978-1-138-07590-0 (pbk)
ISBN 13: 978-1-4398-1973-9 (hbk)

Visit the Taylor & Francis Web site at
http://www.taylorandfrancis.com

and the CRC Press Web site at
http://www.crcpress.com

To Yoriko

Contents

Preface

Artificial materials with atomic/molecular-level tailored structures are the final goal of material/device scientists and engineers. Such tailored structures enable electron wavefunction control, which will improve material performance and generate new photonic/electronic phenomena.

There are a lot of atomic/molecular assembling techniques to build the artificial materials, such as scanning tunneling microscopy (STM), molecular beam epitaxy (MBE), chemical vapor deposition (CVD), atomic layer deposition (ALD), sputtering, and so on. Recently, the molecular layer deposition (MLD) has received attention as the next-generation growth technique for organic thin-film materials. Although MLD was inspired by ALD, a fundamental difference exists between them, that is, growth dimensionality. ALD provides the layer-by-layer growth of conformal ultra-thin films, namely, "one-dimensional growth" to build bulk films. MLD provides the dot-by-dot growth, which means the molecule-by-molecule growth, of polymer wires and molecular wires. The wire locations and orientations can be controlled by seed cores. Therefore, MLD provides "three-dimensional growth" to construct a wire network with designated molecular arrangements. This feature of MLD opens a wide range of applications in thin-film organic photonics and electronics.

This book describes how photonic/electronic properties of thin films can be improved by precise control of atomic and molecular arrangements, and demonstrates the prospect of the artificial materials with atomic/molecular-level tailored structures, especially featuring MLD and conjugated polymers with multiple quantum dots (MQDs) called as polymer MQDs. It also describes other related topics including organic electro-optic materials, optical switches, optical circuits, the self-organized lightwave network (SOLNET), a resource-saving heterogeneous integration process, etc. Some applications of the artificial organic thin films to photonics/electronics are proposed in the fields of optical interconnects within boxes of computers, optical switching systems, solar energy conversion systems, and bio/medical-photonics like the photodynamic therapy.

I would like to thank Prof. K. Asama of Tokyo University of Technology for his helpful advice and encouragement, Drs. K. Kiyota, A. Matsuura, T. Hayano, W. Sotoyama, and S. Tatsuura of Fujitsu Laboratories, Ltd., and students who joined Yoshimura Laboratory in Tokyo University of Technology for their collaboration in the research work that contributes to the writing of this book. I would also like to thank colleagues in Fujitsu Computer Packaging Technologies (FCPT), Inc., San Jose, California. Finally, I would like to express my sincere gratitude to Ms. Ashley Gasque and Ms. Catherine Giacari of CRC Press/Taylor & Francis for giving me the great opportunity to write this book, and Ms. Amy Blalock, Mr. John Edwards, and Mr. Michael Davidson for their help in completing this book.

1 Introduction

Organic functional thin-film materials are expected to be applied to various kinds of photonic/electronic devices, such as nonlinear optical circuits, photovoltaic devices, electroluminescent (EL) devices, thin-film transistors (TFTs), and bio/medical-photonic devices, due to the following four features.

1. **Dimensionality**, which affects optical and electronic properties, is easily controlled in organic thin-film materials, especially between the dimensionality of one and zero. The dimensionality control can be achieved in polymer wires as shown in Figure 1.1. Conjugated polymer wires are regarded as one-dimensional systems, that is, quantum wires, in which π-electrons are widely spread. By inserting molecular units that disconnect the spread of π-electrons, the dimensionality is reduced to somewhere between one and zero to make 0.5-dimensional systems, which correspond to long quantum dots. By increasing the density of the disconnecting molecular units in the polymer wires, lengths of the π-electron spread become shorter, creating zero-dimensional systems, which correspond to short quantum dots. Similar dimensionality control may also be achieved in molecular wires consisting of molecules connected to each other by electrostatic force.

2. **Electron wavefunctions**, which affect optical and electronic properties, are easily controlled in organic thin-film materials. The wavefunction shape control is achieved by placing constituent molecules in designated arrangements as to generate required properties. For example, as shown in Figure 1.2, widely spread wavefunctions, symmetric wavefunctions, and non-centrosymmetric wavefunctions can be formed in polymer wires or molecular wires by controlling the molecular sequences in the wires.

3. **Carrier mobility** is expected to be potentially very high in π-electron systems (Figure 1.3).

4. **The resource problems** will be resolved by using organic materials because they are mainly made of carbon (C), hydrogen (H), nitrogen (N), and oxygen (O), which exist in abundance on the surface of the Earth. At present, as shown in Figure 1.4, energy like electric power and fuel, and materials like semiconductors, metals, and plastics are mostly obtained from resources in the Earth's interior. This destroys energy balance on the surface of the Earth and causes material shortages. In the future, in principle, organic materials for energy, semiconductors, metals, and plastics will be able to be synthesized from C, H, N, and O on the surface with the assistance of solar energy as illustrated in Figure 1.4.

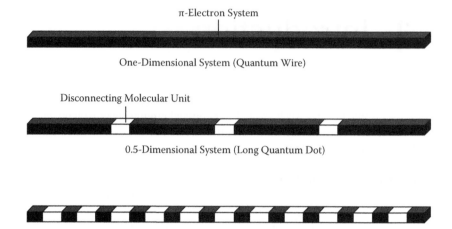

FIGURE 1.1   The dimensionality control achieved in polymer wires.

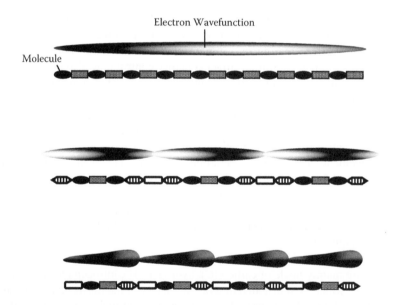

FIGURE 1.2   Control of electron wavefunction shapes by placing constituent molecules in designated arrangements.

In order to realize high-performance thin-film photonic and electronic materials, control of dimensionality and wavefunction shapes by designated arrangements of molecules in the thin films is essential. Molecular layer deposition (MLD) is a promising method for doing this. In MLD, different kinds of molecules are introduced onto a substrate surface with designated sequences to achieve monomolecular step growth of polymer wires or molecular wires using the self-limiting effect.

FIGURE 1.3   π-electron transportation in a conjugated polymer wire.

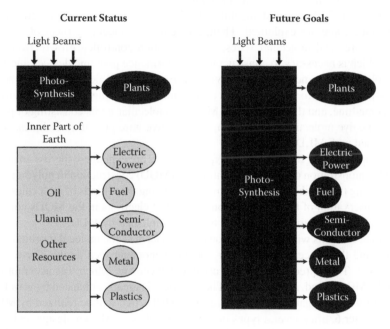

FIGURE 1.4   Solution for the resource problems: organic material production by photosynthesis from C, H, N, and O.

Furthermore, with the assistance of seed cores, MLD enables self-organized growth of polymer/molecular wire networks with designated *molecular sequences, locations*, and *orientations*. Therefore, MLD has the potential to construct artificial organic thin films for various types of photonic/electronic devices like nonlinear optical devices, photovoltaic devices, EL devices, and TFTs, and for molecular nano systems consisting of molecular circuits in large-scale integrated circuits (LSIs). In these applications, π-conjugated systems play especially important roles due to their extended electron wavefunctions. Meanwhile, MLD in the liquid phase is expected to contribute to biomedical fields including selective attack on cancer cells.

In the present book, artificial thin-film photonic materials and devices (in which atomic and molecular arrangements are precisely controlled) and their fabrication methods are reviewed, especially emphasizing MLD. Their applications to optical interconnections within computers, optical switching systems, solar energy conversion systems, and biomedical photonics are discussed.

In Chapter 2, conventional atomic/molecular assembling techniques are shown, including scanning tunneling microscopy (STM), molecular beam epitaxy (MBE), plasma chemical vapor deposition (Plasma CVD), atomic layer deposition (ALD), sputtering, and vacuum deposition polymerization. As examples of the material property control achieved by the atomic arrangement control, direct observation and manipulation of electron wavefunctions, the high electron mobility transistors (HEMT), the multiple quantum well light modulators, photoconduction enhancement in a-Si:H/a-SiN$_x$:H superlattices, and small polaron absorption enhancement in WO$_x$ electrochromic thin films are presented.

In Chapter 3, first, fundamentals of MLD-utilizing chemical reactions and electrostatic force are explained. Then, molecule-by-molecule growth of artificial polymer wires and molecular wires, and orientation-controlled growth from seed cores, which is necessary to construct high-performance photonic devices and polymer/molecular wire networks, are reviewed. Experimental demonstrations include the gas-phase MLD for polymer wires of polyimide and conjugated polymer wires of polyazomethine, and the liquid-phase MLD for molecular wires consisting of p-type and n-type dye molecules. In addition, the selective wire growth techniques, which can be used with MLD, are shown.

In Chapter 4, after fundamentals of quantum dots are explained, there is an experimental demonstration of multiple quantum dot (MQD) construction in polymer wires by arranging two kinds of molecules and by arranging three kinds of molecules using MLD. Observation of the quantum confinement of electrons in the MQDs confirms the prospects of the dimensionality control accomplished by MLD.

Chapter 5 begins with an explanation of the concept of the molecular orbital (MO) method and the concept of nonlinear optical phenomena such as the electro-optic (EO) effect. Then, predictions of EO effect enhancement, which was accomplished using the MO method, are presented. The EO effect is greatly enhanced by wavefunction control in conjugated polymer wires with MQDs that will be realized by MLD.

In Chapter 6, after several types of optical switches are shown, design and simulated switching characteristics are described for waveguide prism deflector (WPD) optical switches and nano-scale optical switches consisting of ring resonators. The impact of EO-conjugated polymers with MQDs on improvement of the optical switch performance is discussed.

In Chapter 7, various examples of organic/polymer photonic materials and devices, whose performance will be improved by utilizing MLD, are presented, including EO molecular crystals and polymers, EO waveguides and switches. As related material/device integration process technologies, nano-scale optical circuits, self-organized lightwave network (SOLNET), resource-saving heterogeneous integration, optical waveguide films with vertical mirrors, and 3-D optical circuits are also described.

Chapter 8 presents possible applications of the thin-film organic photonics to optoelectronic (OE) systems, such as optical interconnects within boxes and three-dimensional micro optical switching systems, are presented.

In Chapter 9, applications of the thin-film organic photonics to solar energy conversion systems are presented. Principle and experimental demonstrations are shown for multidye sensitization and polymer-MQD sensitization, which are realized by

MLD. The chapter also discusses waveguide-type photovoltaic devices, low-cost integrated solar energy conversion systems, and flexible films.

Chapter 10 presents proposals for MLD applications to bio/medical-photonics, including photodynamic therapy for cancer to perform a three-dimensional selective attack on cancer cells by two-photon excitation with different wavelengths utilizing multimolecule stacks constructed by MLD, and a novel therapy in which SOLNET finds cancer cells and guides light beams to them.

2 Atomic/Molecular Assembling Technologies

In Table 2.1, various kinds of atomic/molecular assembling technologies are summarized. The scanning tunneling microscope (STM) is applicable to both atomic and molecular assembling. While sputtering, molecular beam epitaxy (MBE), chemical vapor deposition (CVD), atomic layer apitaxy (ALE), and atomic layer deposition (ALD) are used for atomic assembling, vacuum deposition polymerization and molecular layer deposition (MLD) are used for molecular assembling. In the case of molecular assembling, since the growth is three-dimensional, wire orientations should be controlled. Therefore, selective wire growth has been added in the table.

This chapter begins with an explanation of the similarity of electronic waves to light waves, which is useful to understand the behavior of electron wavefunctions. Then, examples of tailored materials with atomic/molecular assembling by STM, MBE, ALD, plasma CVD, sputtering, and vacuum deposition polymerization are presented. MLD, which is the main theme of the present book, will be precisely described in Chapter 3 and Chapter 4, together with selective wire growth.

2.1   SIMILARITY OF ELECTRONIC WAVES TO LIGHT WAVES

In order to easily understand the behavior of electron wavefunctions, the similarity of electronic waves to light waves is briefly considered in this section. Figure 2.1(a) shows an analogy between the electron confinement and the photon confinement. The wavefunction of electron ψ and the electric field of light waves $\mathbf{E}$ are determined by the following wave equations.

$$\left(-\frac{\hbar^2}{2m}\nabla^2 + V(\mathbf{x})\right)\psi(\mathbf{x}) = E\psi(\mathbf{x})$$

$$\left(\nabla^2 + (n(\mathbf{x})k_0)^2\right)\mathbf{E}(\mathbf{x}) = 0$$

Here, m, V, and E are mass, potential energy, and total energy of an electron, respectively. $\hbar$ is the Plank constant divided by 2π; n is the refractive index, and k_0 is the wavenumber in vacuum. These equations can be written as follows:

$$\left(-\frac{\hbar^2}{2m}\nabla^2 - (E - V(\mathbf{x}))\right)\psi(\mathbf{x}) = 0$$

$$\left(-\frac{\hbar^2}{2m}\nabla^2 - \frac{\hbar^2 k_0^2}{2m}n^2(\mathbf{x})\right)\mathbf{E}(\mathbf{x}) = 0$$

7

TABLE 2.1

Atomic/Molecular Assembling Technologies

Atomic Assembling	Molecular Assembling
Scanning Tunneling Microscope (STM)	
Sputtering	
Molecular Beam Epitaxy (MBE)	
Chemical Vapor Depostion (CVD)	Vacuum Deposition Polymerization, Organic CVD
Atomic Layer Deposition (ALD)	Molecular Layer Deposition (MLD)
Atomic Layer Epitaxy (ALE)	
Selective Wire Growth	

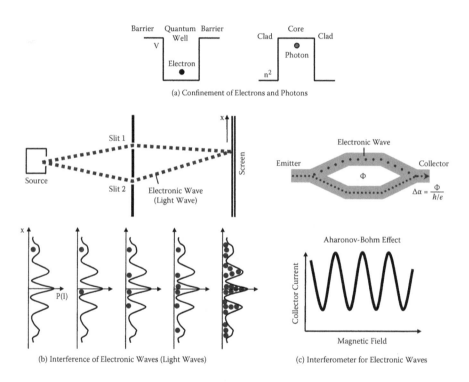

(a) Confinement of Electrons and Photons

(b) Interference of Electronic Waves (Light Waves)

(c) Interferometer for Electronic Waves

FIGURE 2.1 Similarity of electronic waves to light waves.

Comparing these two equations, it is found that the following relationships exist.

$$\psi(\mathbf{x}) \leftrightarrow \mathbf{E}(\mathbf{x})$$

$$E - V(\mathbf{x}) \leftrightarrow \frac{\hbar^2 k_0^2}{2m} n^2(\mathbf{x})$$

It is known that an electron tends to be confined in a region with lower potential energy as shown in Figure 2.1(a). The above-mentioned relationships indicate that

$-V(\mathbf{x})$ corresponds to $n^2(\mathbf{x})$. This suggests that a photon tends to be confined in a region with a larger refractive index. Therefore, by forming lower-potential-energy regions in a media with designated patterns, waveguides for electronic waves can be constructed. Similarly, by forming higher-refractive-index regions, waveguides for light waves can be constructed.

As is well known, light waves passing through two slits make an interference pattern of light intensity I on a screen. Similarly, electronic waves passing through two slits make an interference pattern of probability $P = \psi^*\psi$ that an electron will arrive at x. The interference pattern is schematically illustrated in Figure 2.1(b). Many electrons are attracted to places where $P = \psi^*\psi$ is large, making many spots on the screen.

Figure 2.1(c) shows an interferometer for electronic waves [1]. Waveguides are formed in a shape of the Mach-Zehnder interferometer. Electronic waves are emitted from the emitter. The waves are divided into two paths of the upper arm and the lower arm, and come together in the output waveguide to reach the collector. When magnetic flux exists in the area between the two arms, a phase difference is induced between the two paths via the Aharonov-Bohm effect. Then, the collector current oscillates as a function of the magnetic field due to interference of the electronic waves from the upper arm and the lower arm. Exactly the same phenomena are observed in the Mach-Zehnder interferometer of optical waveguides.

As described, most of the behavior of the electronic waves of ψ can be understood by analogy with the behavior of the light waves of $\mathbf{E}$.

2.2   SCANNING TUNNELING MICROSCOPY (STM)

2.2.1   ATOMIC MANIPULATION PROCESS

Eigler and Schweizer developed an atomic manipulation process that enables atoms to slide across a surface to definite positions as shown in Figure 2.2 [2]. The process utilizes atomic force between the atoms and atoms on the edge of the STM tip.

Figure 2.3 shows the dependence of potential energy on the distance between atoms (molecules). Potential energy arising from Pauli's exclusion principle is extremely large in short distance regions, and it decreases rapidly with distance. The potential energy arising from van der Waals' force, on the other hand, decreases gradually with the distance as compared with the case of Pauli's exclusion principle. As a result, the total potential curve has a minimum. Force $\mathbf{F}$ is given by

$$\mathbf{F} = -\nabla V,$$

using potential energy function V. Then, attractive atomic force arises between atoms (molecules) for distances longer than the minimum point because the slope of the potential curve is positive there. For distances shorter than the minimum point, since the slope of the potential curve is negative, repulsive atomic force arises, giving rise to a stable distance at the minimum point.

Using the atomic force between an atom on a substrate surface and an atom on a tip, the atom on the surface can be carried by the tip as shown in Figure 2.2. When

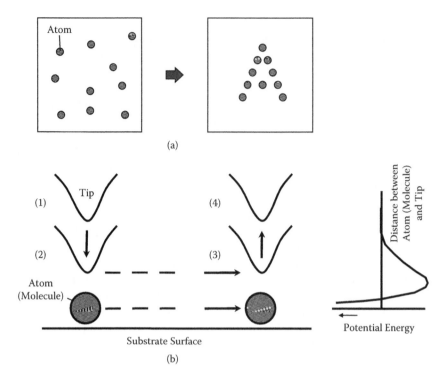

FIGURE 2.2   Schematic illustration of sliding atoms across a surface to definite positions to draw an image, and (b) procedure for sliding atoms by STM.

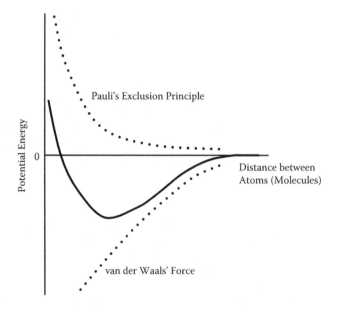

FIGURE 2.3   Dependence of potential energy on distance between atoms (molecules).

the tip is pointing down to the atom and is placed at a distance with a positive potential curve slope, attractive force arises between the atom and the tip. By moving the tip, the atom slides on the surface. After pulling the tip up vertically, the force between the atom and the tip vanishes.

2.2.2 DETECTION OF WAVEFUNCTIONS

Figure 2.4 shows the tunneling process in STM. When a tip is put on a sample surface with a small gap of 1-nm scale, electron wavefunctions penetrate through the gap by the tunneling effect. The tunneling probability is proportional to $e^{-d\sqrt{2m/\hbar^2(V-E)}}$, where, d, m, and E are, respectively, gap length, electron mass, and electron energy. V is the barrier height at the gap and $\hbar$ is Plank's constant divided by 2π. When the tip bias is close to the energy of an electronic state in the sample, a tunnel current increases. When the tip bias is far from the energy of the state, the tunnel current decreases. Therefore, in constant current mode, the distance between the tip and the sample surface becomes large when an electronic state with energy close to the tip bias exists in the sample, and it becomes small when the state does not exist. Using this property of STM, wavefunction shapes can be detected by STM as shown in Figure 2.5. In the figure, ψ represents the wavefunction of the state with energy close to the tip bias. The tip goes away from the surface in the region with large ψ, and gets close to the surface in the region with small ψ, enabling us to draw the shape of $\psi^*\psi$ by the up–down movement of the tip.

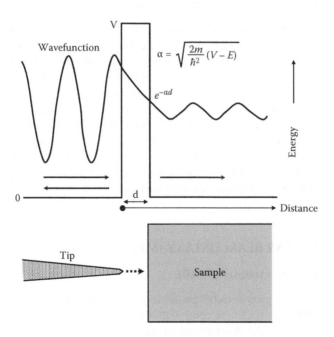

FIGURE 2.4 Tunneling process in STM.

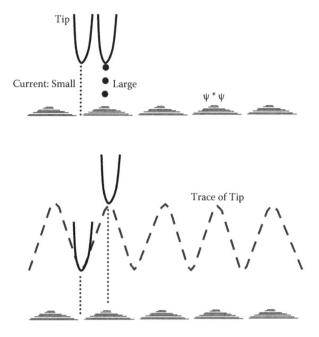

FIGURE 2.5   Detection of wavefunction shapes by STM.

2.2.3   QUANTUM CORRAL AND QUANTUM MIRAGE

Figure 2.6 shows an experiment of electron confinement in a quantum corral, which was reported by Eigler et al. [3]. By STM, 48 Fe atoms are positioned into a ring with a 7.13-nm radius on the Cu surface to construct a quantum corral. Due to the quantum confinement effect by the corral, standing waves of ψ for electrons confined in the corral appear. The shape of $\psi^*\psi$ can be monitored by STM as schematically shown in Figure 2.6. Eigler et al. also succeeded to generate a quantum mirage in an elliptical quantum corral [4]. As Figure 2.7 shows, they put a Co atom at one focus of the corral, and observed a quantum mirage projected around another focus. Due to interference of wavefunctions of electrons scattered by the Co atom, the electronic structures surrounding the Co atom are regenerated near another focus just like the interference of light on an elliptical mirror surface.

2.3   MOLECULAR BEAM EPITAXY (MBE)

2.3.1   GROWTH MECHANISM OF MBE

MBE is an important growth technique for inorganic thin films on flat substrate surfaces. Since MBE has thickness controllability of mono-atomic-layer scales, it has greatly contributed to the development of various kinds of functional devices by the semiconductor industry. The superlattice developed by Esaki and Tsu [5] is the

Number of Fe: 48
Distance between Fe-Fe: 0.88-1.02 nm
Corral Radius: 7.13 nm

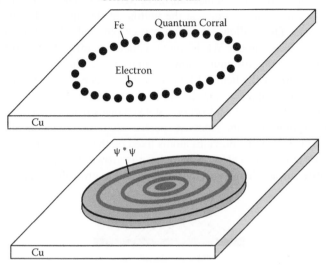

FIGURE 2.6 Electron confinement in a quantum corral.

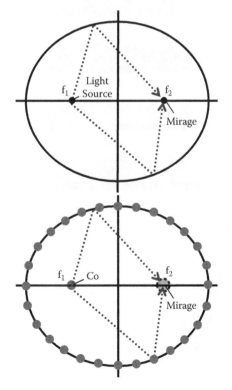

FIGURE 2.7 Quantum mirage.

distinguished innovation achieved by MBE. Transistors, light modulators, laser diodes (LDs) utilizing ultrathin films and quantum wells were also developed by MBE.

Figure 2.8 shows the growth mechanism of MBE. Incident atoms on the surface behave as a potential curve shown at the bottom of the figure. The curve has two minimum points. The shallow minimum causes physical adsorption arising from a weak force such as van der Waals' force without chemical bonds to the surface. This corresponds to the minimum shown in Figure 2.3. The deep minimum causes chemical adsorption with chemical bonds on the surface. Incident atoms, initially, are weakly adsorbed on the surface as physical adsorption. After the atoms migrate on the surface for a while, the adsorption state changes from physical adsorption to chemical adsorption to fix the atoms at specific sites. Next, incident atoms migrating on the surface reach the previous atoms fixed on the surface, and are chemically adsorbed to form growth centers. A monolayer is grown from the growth centers to cover the entire surface. This procedure is repeated to grow thin films.

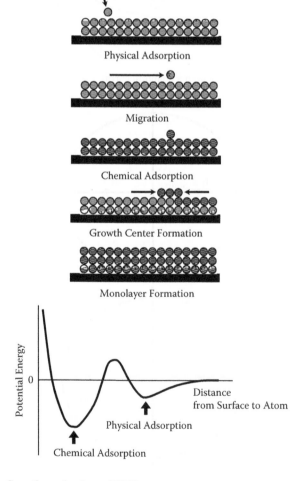

FIGURE 2.8   Growth mechanism of MBE.

2.3.2 High-Electron-Mobility Transistors

One of the functional devices fabricated by MBE is the high-electron-mobility transistor (HEMT) [6], which is basically a field-effect transistor. Figure 2.9 shows an example of the structure. In the $Al_{1-x}Ga_xAs$ layer that has a wide band gap, Si is doped as a dopant for donors. Electrons from the donors are transferred to an adjacent nondoped GaAs layer to provide two-dimensional electron gases that form channels. Thus, in HEMT, the doped layer is separated from the channel layer to reduce scattering of electrons in the channel, increasing carrier mobility.

2.3.3 Multiple Quantum Well Light Modulators

Another functional device is the light modulator, which consists of multiple quantum wells [7] that can be fabricated by MBE. Figure 2.10(a) shows schematic illustrations of band diagrams and electron orbitals in bulk and quantum well structures. When an electron is excited by light absorption, an electron-hole pair, an exciton, is generated. Due to attractive force between an electron and a hole, the energy level of the

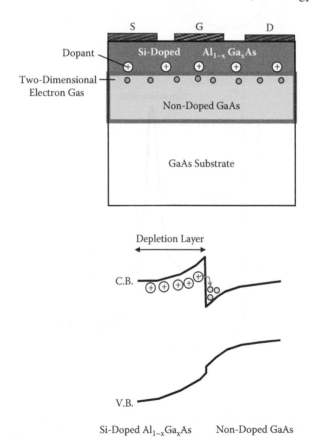

FIGURE 2.9 HEMT.

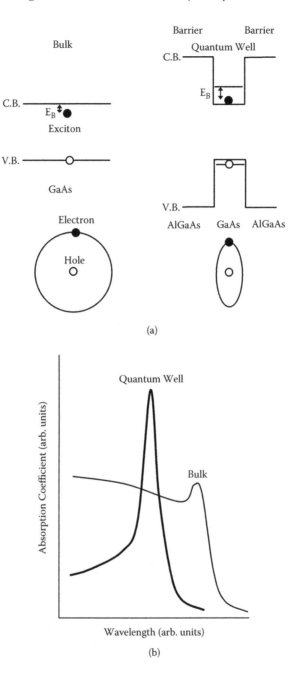

FIGURE 2.10   (a) Schematic illustrations of band diagrams and electron orbitals in bulk and quantum well structures. (b) Sharpening of exciton absorption bands by the quantum confinement in quantum wells.

exciton is located below the conduction band edge. The difference in energy between the band edge and the exciton level gives binding energy of the exciton, denoted by E_B. In quantum wells, excited electrons are confined in narrow spaces; then, deformation in electron orbitals is caused, making the average distance between electrons and holes smaller compared with the bulk case. Consequently, in quantum wells, the binding energy increases, and the dissociation probability of the exciton decreases to increase the life time, τ, of the exciton.

The uncertainty principle tells us that, with an increase in τ, the uncertainty of energy decreases, that is, the width of exciton absorption bands becomes narrow. Thus, as shown in Figure 2.10(b), sharpening of exciton absorption bands is realized in quantum wells. The sharpening is favorable for light modulation and will be described later. It should be noted that a blue shift of the absorption band occurs in quantum wells. This is attributed to an increase in the energy gap due to the electron (hole) confinement effect, which is shown in Figure 2.10(a).

When electric fields are applied to the quantum well, a slope appears in the energy bands as shown in Figure 2.11(a). Then, energy levels for electrons decrease while those for holes increase, resulting in a red shift of the absorption band as shown in Figure 2.11(b). The absorption band shift enables light modulation. With the same amount of wavelength shift of the absorption bands, the sharper the absorption bands become, and the larger the modulation contrast becomes.

In addition, by applying electric fields to the quantum well, the electron wavefunction and the hole wavefunction are moved in opposite directions to reduce wavefunction overlap between the ground state and the excited state. This causes a decrease in transition probability for the electron excitation, inducing a decrease in absorption peak height.

2.3.4 Relationships to Other Growth Techniques

MBE is also applied to fabrication of heterogeneous junctions with thin layers in laser diodes (LDs), superlattice devices, and quantum wires and dots with the assistance of surface treatment. One problem of MBE is lack of conformality. Ultrathin layers can be grown on flat surfaces, but cannot be grown on surfaces with rough structures, porous structures, or three-dimensional structures. Another problem is controllability. Even for flat surfaces, in order to control growth thicknesses with mono-atomic-layer accuracy, high-quality monitoring is required, such as electron diffraction measurements. One more problem is mass productivity. Film growth can be performed only on a limited number of substrates or on a limited area of substrates because MBE utilizes oriented molecular beams in vacuum. For mass production, metal-organic CVD (MOCVD) was developed. A more advanced growth technique is ALD, which has been received much attention recently. The details of ALD are covered in Section 2.4.

2.4 ATOMIC LAYER DEPOSITION (ALD)

ALD is an inorganic film growth method invented by Suntola [8]. In ALD, by introducing two kinds of molecules onto substrate surfaces alternately, ultrathin films are grown with mono-atomic layer steps.

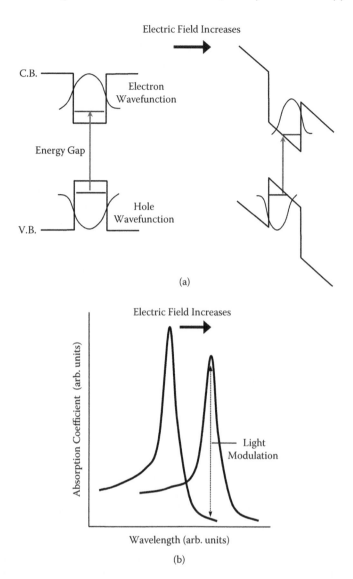

FIGURE 2.11 Effects of applied electric fields on (a) band diagrams of quantum well structures (b) absorption band shifts.

Figure 2.12 shows the ALD process for ZnS thin film growth. On a substrate surface, $ZnCl_2$ is introduced to form a layer that covers the surface. (1) After removing
$ZnCl_2$, H_2S is introduced on the surface. (2) $ZnCl_2$ and H_2S react to form an S layer
on a Zn surface. (3) Once the surface is covered with S, H_2S re-evaporates from the
surface since no chemical reaction occurs between H_2S and S. Thus, the film growth
is terminated. This is called the *self-limiting effect*. (4) Next, after H_2S removal,
$ZnCl_2$ is introduced again. (5) Due to the self-limiting effect, an exact mono layer of

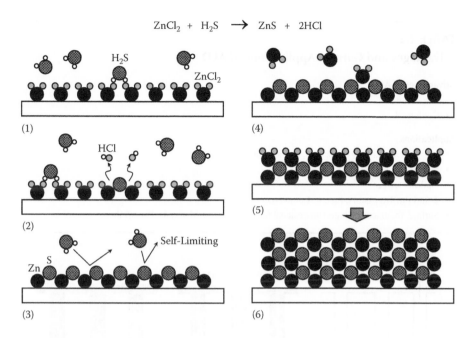

$$ZnCl_2 + H_2S \rightarrow ZnS + 2HCl$$

FIGURE 2.12   ALD process for ZnS thin film growth.

$ZnCl_2$ is formed on the surface. (6) By providing $ZnCl_2$ and H_2S alternately, a ZnS film is formed layer by layer.

ALD was initially developed as atomic layer epitaxy (ALE) for inorganic thin-film AC electroluminescent (EL) devices. Recently ALE and ALD have received much attention for a wide range of applications.

The advantages and current applications of ALD are summarized in Table 2.2. The most outstanding feature of ALD is that it can deposit ultrathin conformal films on surfaces with arbitrary shapes, such as trenches, cylinders, paper, fibers, pours and rough structures, and so on. Typical materials deposited by ALD are ZnS, ZnO, SiO_2, HfO_2, ZrO_2, Al_2O_3, TiO_2, and various types of metals. ALD is applied to ultra-thin/high-k gate insulators of large-scale integrated circuits (LSIs) and capacitors with ultralarge capacitance. It is also applied to solar cells, fuel cells, catalysts, optical devices such as gratings, thin-film transistors, EL devices, and as a surface treatment for various materials including surfaces with three-dimensional structures, porous materials, paper, and fibers.

Figure 2.13 shows examples of the deposition of ultra-conformal films in trenches with high aspect ratios by ALD [9]. The upper picture represents a filling-hole process by ALD. The lower illustration represents ultrathin multilayer formation on a trench surface. This enables capacitors with large capacitance, because the effective area of the capacitors increases. The same structure can be applied to electrostatic or electrochemical batteries. Fuel cells with an ultrathin solid electrolyte, which decreases series electrical resistance, were also reported [10].

TABLE 2.2

Advantages and Current Applications of ALD

Advantages of ALD
- Ultra Thin
- Conformal

Applications
- LSI: Gate Insulators, Capacitors, Barriers for Metal Diffusion
- Solar Cells, Fuel Cells, Catalysts
- Optical Devices
- Thin-Film Transistors, Electro-Luminescent Devices
- Surface Treatments: Three-Dimensional Surfaces, Porous Materials, Paper/Fibers

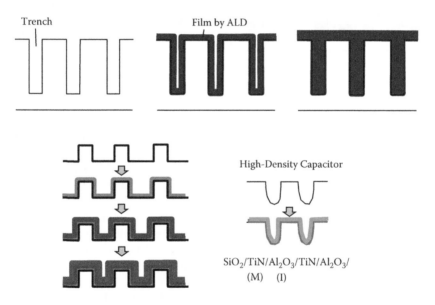

FIGURE 2.13   Examples of deposition of ultra-conformal films in trenches with a high aspect ratio by ALD.

Figure 2.14 shows examples of nano-scale structure fabrication using ALD [11]. On nano-scale cylindrical templates, thin films are deposited by ALD. By removing the templates, nanotubes and vertical walls are obtained. Figure 2.15 shows metal-oxide semiconductor field-effect transistor (MOSFET) with an ultrathin gate insulator layer deposited by ALD. When compared to the conventional deposition method, ALD reduces leakage currents drastically, indicating that conformality of the insulate layer is improved by ALD. ALD applications in other fields, such as hydrophobic surface treatment of cotton fibers and tissue papers, were demonstrated [12].

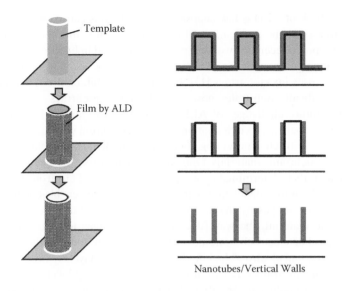

FIGURE 2.14 Examples of nano-scale structure fabrication using ALD.

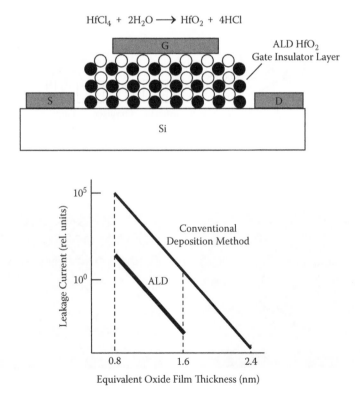

FIGURE 2.15 MOSFET with an ultrathin gate insulator layer deposited by ALD.

One drawback of ALD is low deposition rate, which is caused by gas switching duration. To resolve the problem, novel ALD equipment with improved deposition rates were proposed recently. Two examples are illustrated in Figure 2.16. One example uses rotating gas injectors, in which it takes only a few seconds for one evolution [9]. The other uses injector arrays [13]. Injectors for metal, oxidizer, and purge are arranged periodically. Substrates move through the injector arrays.

Expected future challenges of ALD are shown in Figures 2.17 and 2.18. By arranging more than three kinds of atoms by ALD, quantum wells and superlattices may easily be constructed, providing, for example, artificial nonlinear optical materials and electrochromic materials. Optical nonlinearity like the electro-optic (EO) effect of materials is determined by wavefunction shapes of electrons as described in Chapter 5. Thin film deposition with atomic layer steps enables us to optimize the wavefunction shapes by the arrangement of atoms. When electric fields are applied to the material, the wavefunction shape changes to induce a refractive index change. For electrochromic devices, the electron-donating layer, electron-accepting layer, and electron-blocking layer are stacked layer by layer. By electric fields, electrons are transferred from the electron-donating layer to the electron-accepting layer to induce polarization. Then, each layer changes to mixed valence states to induce coloration due to the intervalence transition, which is explained in Subsection 2.6.2.

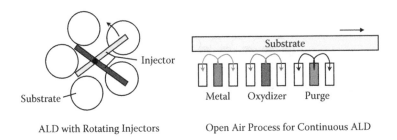

FIGURE 2.16 ALD equipment with high deposition rates.

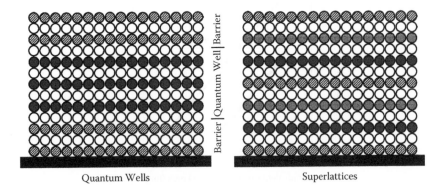

FIGURE 2.17 Proposed quantum wells and superlattices by ALD.

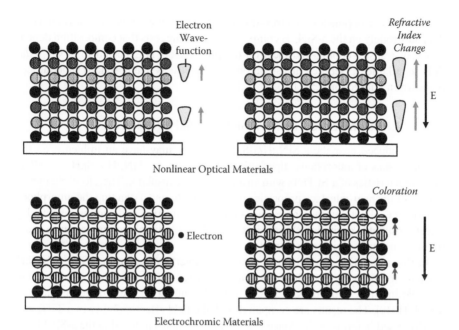

Nonlinear Optical Materials

Electrochromic Materials

FIGURE 2.18 Proposed artificial nonlinear optical materials and electrochromic materials with quantum wells and superlattices by ALD.

2.5 PLASMA CHEMICAL VAPOR DEPOSITION (PLASMA CVD)

For growth of amorphous materials such as a-Si:H, a-SiN$_x$:H, and a-SiO$_x$, plasma CVD is frequently used. Plasma CVD can perform precise composition modulation during film growth by controlling reactive gas composition, enabling fabrication of superlattices as MBE. In this section, as an example of material property control by atomic-level tailored structures, photocurrent enhancement induced by layer composition/thickness optimization in amorphous superlattices is described.

2.5.1 AMORPHOUS SUPERLATTICES

Amorphous superlattices [14] are grown by stacking ultrathin films with different composition sequentially. One of the interesting phenomena observed in amorphous superlattices is the transfer doping effect. The effect has been found in a-SiN$_x$:H/a-Si:H superlattices [14–16], and explained as electron transfer from states in the band gap of the a-SiN$_x$:H layer or interface states to the a-Si:H layer. The electron transfer induces electron accumulation in the a-Si:H layer, giving rise to an increase in the in-plane conductivity by several orders of magnitude. Meanwhile, at an a-SiN$_x$:H/a-Si:H interface, electron trapping by states in the band gap of the a-SiN$_x$:H layer has also been observed. The electron trapping affects the threshold voltage of a-Si thin film transistors (TFTs) with a-SiN$_x$:H gate insulators [17].

The transfer-doping and electron-trapping effects at the a-SiN$_x$:H/a-Si:H interface depend strongly on the a-SiN$_x$:H composition. This suggests that composition dependency of transfer-doping and electron-trapping effects is expected in a-SiN$_x$:H/a-Si:H superlattices. Detailed investigation of these effects in amorphous superlattices is shown in Subsection 2.5.3 [18].

2.5.2   CHARACTERIZATION OF THE A-SiN$_x$:H/A-Si:H INTERFACE

Defects induced near the a-SiN$_x$:H/a-Si:H interface may have an influence on the characteristics of amorphous superlattices consisting of a-SiN$_x$:H/a-Si:H as well as the characteristics of a-Si TFTs with an a-SiN$_x$:H gate insulator. Therefore, characterization of the near-interface properties is important to understanding the amorphous superlattices. Photoluminescence measurement is effective in detecting defects. By selecting suitable excitation energy, it is possible to obtain photoluminescence spectra of a particular region of a thin film.

In this subsection, investigation of the near-interface region in a-SiN$_x$:H/a-Si:H layered structures by photoluminescence measurement is described to show how the near-interface properties of a-Si:H vary depending on the nitrogen content of a-SiN$_x$:H [19]. The results are compared with the field effect mobility in a TFT, and are discussed in terms of the concentration of deep states induced in the a-Si:H layer near the interface.

2.5.2.1   Sample Preparation

Sample structures for photoluminescence measurements are schematically illustrated in Figure 2.19(a). A 300-nm-thick a-SiN$_x$:H layer and a 100-nm-thick a-Si:H layer were successively deposited on a ground quartz substrate by plasma CVD. These layers were deposited in separated chambers to prevent nitrogen contamination of the a-Si:H layer. The a-SiN$_x$:H composition was changed by NH$_3$/SiH$_4$ gas flow ratio.

TFTs of L/W = 20 μm/300 μm were fabricated by depositing a-SiN$_x$:H and a-Si:H layers on a glass substrate as shown in Figure 2.19(b). a-SiN$_x$:H and a-Si:H operate as a gate insulator and an active layer, respectively. The source and drain electrodes were deposited with n$^+$-a-Si:H/Ti/NiCr, and the gate electrode with NiCr.

2.5.2.2   Measurements

For photoluminescence measurements, a He-Cd laser (325 nm), pulsed nitrogen laser (337 nm), and He-Ne laser (633 nm) were used as the excitation light sources as shown in Figure 2.19(a). When He-Cd laser light is introduced to the sample from the a-SiN$_x$:H side, a considerable fraction of the incident light passes through the a-SiN$_x$:H layer and is absorbed in the a-Si:H layer near the interface with a penetration depth of about 10 nm. Consequently, we can obtain photoluminescence spectra of a-Si:H near the a-SiN$_x$:H/a-Si:H interface. Here, a luminescence band that is due to the a-SiN$_x$:H layer is also observed, but it appears on the high-energy side of the band arising from the a-Si:H, so these bands are separated sufficiently. When He-Cd laser light is introduced from the a-Si:H surface side, photoluminescence spectra of a-Si:H near the surface can be obtained. A pulsed nitrogen laser with a pulse width of ~2 ns was used to measure the time-resolved photoluminescence spectra of a-Si:H near the interface. He-Ne laser light, which is uniformly absorbed through a 100-nm-thick a-Si:H layer, was used to obtain photoluminescence spectra of the whole a-Si:H layer. The photoluminescence spectra of a-Si:H and a-SiN$_x$:H were detected by a cooled Ge pin photodiode (response time: ~200 ns) and a photomultiplier, respectively. The cw photoluminescence spectra signals were amplified by a lock-in amplifier and those of the time-resolved photoluminescence spectra by a boxcar averager.

The optical gaps of the a-SiN$_x$:H and a-Si:H layers were obtained from $\sqrt{h v \alpha}$ vs. $h v$ plots according to the relationship $h v \alpha = B(h v - E_g)^2$, where E_g, α, v, and h are, respectively, optical gap, absorption coefficient, frequency of light, and Plank's constant, and B is a material-related constant. The field effect mobility in a TFT was estimated from the slope of I_D vs. V_D $(=V_G)$, where I_D, V_D, and V_G are drain current, drain voltage, and gate voltage, respectively.

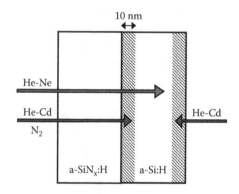

(a) Sample Structure for PL Measurements

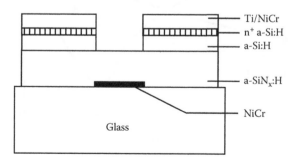

(b) Sample Structure of TFT

FIGURE 2.19 (a) A sample structure for photoluminescence measurements and a schematic illustration of the relationship between the observed region and the excitation light sources. (b) A sample structure of TFT.

2.5.2.3 Optical/Electrical Properties of a-SiN$_x$:H layers

Figure 2.20 shows various optical parameters of a-SiN$_x$:H used in the present experiments as a function of nitrogen content x. The optical gap increases dramatically with increases in x when $x > 0.85$. The peak energy of the photoluminescence band (E_{peak}) also exhibits a high-energy shift. The values of B decrease when x increases from 0 to 0.85, and then, increase with x when $x > 0.85$. It is known that the structure of a-SiN$_x$:H changes from tetrahedral to a Si$_3$N$_4$-like structure with increasing x.[20] Considering this, the results in Figure 2.20 suggest that the principal structural change occurs near the composition $x = 0.85$.

Resistivity of a-SiN$_x$:H is in the order of 10^{15} Ωcm for $x \geq 0.93$. For $x = 0.85$, although the resistivity decreases to 10^{13} Ωcm, the electron blocking property, necessary for TFT operation, is not lost.

2.5.2.4 Photoluminescence Spectra

Figure 2.21 shows the cw photoluminescence spectra of a-Si:H in an a-SiN$_x$:H/a-Si:H layered structure. Each curve corresponds to the result for the observed region in the a-Si:H. The spectra change from region to region. The photoluminescence band of a-Si:H near the interface appears in the low-energy side compared to that of the whole a-Si:H layer. The photoluminescence band of a-Si:H near the surface also exhibits a low-energy shift. These results suggest that the film property varies depending on the region of the a-Si:H layer. So, as described below, we can

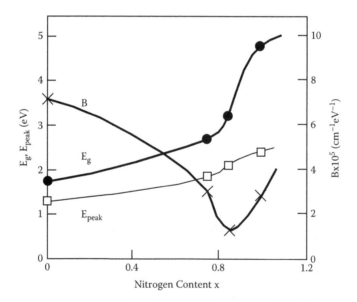

FIGURE 2.20 Optical parameters of a-SiN$_x$:H used in the experiments. From T. Yoshimura, K. Hiranaka, T. Yamaguchi, and S. Yanagisawa, "Influence of a-SiN$_x$:H composition on transfer-doping and electron trapping effects in a-SiN$_x$:H/a-Si:H superlattices," *Philos. Mag.* **B 55**, 409–416 (1987).

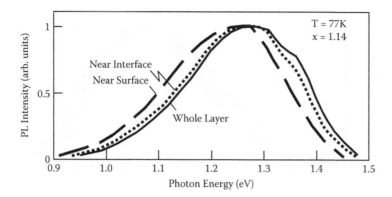

FIGURE 2.21 Cw photoluminescence spectra of a-Si:H in an a-SiN$_x$:H/a-Si:H layered structure. From T. Yoshimura, K. Hiranaka, T. Yamaguchi, S. Yanagisawa, and K. Asama, "Characterization of a-SiN$_x$:H/a-Si:H Interface by Photoluminescence Spectra," *Mat. Res. Soc. Symp. Proc.* **70**, 373–378 (1986).

characterize the near-interface region of a-Si:H by the photoluminescence excited with a He-Cd laser.

Figure 2.22 shows the cw photoluminescence spectra of a-Si:H near the a-SiN$_x$:H/a-Si:H interface for various a-SiN$_x$:H composition. Since the photoluminescence band is extremely broad, the peak position, E_{PL}, is defined as follows.

$$E_{PL} = E_{PL} (-1/2) + (E_{PL} (-1/2) + E_{PL} (+1/2))/2$$

where, $E_{PL} (-1/2)$ and $E_{PL} (+1/2)$ are the energy at the half maxima of the photoluminescence bands for the low-energy side and the high-energy side, respectively. As Figure 2.23 shows, E_{PL} shifts to the low-energy side as the nitrogen content x of the a-SiN$_x$:H layer increases.

To investigate the origin of the spectral shift observed in Figure 2.23, time-resolved photoluminescence spectra were measured with a pulsed nitrogen laser. Figure 2.24 shows the result for a sample with nitrogen content $x = 1.24$, which exhibits the largest low-energy shift of the photoluminescence band. After pulsed excitation, two broad bands are observed. The band on the high-energy side is denoted by the H-band and that on the low-energy side by the L-band. The H-band decays more rapidly than the L-band. The results shown in Figure 2.24 indicate that the cw photoluminescence band in Figure 2.22 contains two different broad bands. In Figure 2.25 the photoluminescence spectra of 2 μs after excitation are compared for $x = 1.24$, which shows the largest low-energy shift of the cw photoluminescence bands, and for $x = 0.93$, which shows the smallest low-energy shift in Figure 2.22. In $x = 1.24$, the L-band is more dominant than when $x = 0.93$. From this it is found that the low-energy shift of the cw photoluminescence bands shown in Figure 2.23 can be attributed to an enhancement of the L-band.

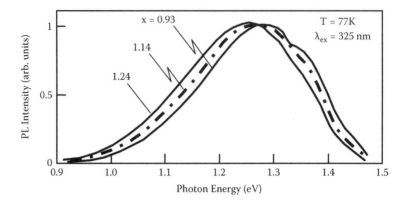

FIGURE 2.22   Cw photoluminescence spectra of a-Si:H near the a-SiN$_x$:H/a-Si:H interface for x = 0.93, 1.14, and 1.24. From T. Yoshimura, K. Hiranaka, T. Yamaguchi, S. Yanagisawa, and K. Asama, "Characterization of a-SiN$_x$:H/a-Si:H interface by photoluminescence spectra," *Mat. Res. Soc. Symp. Proc.* **70**, 373–378 (1986).

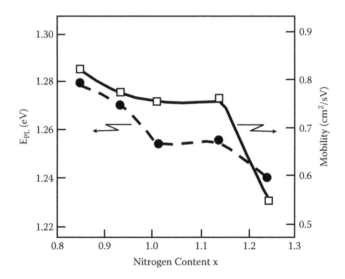

FIGURE 2.23   Dependence of E_{PL} and field effect mobility in a-Si TFT on the nitrogen content of a-SiN$_x$:H layer. From T. Yoshimura, K. Hiranaka, T. Yamaguchi, S. Yanagisawa, and K. Asama, "Characterization of a-SiN$_x$:H/a-Si:H interface by photoluminescence spectra," *Mat. Res. Soc. Symp. Proc.* **70**, 373–378 (1986).

As illustrated in Figure 2.24, the H-band with a peak near 1.35 eV can be ascribed to transitions from band tail states since the photoluminescence peak arising from the inter-tail transition has been known to range between 1.3 and 1.4 eV at the delay of 0.5 − 5 μs [20]. The L-band is supposed to relate to some deep states. These results suggest that the low-energy shift of the cw photoluminescence band shown in Figure 2.23 is caused by an increase in the deep states in the a-Si:H layer. Namely, the more the band shifts to the low-energy side, the more deep states are induced.

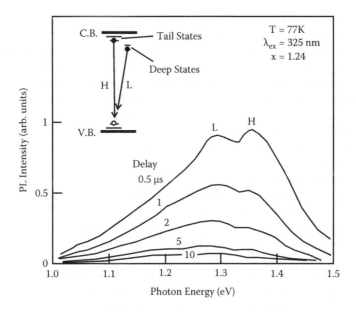

FIGURE 2.24 Time-resolved photoluminescence spectra of a-Si:H near the a-SiN$_x$:H/a-Si:H interface. The insert is a schematic model of the deep states in the a-Si:H. From T. Yoshimura, K. Hiranaka, T. Yamaguchi, S. Yanagisawa, and K. Asama, "Characterization of a-SiN$_x$:H/a-Si:H interface by photoluminescence spectra," *Mat. Res. Soc. Symp. Proc.* **70**, 373–378 (1986).

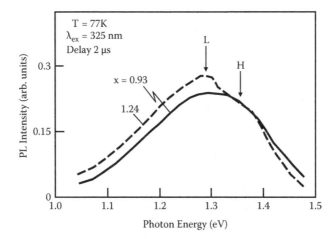

FIGURE 2.25 Photoluminescence spectra of a-Si:H near the a-SiN$_x$:H/a-Si:H interface at 2 μs after pulsed excitation for $x = 0.93$ and 1.24. From T. Yoshimura, K. Hiranaka, T. Yamaguchi, S. Yanagisawa, and K. Asama, "Characterization of a-SiN$_x$:H/a-Si:H interface by photoluminescence spectra," *Mat. Res. Soc. Symp. Proc.* **70**, 373–378 (1986).

Therefore, it is concluded that the deep states in the a-Si:H layer increase with the nitrogen content x of the underlying a-SiN$_x$:H layer.

In Figure 2.26, typical $I_D - V_G$ characteristics are shown for an a-Si TFT with a gate insulator a-SiN$_x$:H of $x = 1.01$. From Figure 2.23, it can be seen that the field effect mobility tends to decrease as the nitrogen content of the a-SiN$_x$:H gate insulator increases, and the decrease in the field effect mobility correlates well with the low-energy shift of E$_{PL}$. This correlation can be explained as follows. The field effect mobility in the TFT is reduced by the deep states in the a-Si:H layer near the interface corresponding to the low-energy shift of the photoluminescence band.

For the origin of the deep states, lattice strain induced in the a-Si:H layer may be considered as below. Roxlo et al. identified lattice strain due to lattice mismatch [22]. Since it is known that the nearest neighbor interatomic distance of a-SiN$_x$:H decreases from 0.24 to 0.17 nm as the nitrogen content increases from 0 to 4/3 [23], the difference in the nearest interatomic distance between a-SiN$_x$:H and a-Si:H becomes large as x increases. This would cause large lattice strain and result in an increase in the deep states. This interpretation is consistent with the experimental results shown in Figure 2.23. In addition, as shown in Figure 2.20, the B value drastically increases with x for $x > 0.9$, suggesting that the lattice structure of a-SiN$_x$:H is ordered to a Si$_3$N$_4$-like structure as x is increased. This would enhance the lattice mismatch between the a-SiN$_x$:H layer

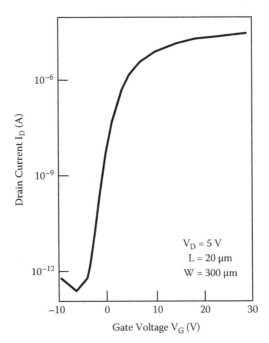

FIGURE 2.26 $I_D - V_G$ characteristics of a-Si TFT. From T. Yoshimura, K. Hiranaka, T. Yamaguchi, S. Yanagisawa, and K. Asama, "Characterization of a-SiN$_x$:H/a-Si:H interface by photoluminescence spectra," *Mat. Res. Soc. Symp. Proc.* **70**, 373–378 (1986).

and the tetrahedral a-Si:H layer. The lattice strain would also be induced by stress effect. It was found that the stress of the a-SiN$_x$:H layer monotonically changes from compressive to strong tensile stress as x increases from 0 to 1.24. This indicates that with increasing x, the stress difference between the a-SiN$_x$:H layer and the a-Si:H layer becomes large. This would also result in an increase in the deep states due to the lattice strain.

It should be noted that with decreasing x, although mobility of a-Si TFTs increases, energy gaps of a-SiN$_x$:H decrease as shown in Figure 2.27, which causes threshold voltage shifts of the TFTs due to electron injection from a-Si:H layer into the a-SiN$_x$:H layer. So, an energy gap of a-SiN$_x$:H should be sufficiently larger than that of a-Si:H in TFTs.

2.5.3 TRANSFER-DOPING AND ELECTRON-TRAPPING EFFECTS IN A-SiN$_x$:H/A-Si:H SUPERLATTICES

2.5.3.1 Fabrication and Measurement Procedures

The a-SiN$_x$:H/a-Si:H superlattice [18], whose structure is schematically illustrated in Figure 2.28, was formed by changing the composition of the reactive gases periodically in plasma CVD on glass or quartz substrates held at 230°C. All the samples have 12 a-Si:H layers and 13 a-SiH$_x$:H layers. For the a-Si: H layers, 20% SiH$_4$ diluted

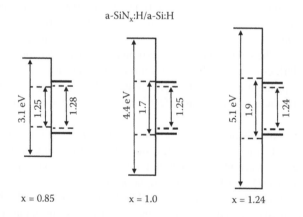

FIGURE 2.27   Band models of a-SiN$_x$:H/a-Si:H layered structures.

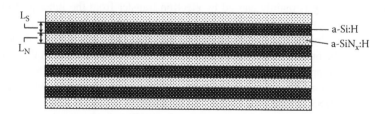

FIGURE 2.28   A structure of a-SiN$_x$:H/a-Si:H superlattice.

with H_2 was used, and the flow rate, r.f. power, and pressure were 50 s.c.c.m., 1.5 W, and 20 Pa, respectively. The deposition rate was 3 nm/min. In the present experiments, the a-Si:H layer thickness, L_S, was fixed to 10 nm. For the deposition of the a-SiN$_x$:H layers, pure NH_3 was added. The flow rate of 20% SiH_4 diluted with H_2 was kept constant at 50 sccm. To obtain a-SiN$_x$:H with various compositions, the flow rate of the pure NH_3 was varied from 25 to 150 sccm. The a-SiN$_x$:H composition was estimated by Auger electron spectroscopy to be $x = 0.75$, 0.85, 1.0, and 1.15 for NH_3 flow rates of 25, 50, 100, and 150 sccm, respectively. The plasma was interrupted for 1.5 min while the gases were changed.

We measured the in-plane conductivity by using slit-type electrodes with a 1-mm gap under 100-V bias. To form the electrodes, the superlattice film was scratched with a diamond scribe forming grooves 1 mm apart and Ag-paste colloidal solution was poured into the grooves and dried. This procedure should result in contact being made to all the layers. The samples were illuminated by a yellow LED (583 nm, 1 W/m^2) to measure the photocurrent. Before the dark current was measured, the samples were kept in the dark for 20 min. The activation energy of the dark current was obtained from Arrhenius plots in the temperature range of 30 to 100°C.

2.5.3.2 Transfer Doping

Figure 2.29(a) shows the influence of the nitrogen content x of a-SiN$_x$:H on the dark conductivity, photoconductivity, and the activation energy of the dark current, E_A, in the superlattices. It can be seen that in the range $0 < x < 0.85$, the dark conductivity increases by three orders of magnitude with increasing x and reaches its maximum of 5×10^{-6} $\Omega^{-1}cm^{-1}$ at $x = 0.85$. For $x > 0.85$, the dark conductivity decreases rapidly with increasing x. The photoconductivity exhibits a similar behavior. The activation energy of the dark conductivity has a minimum of 0.67 eV at $x = 0.85$, corresponding to the peak position of the dark conductivity and the photoconductivity. These results show that the transfer-doping effect is most effective at the nitrogen content $x = 0.85$.

The photocurrent enhancement in the amorphous superlattices is favorable from a viewpoint of applications to photovoltaic devices described in Chapter 9. Since in the superlattice large currents are generated in in-plane directions, planar-type photovoltaic devices with Schottky barriers might be available.

Figure 2.29(b) shows the influence of the a-SiN$_x$:H thickness, L_N, on the dark conductivity, photoconductivity, and the activation energy of the dark current for $x = 0.85$. The dark conductivity increases with increasing a-SiN$_x$:H thickness and, at the same time, the activation energy decreases. This clearly indicates that, as shown in Figure 2.30, the transfer doping is mainly caused by electron transfer from the inner part of the a-SiN$_x$:H layers to the a-Si:H layers rather than just from the interface states.

The reason why the transfer-doping effect becomes large at $x = 0.85$ can be considered from the optical properties of a-SiN$_x$:H shown in Figure 2.20. In the range $0 < x < 0.85$ it is supposed that a-SiN$_x$:H has a tetrahedral structure and hence the concentration of donor-like states, arising from nitrogen atoms, increases with increasing x. For $x > 0.85$, on the other hand, the structure of a-SiN$_x$:H tends toward that of Si_3N_4 in which nitrogen does not act as a donor. This leads to a decrease in the transfer-doping effect.

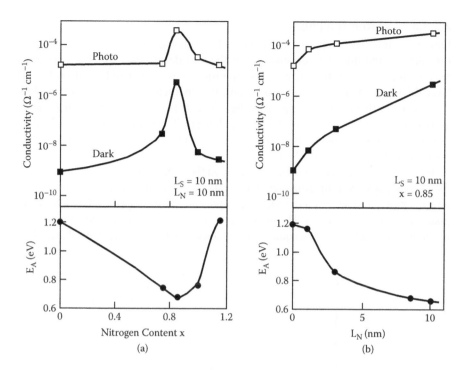

FIGURE 2.29 The dark conductivity, photoconductivity, and the activation energy of the dark current in the superlattices as a function of (a) the nitrogen content of a-SiN$_x$:H and (b) the a-SiN$_x$:H thickness. From T. Yoshimura, K. Hiranaka, T. Yamaguchi, and S. Yanagisawa, "Influence of a-SiNx:H composition on transfer-doping and electron trapping effects in a-SiN$_x$:H/a-Si:H superlattices," *Philos. Mag.* **B 55**, 409–416 (1987).

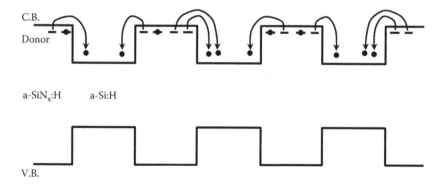

FIGURE 2.30 Schematic model for the transfer doping in a-SiN$_x$:H/a-Si:H superlattices.

2.5.3.3 Electron Trapping

Figures 2.31(a) and (b) show decay curves of the photocurrent for various a-SiN$_x$:H thicknesses, and thickness dependence of the half-decay time, $\tau_{1/2}$. The horizontal axis of (a) corresponds to time after the illumination is interrupted. As the a-SiN$_x$:H layer thickness increases from 1 to 10 nm, the photocurrent decay becomes slow. This indicates that electron trapping by the a-SiN$_x$:H layers occurs in the superlattice and extends to the inner part of the a-SiN$_x$:H layers. The tendency for $\tau_{1/2}$ to saturate around $L_N \sim$ 10 nm indicates that the interaction depth for the electron-trapping phenomena is ~10 nm in the a-SiN$_x$:H layer.

Figure 2.32 shows the photocurrent decay in the superlattice with $L_S = L_N = 10$ nm for various x at various measurement temperatures. A fast initial decay is observed in all the samples, followed by a slow decay component. The decay speed of the slow component depends on the a-SiN$_x$:H composition. The slow component is greatly affected by temperature, while the fast component exhibits little temperature dependence. This

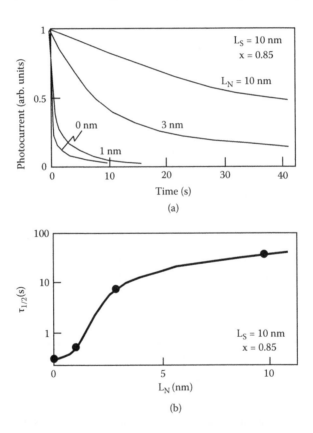

FIGURE 2.31   (a) Photocurrent decay curves for various a-SiN$_x$:H thicknesses, and (b) thickness dependence of the half-decay time of the photocurrent decay curves, in a-SiN$_x$:H/a-Si:H superlattices. From T. Yoshimura, K. Hiranaka, T. Yamaguchi, and S. Yanagisawa, "Influence of a-SiNx:H composition on transfer-doping and electron trapping effects in a-SiN$_x$:H/a-Si:H superlattices," *Philos. Mag.* **B 55**, 409–416 (1987).

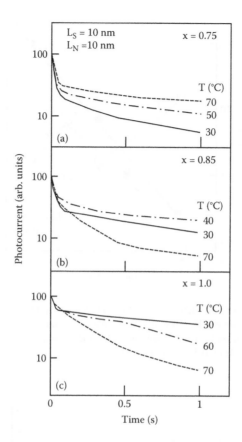

FIGURE 2.32 Photocurrent decay in a-SiN$_x$:H/a-Si:H superlattices at various measurement temperatures for (a) $x = 0.75$, (b) $x = 0.85$, and (c) $x = 1.0$. From T. Yoshimura, K. Hiranaka, T. Yamaguchi, and S. Yanagisawa, "Influence of a-SiNx:H composition on transfer-doping and electron trapping effects in a-SiN$_x$:H/a-Si:H superlattices," *Philos. Mag.* **B 55**, 409–416 (1987).

indicates that the slow component is attributed to deep electron-trapping states in the a-SiN$_x$:H layers and the process of the electron release from the deep states limits the decay time. Figure 2.33 shows the temperature dependence of the photocurrent decay time τ of the slow component for various x. Here, τ is the time for the slow component to decay to $1/e$ of the initial value. For $x = 0.75$, τ increases with increasing T, whereas for $x = 1.0$, τ decreases with increasing T. For $x = 0.85$, τ increases with T if $1000/T <$ 3.2, and then decreases when $1000/T > 3.2$. Thus electron trapping by the a-SiN$_x$:H layers varies strongly with the a-SiN$_x$:H composition.

As reported by Rose [24], τ versus T characteristics are determined by the relationship between the position of the Fermi level and the distribution of the electron-trapping states. When the Fermi level is located in the middle of a continuous distribution of electron-trapping states, τ increases with T. Conversely, when the states are located above the Fermi level, τ decreases with T. According to this, qualitative models for the electron-trapping states are schematically illustrated as in Figure 2.34. For $x = 0.75$, the Fermi level is in the middle of the electron-trapping

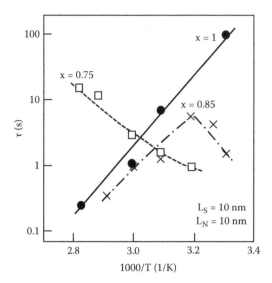

FIGURE 2.33 Temperature dependence of the photocurrent decay time of the slow component. From T. Yoshimura, K. Hiranaka, T. Yamaguchi, and S. Yanagisawa, "Influence of a-SiNx:H composition on transfer-doping and electron trapping effects in a-SiN$_x$:H/a-Si:H superlattices," *Philos. Mag.* **B 55**, 409–416 (1987).

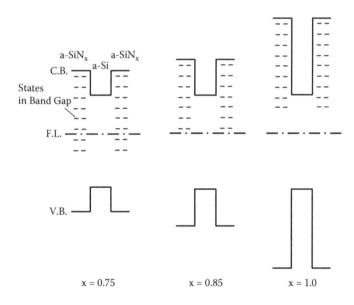

FIGURE 2.34 Schematic model for the electron-trapping-state distribution in the a-SiN$_x$:H layer. From T. Yoshimura, K. Hiranaka, T. Yamaguchi, and S. Yanagisawa, "Influence of a-SiNx:H composition on transfer-doping and electron trapping effects in a-SiN$_x$:H/a-Si:H superlattices," *Philos. Mag.* **B 55**, 409–416 (1987).

states. For $x = 1.0$, the Fermi level is below the electron-trapping states. For $x = 0.85$, the Fermi level should be near the bottom of the electron-trapping states, since τ decreases with T in the range of $1000/T > 3.2$.

2.6 SPUTTERING

Sputtering is one of the popular thin-film deposition methods. It enables us to control crystallinity and composition, that is, atomic arrangements in thin films, by adjusting sputtering power, gas composition and pressure, substrate temperature, and source material composition. This section describes an example of material property control by film structures in an atomic level, and enhancement of coloration efficiency in WO_x electrochromic thin films by sputtering [25, 26].

2.6.1 ELECTROCHROMISM IN WO_x THIN FILMS

WO_x thin films exhibit electrochromism [27]. The coloration is attributed to color centers generated by electrons injected in the films. Since the absorption spectra of the color centers cover the visible and near infrared regions, WO_x thin films are useful for displays and heat-cut windows.

Figure 2.35 shows the coloring/bleaching mechanism in electrochromic devices (ECDs). ECD consists of an electrochromic layer of WO_3 and electrolyte that are sandwiched by an indium tin oxide (ITO) electrode and a counter electrode. By applying negative voltage to the ITO electrode, electrons are injected into the WO_3 thin film, and at the same time, protons are injected to the film from the electrolyte to cancel the negative charge of the injected electrons. Due to the neutralization, many electrons can be injected into the WO_3 thin film, say, several

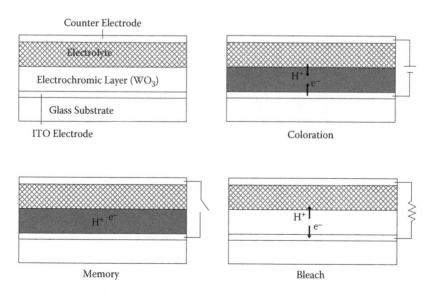

FIGURE 2.35 Coloring/bleaching mechanism in electrochromic devices (ECDs).

tenths of the number of W atoms in the film. The injected electrons are trapped at W^{6+} sites to form color centers. The trapped electrons hop to the neighboring W^{6+} sites by absorbing light. The light absorption contributes to the coloration of WO_3. By making the external circuits open, the injected charge is kept in the film, giving rise to memory function as well as battery function. By shorting the ITO and counter electrodes, the electrons and protons are ejected from the WO_3 thin film to bleach the film.

Since WO_3 is an n-type semiconductor, it is expected to be applied to photovoltaic devices having battery function, which is described in Chapter 9, Section 9.4.

2.6.2 ENHANCEMENT OF COLORATION EFFICIENCY IN WO_x WITH CONTROLLED FILM STRUCTURES

The coloration property of WO_3 is affected by the film structures, that is, atomic arrangements in the film. We fabricated two kinds of ECDs with a WO_3 thin film deposited from WO_3 powder and a "WO_2" thin film from WO_2 powder using vacuum evaporation [26]. Figure 2.36 shows the absorption spectra of the color canters in WO_3 and "WO_2" thin films for the same injected charge (~3 mC/cm^2). The "WO_2" thin film has much larger absorption than does the WO_3 thin film.

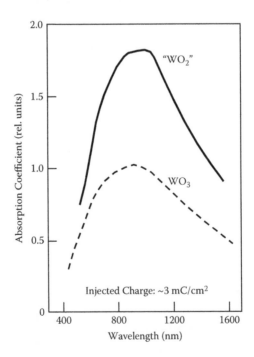

FIGURE 2.36 Absorption spectra of the color centers in WO_3 and "WO_2" thin films for the same injected charge. From T. Yoshimura, M. Watanabe, Y. Koike, K. Kiyota, and M. Tanaka, "Enhancement in oscillator strength of color centers in electrochromic thin films deposited from WO_2," *J. Appl. Phys.* **53**, 7314–7320 (1982).

In order to reveal the reason for the difference in the absorption property, precise investigation was carried out using rf reactive magnetron sputtering [25]. Figure 2.37 shows the coloration efficiency spectra of WO_x thin films deposited at an oxygen gas pressure of 3 Pa at various rf powers. The efficiency is drastically enhanced as rf power increases. At rf power of 1000 W, the peak height is seven times greater than that of the conventional WO_x thin film prepared by vacuum evaporation. In Figure 2.37, it is noted that, with an increase in the efficiency, the peak position of the efficiency spectra shifts to the low-energy side.

Table 2.3 lists the lattice constants obtained by x-ray diffraction measurements on the film prepared by rf reactive magnetron sputtering, reactive ion plating, and vacuum evaporation. In the left-hand columns of the table, the powder data for monoclinic WO_3, $W_{20}O_{58}$, and $W_{18}O_{49}$ are shown for reference.

Figure 2.38 shows typical x-ray photoelectron spectroscopy (XPS) spectra of W_{4f} and O_{1s} levels in WO_x thin films deposited by rf reactive magnetron sputtering. The intensity of the XPS spectrum for O_{1s} level decreases with an increase in rf power, while that for the W_{4f} level shows a slight difference among the three films. Table 2.4 lists the counting ratio of photoelectrons from the O_{1s} level to those from the W_{4f}

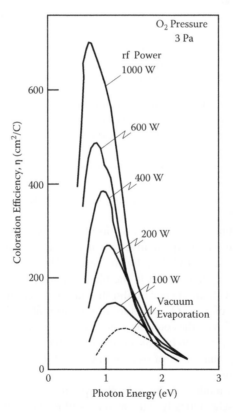

FIGURE 2.37 Coloration efficiency spectra of WO_x thin films deposited by rf reactive magnetron sputtering for various rf powers. From T. Yoshimura, "Oscillator strength of small-polaron absorption in WO_x ($x \leq 3$) electrochromic thin films," *J. Appl. Phys.* **57**, 911–919 (1985).

TABLE 2.3

Lattice Constants Obtained from X-Ray Diffraction Patterns for Various WO$_x$ Thin Films

WO$_3$	W$_{20}$O$_{58}$	W$_{18}$O$_{49}$	Sputtering (3 Pa, 1000 W)	Ion Plating Film 1	Vacuum Evaporation (annealed at 400°C)
3.84 (001)			3.84	3.84	3.84
		3.78 (010)			
	3.77 (010)			3.77	3.77
3.76 (020)					
		3.73 (103)			
	3.70 (303)		3.70		
				3.66	
3.64 (200)	3.64 (106)				3.65
		3.63 (011) (502)			

Source: T. Yoshimura, "Oscillator strength of small-polaron absorption in WO$_x$ ($x \leq 3$) electrochromic thin films," *J. Appl. Phys.* **57**, 911–919 (1985).

Note: Lattice constants from the powder data are also listed with Miller indices (hkl) for monoclinic WO$_3$, W$_{20}$O$_{58}$, and W$_{18}$O$_{49}$.

level for the WO$_x$ thin films. The counting ratio was calculated by integrating the line profiles of the XPS spectra.

Although the counting ratio of O$_{1s}$/W$_{4f}$ itself does not give the O/W ratio in the film directly, we can obtain qualitative information about the O/W ratio by comparing the counting ratio of O$_{1s}$/W$_{4f}$ for the various WO$_x$ thin films. It can be seen from Table 2.4 that oxygen content in the film decreases with increasing rf power, which is consistent with the change in film structure induced by increasing rf power, that is, development of a W$_{20}$O$_{58}$ structure in WO$_x$ thin film. Comparing the results shown in Figure 2.37 with the counting ratio of O$_{1s}$/W$_{4f}$ and the x-ray diffraction patterns described above, we can find the clear correlation between the enhancement of coloration efficiency and a lack of oxygen in the WO$_x$ thin film, which induces the W$_{20}$O$_{58}$-like structure.

In Figure 2.39, a model for the light absorption in WO$_3$, called *small-polaron absorption* [27], is schematically illustrated as well as the lattice structure of WO$_3$. The crystal structure of WO$_3$ is a corner-shared WO$_6$ octahedral. Each W site is surrounded by six O sites. When an injected electron is trapped at the W$_A$ site, the electron distorts the surrounding lattice, forming a small polaron. When the electron

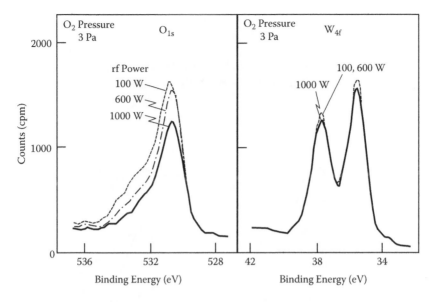

FIGURE 2.38 Typical XPS spectra of the W_{4f} and O_{1s} levels in WO_x thin films deposited by rf reactive magnetron sputtering. From T. Yoshimura, "Oscillator strength of small-polaron absorption in WO_x (x≤3) electrochromic thin films," *J. Appl. Phys.* **57**, 911–919 (1985).

TABLE 2.4

Counting Ratio of Photoelectrons from O_{1s} Level to Those from the W_{4f} Level for WO_x Thin Films

Oxygen Pressure (Pa)		3	
rf Power (W)	100	600	1000
O_{1s}/W_{4f} Ratio	0.95	0.8	0.7

Source: T. Yoshimura, "Oscillator strength of small-polaron absorption in WO_x (x ≤ 3) electrochromic thin films," *J. Appl. Phys.* **57**, 911–919 (1985).

---

is transferred to the neighboring W_B site by absorbing light, the distortion is also transferred to the W_B site. This means that the small polaron hops from the W_A site to the W_B site. So, the light absorption caused by the trapped electron in WO_3 is called small-polaron absorption.

Faughnan et al. determined that the oscillator strength of the absorption band in the evaporated WO_x thin film is $f_0 = 0.1$ [28]. Then, oscillator strength of the small-polaron absorption is estimated for various WO_x thin films using the following expression:

$$f = f_0 (I / I_0).$$

Here, I and I_0 are respectively the integrated intensity of the coloration efficiency spectra of WO_x thin films and that of a WO_x thin film prepared by vacuum evaporation.

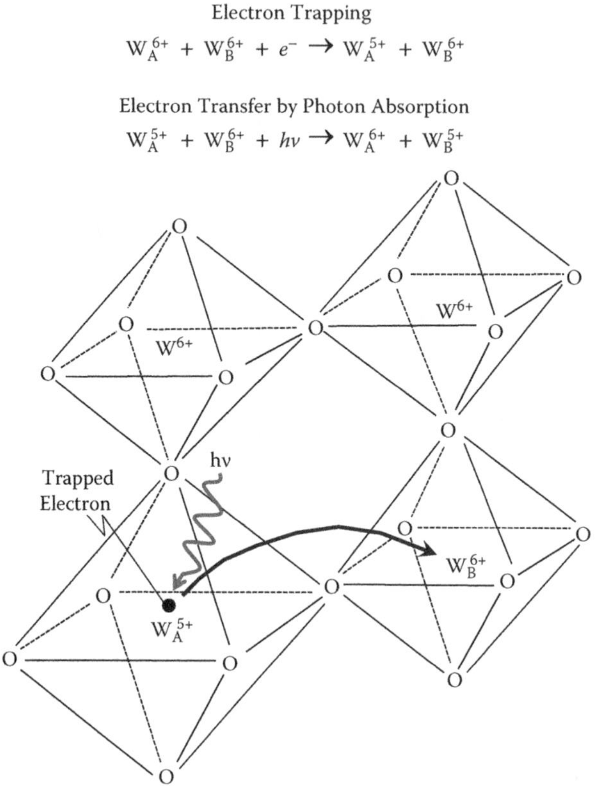
Electron Trapping
$W_A^{6+} + W_B^{6+} + e^- \rightarrow W_A^{5+} + W_B^{6+}$
Electron Transfer by Photon Absorption
$W_A^{5+} + W_B^{6+} + h\nu \rightarrow W_A^{6+} + W_B^{5+}$
W^{6+}
W^{6+}
hv
Trapped
Electron
W_B^{6+}
W_A^{5+}

FIGURE 2.39 Schematic illustration of a lattice structure, and a model for the small-polaron absorption in WO_3. From T. Yoshimura, "Oscillator strength of small-polaron absorption in WO_x ($x \leq 3$) electrochromic thin films," *J. Appl. Phys.* **57**, 911–919 (1985).

Table 2.5 lists the oscillator strength f and the peak position E_0 of absorption band for various WO_x thin films. Oscillator strength varies widely depending on the procedure used for the film preparation. An extremely large oscillator strength is obtained for the film deposited by rf reactive magnetron sputtering at 1000 W. The value is five times larger than that of a conventional evaporated WO_x thin film, and one order of magnitude larger than that of the evaporated film annealed at 200°C in air.

The mechanism of the oscillator strength enhancement can be discussed from the viewpoint of the degree of polaron extension. Faugnan et al. reported that f, E_0, in eV, and the dipole strength D in nm are related as follows [28].

$$f = 9E_0D^2 \tag{2.1}$$

Considering that each W^{5+} site is surrounded by six neighboring W^{6+} sites, D is given by

$$D = \sqrt{6}\xi r, \tag{2.2}$$

TABLE 2.5
Properties of Small-Polaron Absorption for Various WO_x Thin Films

(Sputtering)				
rf Power (W)	Oxygen Pressure (Pa)	E_0 (eV)	f	ξ
1000	3	0.7	0.51	0.22
600	3	0.85	0.42	0.18
400	3	1.0	0.32	0.14
200	3	1.1	0.26	0.12
100	3	1.2	0.16	0.09
600	7	0.95	0.35	0.15
600	0.7	0.95	0.2	0.12
600	0.08	1.05	0.14	0.09
	(Ion Plating)			
	Film 1	0.75	0.31	0.16
	Film 2	1.3	0.08	0.06
	(Vacuum Evaporation)			
WO_3	As deposited	1.3	0.1	0.07
	Annealed at 200°C	1.45	0.04	0.04
	Annealed at 400°C	0.7	0.24	0.15
"WO_2"	As deposited	1.2	0.18	0.09

Source: T. Yoshimura, "Oscillator strength of small-polaron absorption in WO_x ($x \le 3$) electro-chromic thin films," *J. Appl. Phys.* **57**, 911–919 (1985).

where r is nearest W–W distance in nm, and ξ is the delocalization parameter of the polaron wavefunction. ξ has the following physical meaning: when an electron is trapped at a W site, the wavefunction of the trapped electron is not completely localized at the site, but a fraction of the wavefunction ξ penetrates to the neighboring W site. Substituting Equation (2.2) into Equation (2.1), f is expressed in terms of E_0 and ξ as follows:

$$f = 54E_0\xi^2r^2 \tag{2.3}$$

In this equation, it must be noted that E_0 is also a function of ξ as will be qualitatively explained by the configuration coordinate diagram in Figure 2.40. ξ can be determined by substituting E_0, f, and r into Equation (2.3). The nearest W–W distance in the crystalline WO_3, 0.54 nm, is used for r as a zeroth approximation. The obtained values of ξ are listed in Table 2.5.

It should be noted in Table 2.5 that f increases from 0.04 to 0.51 as ξ increases from 0.04 to 0.22. This relationship between f and ξ can be explained as follows. As can be seen from Figure 2.39, the oscillator strength of the absorption band is

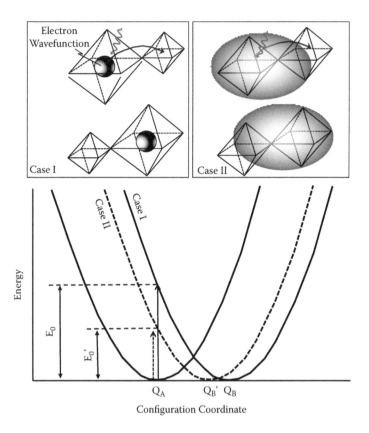

FIGURE 2.40   Configuration coordinate diagrams for small-polaron absorption. From T. Yoshimura, "Oscillator strength of small-polaron absorption in WO_x ($x \leq 3$) electrochromic thin films," *J. Appl. Phys.* **57**, 911–919 (1985).

expected to increase with an increase in the transition probability for the electron to hop from the W_A site to the W_B site. If the trapped electron is completely localized, that is, $\xi = 0$, transition probability is reduced to zero since the initial state wavefunction (trapped state at the W_A site) does not overlap with the final state (trapped state at the W_B site). With an increase in penetration of wavefunction to the neighboring W_B site, that is, an increase in ξ, the overlap of the wavefunction between the initial and the final states increases, causing the transition probability to become larger.

The previously descrbed behavior of the oscillator strength is explained by a configuration coordinate diagram for small-polaron absorption, proposed by Faughnan et al., [28]. In Figure 2.40, Q_A or Q_B is the equilibrium position of the lattice surrounding the electron, which is trapped at the W_A site or the W_B site, respectively. When an electron trapped at the W_A site is excited by absorbing a photon with energy E_0, it transfers from the initial state to the final state (trapped at the W_B site). After the transition of the electron, a relaxation of the surrounding lattice occurs along the curve with a minimum point at Q_B. So, $\Delta Q = Q_B - Q_A$ represents the change of lattice distortion, accompanied by the electron hopping from the W_A site to the W_B site.

Since $\Delta Q = Q_B - Q_A$ corresponds to the change of lattice distortion induced by the electron hopping, ΔQ depends on the degree of localization of the trapped electron. When the trapped electron is highly localized (Case I in Figure 2.40), the lattice distortion is large because, at first, lattice distortion is well localized around the W_A site, then, it shifts around the W_B site following the transfer of the electron from W_A to W_B. With delocalization of the trapped electron (Case II in Figure 2.40), the lattice distortion becomes small, indicating that ΔQ decreases. Since ξ represents the penetrating fraction of the trapped electron into the neighboring site, the degree of delocalization of the trapped electron increases with an increase in ξ. Therefore, it is suggested that ΔQ will decrease with an increase in ξ, implying that the energy curve for the W_B site in the configuration coordinate diagram shifts from the solid curve to the dotted curve with increasing ξ. In this case, then, it is expected that the peak position of the absorption band shifts to the low-energy side with an increase in ξ.

Figure 2.41 shows the relationship between experimentally obtained E_0 and ξ. It is found that E_0 tends to decrease with increasing ξ as expected from the previously mentioned configuration coordinate model. Thus, it is confirmed that the enhancement of the oscillator strength of small-polaron absorption arises from a development of delocalization of the polaron wavefunction.

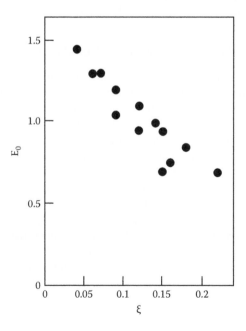

FIGURE 2.41   Relationship between experimentally obtained E_0 and ξ. From T. Yoshimura, "Oscillator strength of small-polaron absorption in WO_x ($x \leq 3$) electrochromic thin films," *J. Appl. Phys.* **57**, 911–919 (1985).

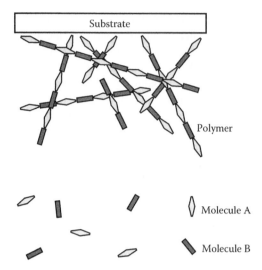

FIGURE 2.42   Mechanism of the vacuum deposition polymerization.

As described, by controlling sputtering conditions, the atomic arrangements in the films are controlled, which enables control of optical properties of the film, such as oscillator strength, namely, coloration efficiency.

2.7   VACUUM DEPOSITION POLYMERIZATION

Vacuum deposition polymerization, shown in Figure 2.42, is a method for polymer film deposition in vacuum using vacuum evaporation [29–32]. Two kinds of reactive molecules A and B are evaporated in a vacuum chamber. The molecules are combined to deposit polymer films on the substrate surface. Vacuum deposition polymerization is regarded as CVD for polymer film fabrication.

Various applications of vacuum deposition polymerization have been reported, such as polyimide films for nonlinear optical materials, organic EL devices, insulators in multilayer circuit substrates and LSIs, surface coating, and so on.

Weaver and Bradley demonstrated that the chemical vapor deposition polymerization is an effective method to fabricate conjugated polymer films of polyazomethine (poly-AM) in the organic EL [33]. The result confirms that vacuum deposition polymerization is useful to grow functional polymer films, and at the same time, confirms the prospect of applications of poly-AM to thin-film organic photonic/electronic devices.

REFERENCES

1. E. Buks, R. Schuster, M. Heiblum, D. Mahalu, and V. Umansky, "Dephasing in electron interference by a 'which-path' detector," *Nature* **391**, 871–874 (1998).
2. D. M. Eigler and E. K. Schweizer, "Positioning single atoms with a scanning tunneling microscope," *Nature* **344**, 524–526 (1990).

3. M. F. Crommie, C. P. Lutz, and D. M. Eigler, "Confinement of electrons to quantum corrals on a metal surface," *Science* **262**, 218–220 (1993).

4. H. C. Manoharan, C. P. Lutz, and D. M. Eigler, "Quantum mirages formed by coherent projection of electronic structures," *Nature* **403**, 512–515 (2000).

5. L. Esaki and R. Tsu, "Superlattice and negative differential conductivity in semiconductors," *IBM Journal of Research and Development* **14**, 61–65 (1970).

6. T. Mimura, S. Hiyamizu, T. Fujii, and K. Nanbu, "A new field-effect transistor with selectively doped GaAs/n-$Al_xGa_{1-x}As$ heterojunctions," *Jpn. J. Appl. Phys.* **19**, L225–L227 (1980).

7. D. S. Chemla, D. A. B. Miller, and P. W. Smith, "Nonlinear optical properties of GaAs/GaAlAs multiple quantum well material: Phenomena and applications," *Opt. Eng.* **24**, 556–564 (1985).

8. T. Suntola, "Atomic layer epitaxy," *Material Science Reports* **4** (7), 1989, Elsevier Science Publishers.

9. F. Roozeboom, W. Dekker, K. B. Jinesh, J. H. Klootwijk, M. A. Verheijen, H.-D. Kim, and D. Blin, "Ultrahigh-density trench decoupling capacitors comprising multiple MIM layer stacks grown by atomic layer deposition," AVS, 8th International Conference on Atomic Layer Deposition, MonA1-1, Bruges, Belgium (2008).

10. J. H. Shim, J. S. Park, J. An, T. M. Gur, and F. B. Prinz, "Atomic layer deposition of proton conducting Y:$BaZrO_3$," AVS, 9th International Conference on Atomic Layer Deposition, Monterey, California, 76 (2009).

11. H. Shin and C. Bae, "Nanoscale tubular structures and ampoules of oxides," AVS, 8th International Conference on Atomic Layer Deposition, TueM2-1, Bruges, Belgium (2008).

12. G. N. Parsons, Q. Peng, J. C. Spagnola, G. Scarel, G. K. Hyde, B. Gong, C. Devine, K. Lee, J. Jur, K. Roberts, and J. S. Na, "Modification of fibers and nonwoven fiber mats using atomic layer deposition," AVS, 9th International Conference on Atomic Layer Deposition, Monterey, California, 29 (2009).

13. D. H. Levy, D. C. Freeman, S. F. Nelson, and P. J. Cowdery-Corvan, "A high-speed continuous process for ALD depositions," AVS, 8th International Conference on Atomic Layer Deposition, WedA2b-3, Bruges, Belgium (2008).

14. C. B. Roxlo, B. Abeles, and T. Tiedje, "Amorphous semiconductor superlattices," *Phys, Rev. Lett.* **51**, 2003–2006 (1983).

15. T. Tiedje and B. Abeles, "Charge transfer doping in amorphous semiconductor superlattice," *Appl. Phys. Lett.* **45**, 179–181 (1984).

16. J. Kakalios, H. Fritzsche, N. Ibaraki, and S. R. Ovshinsky, "Properties of amorphous semiconducting multilayer films" *J. Non-Crystalline Solids* **66**, 339–344 (1984).

17. M. J. Powell, "Charge trapping instabilities in amorphous silicon-silicon nitride thin-film transistors," *Appl. Phys. Lett.* **43**, 597–599 (1983).

18. T. Yoshimura, K. Hiranaka, T. Yamaguchi, and S. Yanagisawa, "Influence of a-SiN_x:H composition on transfer-doping and electron trapping effects in a-SiN_x:H/a-Si:H superlattices," *Philos. Mag.* B **55**, 409–416 (1987).

19. T. Yoshimura, K. Hiranaka, T. Yamaguchi, S. Yanagisawa, and K. Asama, "Characterization of a-SiN_x:H/a-Si:H interface by photoluminescence spectra," in *Materials issues in amorphous-semiconductor technology*, edited by D. Adler, Y. Hamakawa, and A. Madan (*Mater. Res. Soc. Symp. Proc.* **70**, 373–378, (1986).

20. T. V. Herak, R. D. Meleod, K. C. Kao, H. C. Card, H. Watanabe, K. Katoh, M. Yasui, and Y. Shibata, "Undoped amorphous SiNx:H alloy semiconductors: Dependence of electronic properties on composition," *J. Non-Crystalline Solids* **69**, 39–48 (1984).

21. C. Tsang and R. A. Street, "Recombination in plasma-deposited amorphous Si:H luminescence decay," *Phys. Rev.* B **19**, 3027–3040 (1979).

22. C. B. Roxlo, B. Abeles, and T. Tiedje, "Evidence for lattice-mismatch-induced defects in amorphous semiconductor hetero-junctions," *Phys. Rev. Lett.* **52**, 1994–1997 (1984).

23. K. Tanaka, *Fundamentals of amorphous semiconductors*, (Ohmusha, Tokyo, 1982) [in Japanese].
24. A. Rose, "Outline of some photoconductive processes," *RCA Rev.* **12**, 362–414 (1951).
25. T. Yoshimura, "Oscillator strength of small-polaron absorption in WO_x ($x \leq 3$) electrochromic thin films," *J. Appl. Phys.* **57**, 911–919 (1985).
26. T. Yoshimura, M. Watanabe, Y. Koike, K. Kiyota, and M. Tanaka, "Enhancement in oscillator strength of color centers in electrochromic thin films deposited from WO_2," *J. Appl. Phys.* **53**, 7314–7320 (1982).
27. S. K. Deb, "Optical and photoelectric properties and colour centres in thin films of tungsten oxide," *Pholos. Mag.* **27**, 801–822 (1973).
28. B. W. Faughnan, R. S. Crandall, and D. Heyman, "Electrochromism in WO_3 amorphous films," *RCA Rev.* **36**, 177–197 (1975).
29. M. Iijima, Y. Takahashi, K. Inagawa, and A. Itoh, *J. Vac. Soc. Jpn.* **28**, 437 (1985) [in Japanese].
30. Y. Takahashi, M. Iijima, K. Iinagawa, and A. Itoh, "Synthesis of aromatic polyimide film by vacuum deposition polymerization," *J. Vac. Sci. Technol.* **A 5**, 2253–2256 (1987).
31. M. Iijima, Y. Takahashi, and E. Fukuda, "Vacuum deposition polymerization," *Nikkei New Materials*, December 11, 93–100 (1989) [in Japanese].
32. A. Kubono, N. Okui, K. Tanaka, S. Umemoto, and T. Sakai, "Highly oriented polyamide thin films prepared by vapor deposition polymerization," *Thin Solid Films* 199, 385–393 (1991).
33. M. S. Weaver and D. D. C. Bradley, "Organic electroluminescence devices fabricated with chemical vapour deposited polyazomethine films," *Synthetic Metals* **83**, 61–66 (1996).

3 Fundamentals of Molecular Layer Deposition (MLD)

As described in Chapter 1, to realize high-performance photonic/electronic devices consisting of organic functional thin films, growth methods for polymer wires and molecular wires with designated (1) molecular sequences, (2) locations, and (3) orientations are essential. Molecular layer deposition (MLD) [1–6] is a method to grow polymer wires and molecular wires with designed molecular sequences, enabling precisely controlled growth of artificial polymer wires and molecular wires. MLD for growth of polymer wires utilizes chemical reactions between source molecules. MLD for growth of molecular wires, in which different kinds of molecules are aligned in designated sequences, utilizes electrostatic force between source molecules.

In the present chapter, the concept of MLD is reviewed, and the proof of concept is experimentally demonstrated. Selective wire growth for location and orientation control of polymer/molecular wires, which can be combined with MLD, is also described.

3.1 CONCEPT OF MLD

3.1.1 MLD UTILIZING CHEMICAL REACTIONS

The concept of MLD utilizing selective chemical reactions between source molecules [1] is shown in Figure 3.1 for a case in which four kinds of molecules are used: molecules A, B, C, and D. These molecules are prepared under the following guidelines.

- A molecule has two or more reactive groups.
- Reactive groups in a molecule cannot react with reactive groups in the same kind of molecules. For example, reactive groups in molecule A cannot react with reactive groups in molecule A.
- Reactive groups in a molecule can react with reactive groups in other kinds of molecules. For example, reactive groups in molecule A can react with reactive groups in molecule B.

Thus, the same kinds of molecules cannot be connected, and different kinds of molecules can be connected through the reactions between groups. The molecular connections are achieved by chemical force. In the case shown in Figure 3.1, strong chemical reactions occur between molecules A–B, B–C, C–D, and D–A, and weak reactions between molecules A–A, B–B, C–C, D–D, A–C, and B–D.

In MLD, there are two typical growth configurations—*one-sided growth* and *two-sided growth*. For the one-sided growth, as shown in Figure 3.1(a), when molecule A is provided to a surface, a monomolecular layer of molecule A is formed. Once the surface is covered with molecule A, the deposition of the molecules is automatically terminated due to the small reactivity between the same kinds of groups, giving rise to thickness saturation. When molecules are switched from A to B, molecule B is connected to molecule A to induce a rapid thickness increase followed by saturation, exhibiting monomolecular step growth. This enables us to carry out sequential self-limiting reactions, like atomic layer deposition (ALD) [7,8], by switching molecules. When molecules are switched from B to C, molecule C is connected to molecule B. Similarly, when molecules are switched from C to D, molecule D is connected to molecule C. By repeating the switching with designated sequences, artificial polymer wires with a molecular sequence like ABCD … can be obtained.

For two-sided growth, as shown in Figure 3.1(b), molecule A is adsorbed on a surface. When molecule B is provided to the surface, molecules B is connected to both sides of molecule A. Once both sides of molecule A are covered with molecules B, the deposition of the molecules is automatically terminated due to the self-limiting effect, forming a BAB structure. By switching molecules from B to C, molecule C is connected to both sides of the BAB structure to form a CBABC structure. Similarly, by switching molecules from C to D, molecule D is connected to both sides of the CBABC structure to form a DCBABCD structure. By repeating the switching with designated sequences, artificial polymer wires with a molecular sequence like … DCBABCD … can be obtained with monomolecular step growth.

If a molecule with three reactive groups is used as shown in Figure 3.2(a), three-way branching polymer wires can be constructed. Similarly, more than four-way branching polymer wires can be constructed by using a molecule with more than four reactive groups. Bending of polymer wires is also possible by using a molecule with two reactive groups substituted in appropriate configurations.

An additional advantage of MLD is that it can deposit ultrathin conformal films on three-dimensional surfaces with arbitrary shapes, such as trenches, cylinders, pours/rough surfaces, and so on, as shown in Figure 3.2(b), in a manner similar to ALD.

3.1.2 MLD UTILIZING ELECTROSTATIC FORCE

The concept of MLD utilizing repulsive and attractive electrostatic force between source molecules [5,6] is shown in Figure 3.3 for one-sided growth. When four kinds of molecules are used, attractive force is generated in the combination of molecules A–B, B–C, C–D, and D–A, and repulsive force is generated in the combinations of molecules A–A, B–B, C–C, D–D, A–C, and B–D.

When molecule A is provided to a surface, a monomolecular layer of molecule A is formed. Once the surface is covered with molecule A, the deposition of the molecules is automatically terminated due to the repulsion between the same kinds of molecules, giving rise to thickness saturation. When molecules are switched from A to B, molecule B is connected to molecule A to induce a rapid thickness increase followed by saturation, exhibiting monomolecular step growth. When molecules are switched from B to C, molecule C is connected to molecule B. Similarly, when

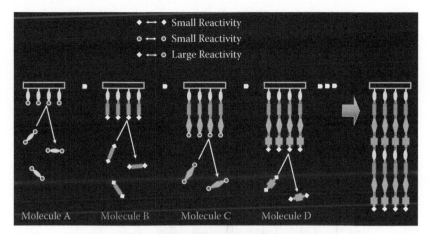

(a) One-Sided Growth

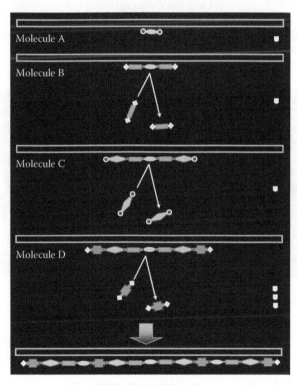

(b) Both-Sided Growth

FIGURE 3.1 Concept of MLD utilizing chemical reactions. (a) One-sided growth and (b) Both-sided growth. From T. Yoshimura, Trends in Thin Solid Films Research (ed. A. R. Jost), Ch. 5 "Self-Organized Growth of Polymer Wire Networks with Designed Molecular Sequences for Wavefunction-Controlled Nano Systems." Nova Science Publishers, New York (2007).

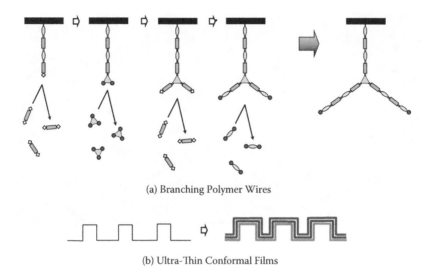

(a) Branching Polymer Wires

(b) Ultra-Thin Conformal Films

FIGURE 3.2   Growth of (a) branching polymer wires and (b) ultra-thin conformal films on three-dimensional surfaces by MLD.

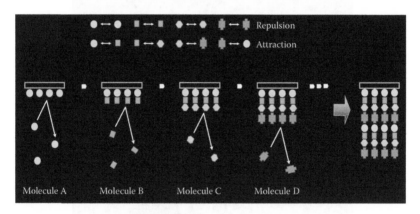

FIGURE 3.3   Concept of MLD utilizing electrostatic force.

molecules are switched from C to D, molecule D is connected to molecule C. By repeating the switching with designated sequences, artificial molecular wires with a molecular sequence like ABCD … can be obtained. As in MLD for polymer wire growth, two-sided growth configurations might be possible.

3.1.3   MLD WITH MOLECULE GROUPS

Usually MLD is carried out by introducing one kind of molecule onto substrates at a molecular supplying step. In MLD with molecule groups, several kinds of molecules are introduced at the same molecular supplying step.

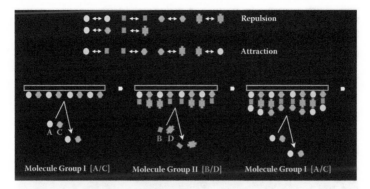

FIGURE 3.4   Concept of MLD with molecule groups.

In the example depicted in Figure 3.4, Molecule Group I containing molecules A and C and Molecule Group II containing molecules B and D are prepared. By providing molecules onto the substrate in the order of Molecule Group I, Molecule Group II, Molecule Group I, molecular wires of ABA, ABC, ADA, ADC, CBC, CBA, CDC, and CDA are formed in parallel.

In some cases, such feature that molecular wires or polymer wires with different molecular arrangements are grown simultaneously is useful, as described in Section 9.1.1.

3.2   MLD EQUIPMENT

In MLD, source molecules are supplied onto a substrate surface in gas phase or liquid phase. The former is the gas-phase MLD like K cell–type MLD and carrier gas–type MLD, and the latter is the liquid-phase MLD like fluidic circuit–type MLD.

3.2.1   Gas-Phase MLD

3.2.1.1   K Cell–Type MLD

Figure 3.5 shows a schematic illustration and a photograph of the K cell–type MLD [1]. Molecular gases are provided onto a surface in vacuum from low-temperature K cells, in which source molecules are loaded. Molecular gas switching is performed by shutters. When the vapor pressure of the source molecules is high, molecular gases leak from the gap between the K cells and the shutters. In such cases, valves are used for the molecular gas switching.

3.2.1.2   Carrier Gas–Type MLD

Figure 3.6 shows a schematic illustration and a photograph of the carrier gas–type MLD [9,10], in which carrier gas is employed to introduce molecular gases in temperature-controlled molecular cells onto a substrate surface. Molecular gas switching is performed by valves. The molecules are combined through chemical reactions on the surface, while excess gases are removed with the carrier gas to a pump.

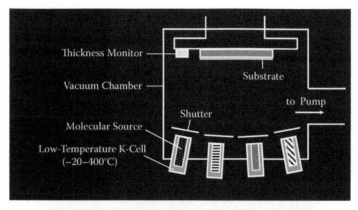

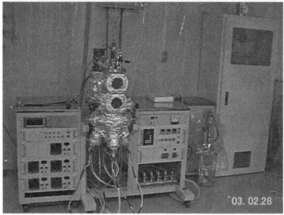

FIGURE 3.5 K cell–type MLD.

To ensure removal of remaining molecules and by-products generated by the chemical reactions used in the previous steps, gas purges with inert gases before the next molecular gas injection is effective.

The carrier gas–type MLD has the following advantages.

1. By-products generated by the chemical reactions are swept away from the surfaces by carrier gas flows.
2. The carrier gas prevents contaminations in the surrounding region from reaching the substrate surface.
3. Stable supply rates of molecular gases can be easily maintained.

Items 1 and 2 can keep the surface clean, which is especially important to carry out MLD. If by-products remain near the surface and molecules are adsorbed on them, polymer wires will grow randomly from the by-product sites. This will prevent polymer wires from growing with the designated orientations and locations. Furthermore, by-products may terminate polymer wire growth when their presence

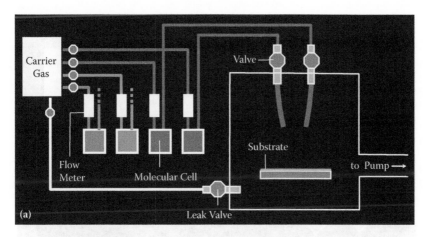

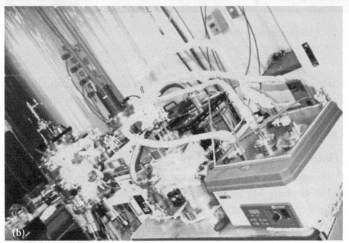

FIGURE 3.6   Carrier gas–type MLD.

is near the wire edges. Contaminations such as remaining molecules used in the previous steps also prevent MLD from performing normal molecule-by-molecule growth. Therefore, the removal of contamination is essential in MLD of orientation-controlled long wires with designated locations.

The carrier gas–type MLD does not always need to be carried out in high vacuum since, as described previously, the carrier gas flow isolates the substrate surface from outside to maintain desirable reactions between reactive groups of the molecules on the surface and reactive groups of the supplied molecules.

The advantages for the carrier gas–type MLD can similarly be found in the carrier gas–type organic chemical vapor deposition (CVD), where carrier gas is employed to introduce two or more kinds of molecular gases onto a substrate surface simultaneously. The molecules are combined through chemical reactions on the surface. Although the growth mechanism of the carrier gas–type organic CVD is similar to

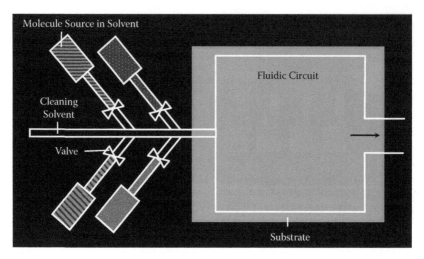

FIGURE 3.7   Fluidic circuit–type MLD.

that of vacuum deposition polymerization, the former has the advantage that clean surfaces and stable gas flows are easily maintained during growth.

3.2.2   LIQUID-PHASE MLD

3.2.2.1   Fluidic-Circuit Type MLD

Figure 3.7 shows a schematic illustration of the fluidic circuit–type MLD as an example of liquid-phase MLD. Source molecules are carried with solvent to be supplied on a surface. The molecules supplied to substrate surfaces are switched by valves. To ensure removal of remaining molecules, cleaning solvent injection during molecular switching is necessary. Fluidic processes using capillary circuits are effective to reduce the volume of solutions required for the MLD and solution switching duration. The liquid-phase MLD makes it easy to add catalysts for polymer wire or molecular wire growth.

It should be noted that liquid-phase MLD can be performed in human bodies for medical applications. In this case, for example, the human body corresponds to the chamber and the substrates correspond to cancer cells, and molecules are introduced from outside of the bodies by injection. By introducing different kinds of molecules into the bodies with designated sequences by injections, polymer wires or molecular wires will be grown on the cancer cells.

3.3   PROOF OF CONCEPT OF MLD

3.3.1   MLD UTILIZING CHEMICAL REACTIONS

MLD utilizing chemical reactions can be applied to a variety of polymers with covalent bonds arising from various kinds of reactive groups such as carbonyl-oxy-carbonyl groups, amino groups, −CHO groups, epoxy ring, −NCO groups, and so on. In

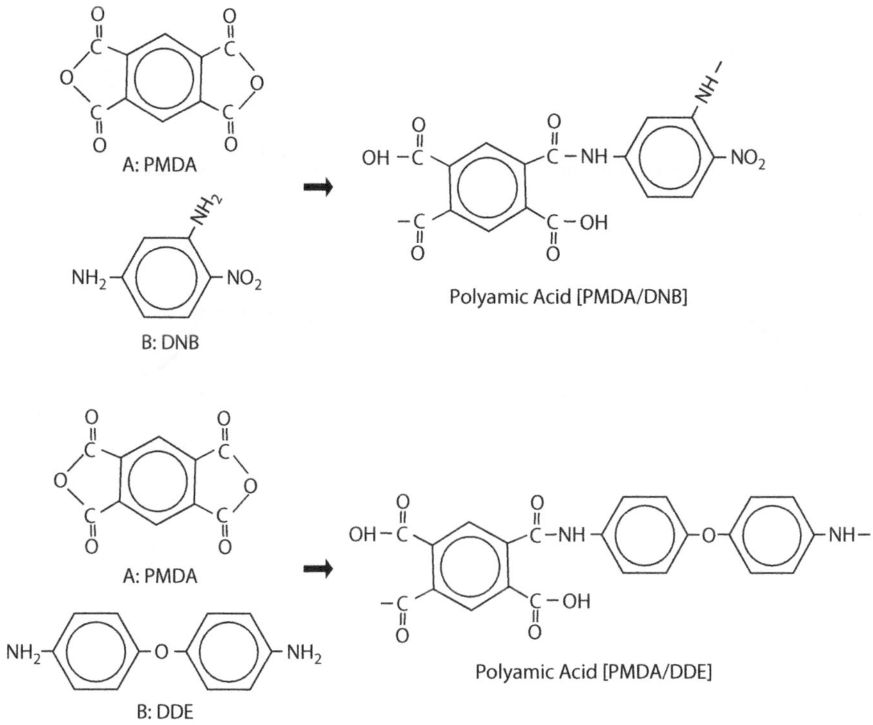

FIGURE 3.8 Molecules and reactions for polyamic acid wire growth.

the present section, in order to present proofs of concept of MLD for polymer wires, MLD of polyimide and polyazomethine (poly-AM) is shown using the K cell–type MLD.

3.3.1.1 Polyimide

To demonstrate MLD, growth dynamics of polymer wires on a surface were investigated using molecule A and molecule B [1]. For the experiments, two sets of combinations of source molecules were considered. One is a set of A: pyromellitic dianhydride (PMDA) and B: 2,4-diaminonitrobenzene (DNB), and the other is a set of A: PMDA and B: 4,4'-diaminodiphenyl ether (DDE). The molecular structures and reactions are shown in Figure 3.8 [11,12]. The molecules are connected by reactions between the carbonyl-oxy-carbonyl group in molecule A and the amino group in molecule B to form alternately connected polymer wires. The reactions between carbonyl-oxy-carbonyl and carbonyl-oxy-carbonyl, and between amino and amino are weak, forming no bonds between molecules. From PMDA and DNB, polymer wires of polyamic acid [PMDA/DNB] are formed, and from PMDA and DDE, polymer wires of polyamic acid [PMDA/DDE]. By heating the polyamic acid in vacuum, polyimide is obtained.

Molecules A and B were evaporated separately from temperature-controlled K cells. The supplied molecules were selected by shutters on the K cells. The background

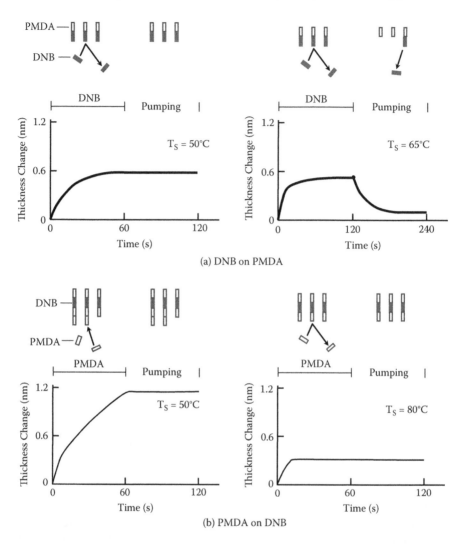

FIGURE 3.9   Thickness change versus shutter operations in MLD of polyamic acid [PMDA/DNB]. From T. Yoshimura, S. Tatsuura, and W. Sotoyama, "Polymer films formed with monolayer growth steps by molecular layer deposition," *Appl. Phys. Lett.* **59**, 482–484 (1991).

pressure was 5×10^{-6} Pa and molecular gas pressures $1–5 \times 10^{-3}$ Pa. The thickness of deposited films was measured using a quartz oscillator thickness monitor. The monitor head contacted the heated holder and was secured by an aluminum belt. The temperature was measured using a thermocouple on the holder. It is suspected that the actual temperature of the thickness monitor head was lower than that displayed. In this report, we used the displayed value as substrate temperature T_s for the sake of convenience.

Figure 3.9(a) shows the thickness change versus shutter operations for DNB growth on PMDA. Here, shutter A is for PMDA and shutter B for DNB. First, PMDA molecules are introduced on the surface by opening shutter A to form the PMDA layer. For duration t_d, both shutters A and B are closed. DNB molecules are then supplied onto the PMDA layer by opening shutter B. For $T_s = 50°C$, the film thickness rapidly increases and saturates in 40 s after shutter B opens. Thickness is maintained after shutter B closes. The change in thickness is 0.6 nm, close to the molecular size of DNB. The time dependence of the change in thickness is not affected by t_d from 5 to 60 s, which excludes the possibility that the thickness changes due to PMDA molecules remaining just after shutter A is closed.

These results suggest that a monomolecular step growth of DNB is achieved on the PMDA as follows. The binding energy between PMDA and DNB, that is, between carbonyl-oxy-carbonyl and amino, is larger than that between DNB and DNB (amino and amino). So the DNB molecules can be connected to the PMDA molecules. Once the surface is covered with DNB molecules, additional DNB molecules cannot be connected onto the surface to terminate the growth automatically.

When DNB is supplied to the PMDA layer at $T_s = 65°C$, a similar thickness increase and saturation as at $T_s = 50°C$ is observed. When both shutters are closed after DNB layer growth, however, the film thickness decreases to the initial value. This means that DNB molecules re-evaporate from the surface before strong bonds between DNB molecules and PMDA molecules are formed. When we consider that the average time for DNB to remain at the surface corresponds to the time constant of the thickness decay, the average time is about 30 s at $T_s = 65°C$.

Figure 3.9(b) shows the thickness change versus shutter operations for PMDA growth on DNB. When PMDA is supplied to a DNB layer at $T_s = 50°C$, an initial rapid increase in thickness is observed. Thereafter, the thickness continues to gradually increase and does not saturate. This indicates that PMDA molecules condense on the surface to form a multimolecular layer. When the substrate temperature is raised to 80°C, the thickness saturates, indicating that PMDA condensation is avoided at this temperature and mono molecular step growth of PMDA on DNB is achieved.

As described, the optimum temperature for monomolecular step growth differs for DNB on PMDA and for PMDA on DNB. This may be due to the difference in vapor pressure between PMDA and DNB. To fabricate polymer wires by repeatedly stacking with monomolecular steps, the substrate temperature must be elevated synchronously with molecule switching.

Figure 3.10 shows an example of MLD using PMDA for molecule A and DDE for molecule B. Mono molecular step growth is observed at the same temperature of $T_s = 80°C$ for both DDE on PMDA and PMDA on DDE. Therefore, MLD can be done without elevating the substrate temperature. This might be attributed to the fact that vapor pressures of PMDA and DDE is not largely different. A 10-nm-thick polymer film of polyamic acid was deposited by MLD with 15 steps of alternate molecule supply. By annealing in vacuum, the polyamic acid is converted into polyimide, as shown in Figure 3.11.

It should be noted that recently the gas-phase MLD utilizing chemical reactions was drastically advanced. Putkonen et al. succeeded in growing thick polyimide films of several hundred nm by MLD with over 500 steps [13]. Salmi et al. reported inorganic-organic nanolaminates consisting of polyimide and Ta_2O_5 that

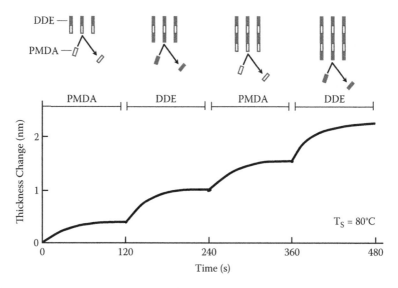

FIGURE 3.10 Thickness change versus shutter operations in MLD of polyamic acid [PMDA/DDE]. From T. Yoshimura, S. Tatsuura, and W. Sotoyama, "Polymer films formed with monolayer growth steps by molecular layer deposition," *Appl. Phys. Lett.* **59**, 482–484 (1991).

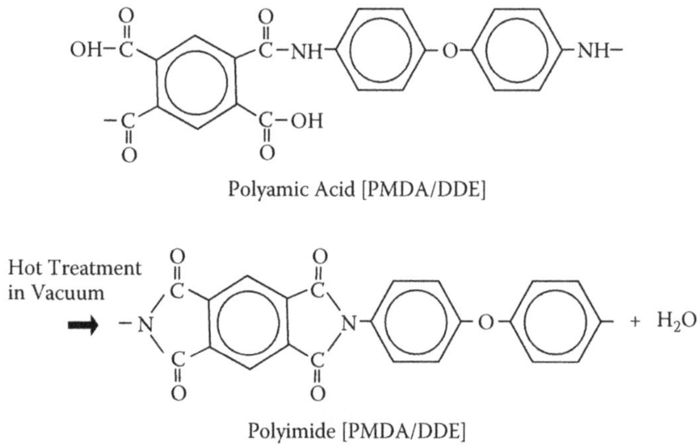

FIGURE 3.11 A conversion reaction from polyamic acid to polyimide.

are applicable to photonic gratings [14]. George et al. clarified surface reactions for MLD and developed new approaches in MLD [15], including the three-step ABC process using ring-opening and heterobifunctional reactants. The process suppresses double reactions, enabling long-wire growth. Liang and Weimer achieved the conformal coating of ultrathin aluminum alkoxide (alucone) polymer films on silica and titania nanoparticles as well as fabrication of porous films having voids with designed sizes [16]. Loscutoff et al. applied rigid and thermally stable nano-scale oligourea films, in which hydrogen bonds exist between neighboring oligomer wires, grown by MLD to copper diffusion barriers in nanoelectronics [17].

3.3.1.2 Conjugated Polymers

MLD can be applied to molecule-by-molecule growth of conjugated polymers. As shown in Figure 3.12, by using terephthalaldehyde (TPA) for molecule A and *p*-phenylenediamine (PPDA) for molecule B, conjugated polymers of poly-AM [TPA/PPDA] are grown. TPA and PPDA react through –CHO in TPA and –NH$_2$ in PPDA to form –C=N– inducing π-conjugated systems [12]. When oxalic acid (OA) and oxalic dihydrazide (ODH) react, conjugated polymers of poly-oxadiazole (OXD) are synthesized with the assistance of heating in a vacuum [18].

We demonstrated MLD of poly-AM using TPA and PPDA [19]. Figure 3.13 shows the results. When PPDA is provided on TPA with a molecular gas pressure of 1–8 x 10^{-2} Pa, steplike growth of ~0.7 nm that is close to the molecule size is observed at a substrate temperature $T_S = 25°C$. After removing PPDA from the vacuum chamber, by providing TPA on PPDA existing at the top surface with a molecular gas pressure of 10^{-1} Pa or more, steplike growth is observed again. These results indicate that monomolecular step growth with self-limiting is achieved by MLD. It is also observed that the thickness is proportional to the molecular gas switching count, just like ALD. The molecule-by-molecule growth characteristics of MLD enable us to obtain desired properties of polymers with controlled π-electron wavefunction shapes.

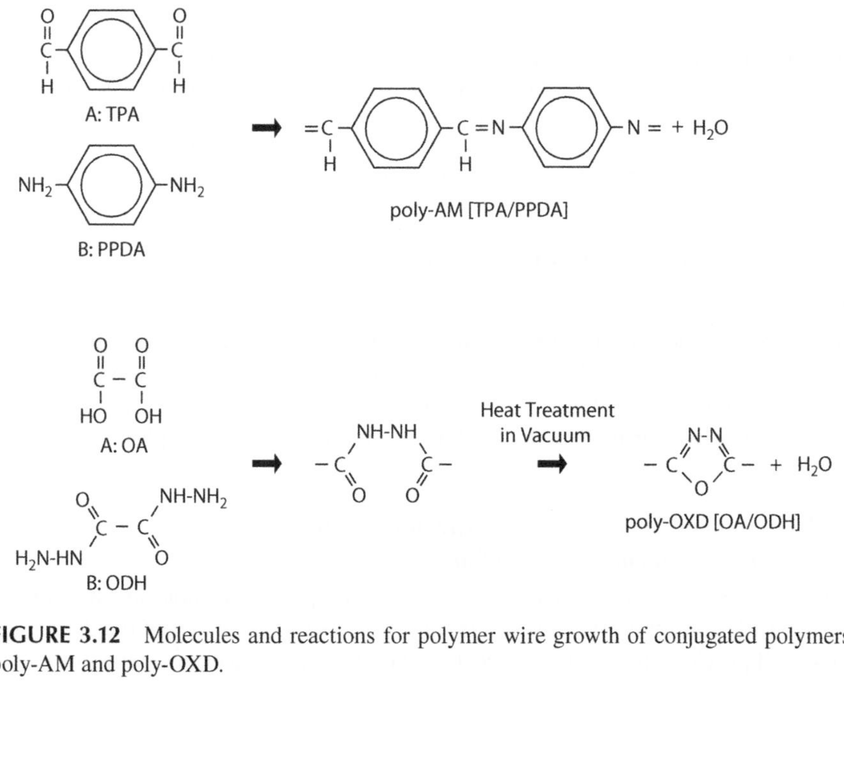

FIGURE 3.12 Molecules and reactions for polymer wire growth of conjugated polymers: poly-AM and poly-OXD.

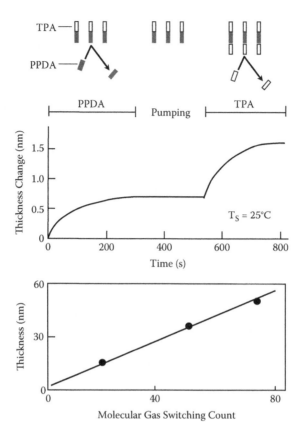

FIGURE 3.13 (a) Thickness change versus shutter operations, and (b) thickness versus molecular gas switching count characteristics in MLD of poly-AM [TPA/PPDA]. From T. Yoshimura, S. Tatsuura, W. Sotoyama, A. Matsuura, and T. Hayano, "Quantum wire and dot formation by chemical vapor deposition and molecular layer deposition of one-dimensional conjugated polymer," *Appl. Phys. Lett.* **60**, 268–270 (1992).

3.3.2 MLD Utilizing Electrostatic Force

MLD can be applied to molecular wires, in which different kinds of molecules are aligned by the electrostatic force with designated sequences. In this subsection, as a proof of concept for MLD utilizing electrostatic force, formation of stacked structures of p-type and n-type dye molecules on ZnO surfaces is presented. In addition, MLD in molecular crystals of charge-transfer complexes is briefly described.

3.3.2.1 Stacked Structures of p-Type and n-Type Dye Molecules on ZnO Surfaces

In the 1970s, K. Kiyota et al. were able to self-assemble p-type dye molecules and n-type dye molecules on semiconductor surfaces [20]. The self-assembling of the dye molecules itself provides the proof of concept for the MLD utilizing electrostatic force.

To analyze the molecular assembling structures on semiconductor surfaces, surface potential measurements were used [21]. Samples used for the measurement were prepared as follows. For the semiconductor, a ZnO powder layer was spread on a glass substrate with a SnO_2 electrode. For the first dye molecule adsorption, 0.01 mL of an alcohol solution of the first dye molecule with a concentration of 1.6 mol/L was dropped on the ZnO layer (3 cm^2 area). For the second dye molecule adsorption, an alcohol solution of the second dye molecule was similarly dropped on the surface, where the first dye molecules were already adsorbed. Structures of various p-type and n-type dye molecules are illustrated in Figure 3.14. In the present experiment, rose bengal (RB) and bromphenol blue (BPB) were used for the p-type dye molecules, and malachite green (MG) and crystal violet (CV) for the n-type dye molecules.

Figure 3.15 shows the surface potential of a ZnO layer and ZnO layers adsorbing the first dye molecules. The surface potential of the ZnO layer without dye molecule adsorption is about −200 mV. When p-type dye molecules are adsorbed on the ZnO layer, the surface potential becomes more negative than the ZnO layer without dye molecule adsorption. Conversely, when n-type dye molecules are adsorbed, the surface potential becomes less negative.

This result can be explained as follows. In ZnO, zinc atoms in interstitial positions donate their electrons to oxygen adsorbed on the ZnO surface to generate electric dipoles, causing a negative surface potential for the ZnO layer. When p-type dye molecules are adsorbed on the ZnO surface, electrons in the ZnO layer are transferred to the dye molecules to generate additional electric dipoles with the same direction as that of the dipoles arising from the oxygen, enhancing the total electric dipole moment at the ZnO surface. In the case of an n-type dye molecule, conversely, the electrons in the dye molecules are transferred to the ZnO layer to generate electric dipoles with opposite direction, which suppresses the total electric dipole moment at the ZnO surface. Kokado et al. found that conductivity of the ZnO layer was raised by adsorption of n-type dye molecules and reduced by adsorption of p-type dye molecules, and they attributed the phenomenon to the transfer of electrons between the dye molecules and the ZnO [22]. The results of the surface potential measurements are consistent with their findings.

Figure 3.16 shows the surface potential of ZnO layers adsorbing the first dye molecules and the second dye molecules successively. Here, MG→RB indicates that the first dye molecule is MG and the second dye molecule is RB. As can be seen in the figure, variation of the surface potential arising from the adsorption of two dye molecules mainly depends on the second dye molecule. This indicates that the two-dye-molecule-stacked structure is definitely constructed and the surface potential is determined by differences in Fermi levels between the second dye molecules (the top molecules) and the ZnO layer.

The surface potential of the samples, in which p-type and n-type dye molecules are coadsorbed on the ZnO layer simultaneously, is also shown in Figure 3.16. RB + MG indicates the coadsorption of RB and MG. For the coadsorption of p-type and n-type dye molecules, 0.01 mL of an alcohol solution containing both p-type dye molecules and n-type dye molecules at equal concentrations of 1.6 mol/L is dropped on the ZnO layer. The surface potential of the ZnO layer coadsorbing the p-type and n-type dye molecules is very similar to that of the ZnO layer adsorbing p-type

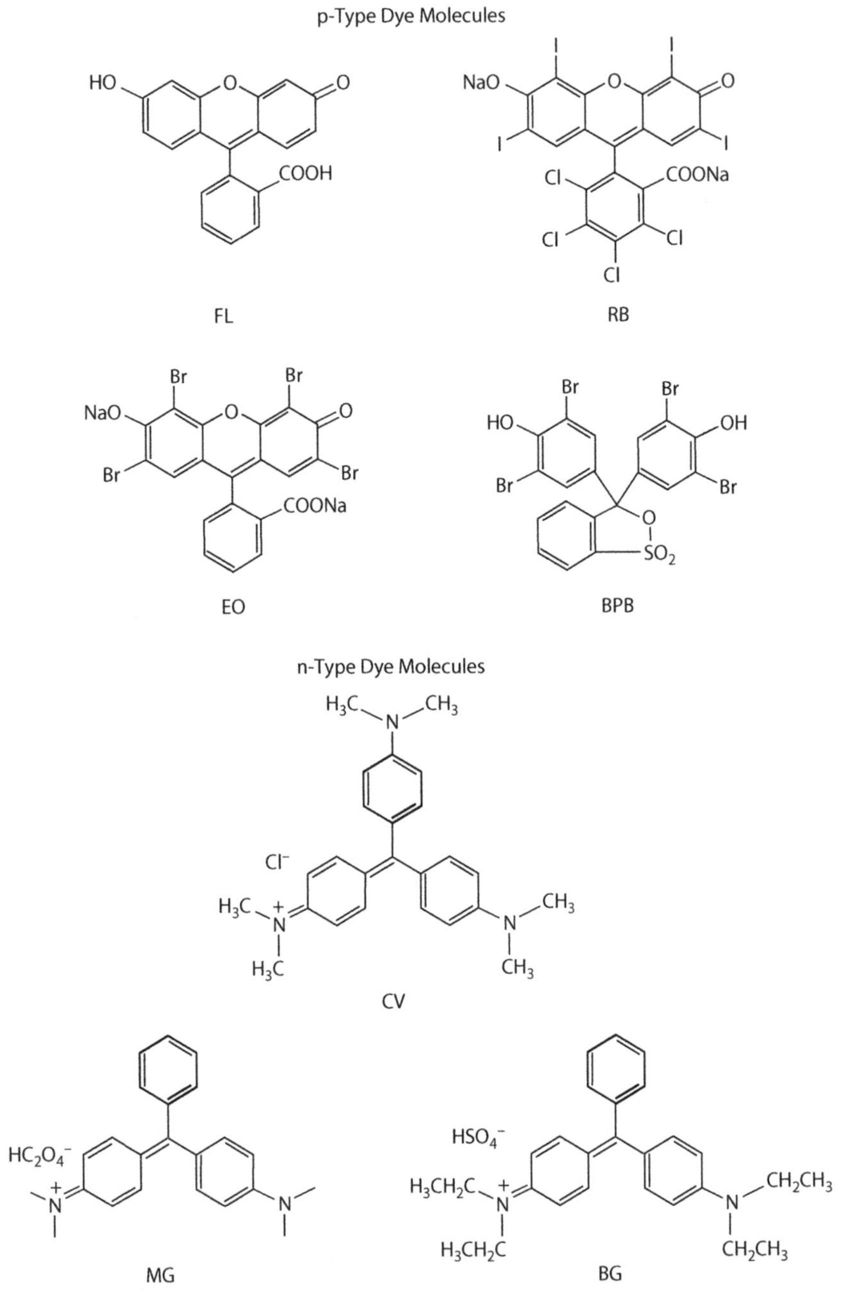

FIGURE 3.14 Structures of p-type and n-type dye molecules. Fluorescein: FL, Rose bengal: RB, Eosine: EO, Bromphenol blue: BPB, Crystal violet: CV, Malachite green: MG, and Brilliant green: BG.

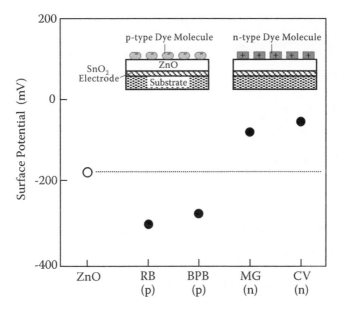

FIGURE 3.15 Surface potential of a ZnO layer and ZnO layers adsorbing dye molecules. From T. Yoshimura, K. Kiyota, H. Ueda, and M. Tanaka, "Contact potential difference of ZnO layer adsorbing p-type dye and n-type dye," *Jpn. J. Appl. Phys.* **18**, 2315–2316 (1979).

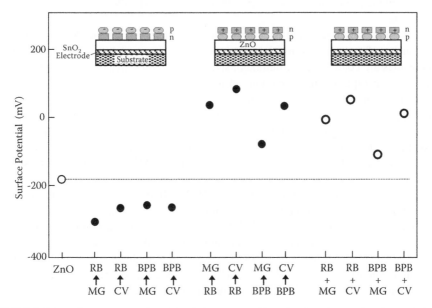

FIGURE 3.16 Surface potential of ZnO layers adsorbing the first dye molecules and the second dye molecules successively, and surface potential of ZnO layer coadsorbing p-type dye molecules and n-type dye molecules simultaneously. From T. Yoshimura, K. Kiyota, H. Ueda, and M. Tanaka, "Contact potential difference of ZnO layer adsorbing p-type dye and n-type dye," *Jpn. J. Appl. Phys.* **18**, 2315–2316 (1979).

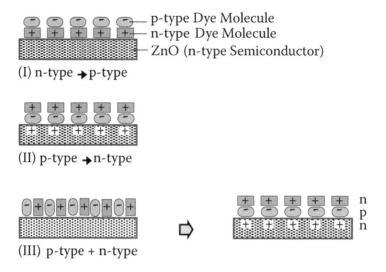

(I) n-type → p-type

(II) p-type → n-type

(III) p-type + n-type

FIGURE 3.17 Assembled structures of p-type and n-type dye molecules stacked on ZnO surfaces.

dye molecules as the first dye molecule and n-type dye molecules as the second dye molecule. This suggests that p-type dye molecules and n-type dye molecules are self-assembled to form a stacked structure of a ZnO/p-type dye molecule/n-type dye molecule.

In Figure 3.17, assembled structures of p-type and n-type dye molecules stacked on ZnO surfaces are summarized. When p-type dye molecules and n-type dye molecules are coadsorbed simultaneously, dye molecules are self-assembled in the order of a p-type dye molecule and an n-type dye molecule on an n-type ZnO to form npn structures. The self-assembling arises from electron transfers from ZnO to p-type dye molecules and from n-type dye molecules to p-type dye molecules, resulting in a strong connection due to electrostatic force between ZnO and p-type dye molecules, and n-type dye molecules and p-type dye molecules. The self-assembling effect strongly suggests the viability of MLD utilizing electrostatic force.

Actual experimental demonstration of MLD utilizing electrostatic force is presented in Chapter 9, Section 9.1.3, where, multidye sensitization is described.

The concept of the self-assembled structures of the p-type and n-type molecules provides the base for the self-assembled monolayers (SAM) developed in 1980s. As Figure 3.18 shows, in the case of SAM, two different molecules are combined into one molecule to achieve the self-assembling. The SAM is currently applied to surface treatment for various kinds of organic thin-film devices.

3.3.2.2 Molecular Crystals

It is known that tetrathiafulvalene (TTF) and tetracyanoquinodimethane (TCNQ) combine through charge transfers between them to form molecular crystals of charge transfer complexes. From this property, it is expected that MLD utilizing

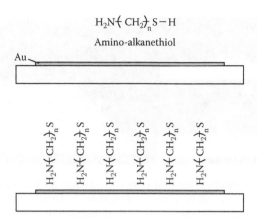

FIGURE 3.18   Self-assembled monolayer.

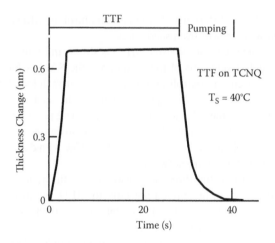

FIGURE 3.19   MLD of TTF on TCNQ.

electrostatic force might be accomplished using these molecules. Figure 3.19 shows MLD of TTF on TCNQ. As in the PMDA/DNB, PMDA/DDE, and TPA/PPDA systems, monomolecular adsorption occurs. TTF re-evaporates when shutter A is closed, however, suggesting that the connecting force between TTF and TCNQ is not sufficient. Since the time constant for thickness decay increases with decreasing substrate temperature, by optimizing the substrate temperature, we might be able to apply MLD to charge transfer complexes.

3.4   MLD WITH CONTROLLED GROWTH ORIENTATIONS AND LOCATIONS

In order to construct thin-film organic photonic/electronic devices and molecular nano systems consisting of self-organized polymer/molecular wire networks, three-

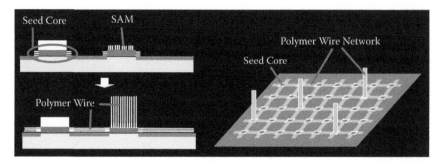

FIGURE 3.20 The concept of the seed-core-assisted MLD for three-dimensional growth. From T. Yoshimura, Y. Suzuki, N. Shimoda, T. Kofudo, K. Okada, Y. Arai, and K. Asama, "Three-dimensional chip-scale optical interconnects and switches with self-organized wiring based on device-embedded waveguide films and molecular nanotechnologies," *Proc. SPIE* **6126**, 612609-1-15 (2006).

dimensional growth with location and orientation control by MLD is required. To do this we proposed wire growth from seed cores.

The concept of the seed-core-assisted MLD is shown in Figure 3.20 [3,4,23]. On the top surfaces and/or sidewalls of the seed cores, the SAM is formed selectively. The seed cores are used for controlling polymer wire growth locations and orientations. For vertical growth, the SAM is put on the top of the seed cores, while for horizontal growth it is put on the sidewalls. The regions, where polymer wires should not grow are covered with, for example, SiO_2 films. By distributing the seed cores with designated patterns, polymer wires are expected to grow with designated configurations, constructing polymer wire networks.

It should be noted that the three-dimensional growth orientation control technique using seed cores can be applied not only to MLD, but also to other growth methods such as vacuum deposition polymerization and organic CVD.

In this section, using the carrier gas–type MLD, we demonstrate vertical growth of poly-AM wires from seed cores with amino-alkanethiol SAM on Au films [9,10].

3.4.1 GROWTH CONTROL BY SEED CORES

3.4.1.1 MLD from SAM

Wires of poly-AM, a conjugated polymer, were grown from seed cores by MLD using source molecules of TPA and PPDA. The carrier gas was nitrogen (N_2), and the molecular cell temperature was 50°C. Figure 3.21 shows the schematic illustration for the process of molecule-by-molecule growth of poly-AM wires by the carrier gas–type MLD. In step 0, in order to anchor seed core molecules to a substrate surface, a SAM of amino-alkanethiol (11-amino-1-undecanethiol) was formed on patterned Au films deposited on a glass substrate. The SAM formation was carried out by putting the substrate in an ethanol solution of 11-amino-1-undecanethiol, hydrochloride, and then rinsing and drying it. In step 1, TPA molecules were provided onto the surface by opening the valve for TPA. The surface was exposed to TPA molecular gas for 5 min, to connect TPA to the SAM. In step 2, after removing TPA molecular gas by

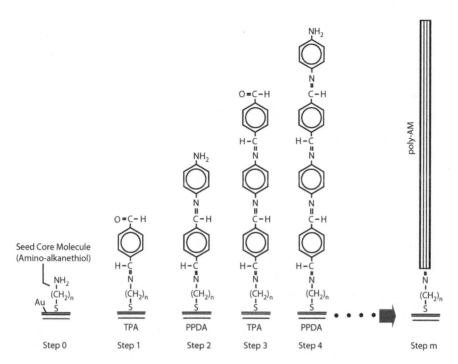

FIGURE 3.21 Process of molecule-by-molecule growth of poly-AM wires. From T. Yoshimura, S. Ito, T. Nakayama, and K. Matsumoto, "Orientation-controlled molecule-by-molecule polymer wire growth by the carrier-gas-type organic chemical vapor deposition and the molecular layer deposition," *Appl. Phys. Lett.* **91**, 033103-1-3 (2007).

closing the valve for TPA and pumping for 5 min, the valve for PPDA was opened. The surface was exposed to PPDA molecular gas for 5 min, to connect PPDA to TPA. TPA and PPDA were alternately introduced onto the surface by switching the valves, reaching step m. In the present case, TPA and PPDA were respectively provided onto the surface 10 times, that is, m = 20. Thus, poly-AM grown on a substrate with SAM was obtained. For the purpose of comparison, poly-AM grown on a substrate without SAM was also prepared by the same procedure described above.

Figure 3.22 shows the Fourier transform infrared reflection absorption spectroscopy (FTIR-RAS) spectra of TPA-adsorbed surfaces (step 1). The absorption peaks attributed to TPA in wavenumber regions around 820 cm^{-1} and 780 cm^{-1} are larger for the sample that was prepared using a surface with SAM (denoted by "SAM surface") than those for the sample that was prepared using a surface without SAM (denoted by "non-SAM surface"). When the sample temperature was raised to 45°C, the peak height in the non-SAM surface decreased while that in the SAM surface did not decrease. These results indicate that more TPA molecules exist on the surface with stronger adsorption strengths in the case of the SAM surface than in the case of the non-SAM surface, suggesting that TPA–SAM connections are constructed in the SAM surface.

Figure 3.23 shows the FTIR-RAS spectra of poly-AM grown by the carrier gas–type MLD (step m = 20). It can be seen that both the SAM surface and the non-SAM

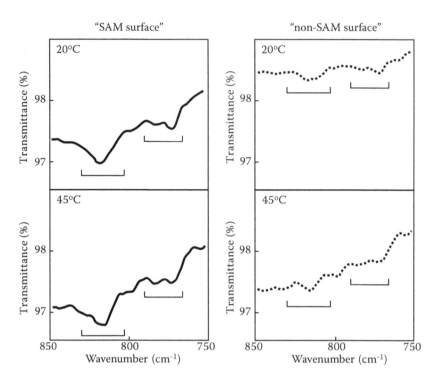

FIGURE 3.22   FTIR-RAS spectra of TPA-adsorbed surfaces (step 1). From T. Yoshimura, S. Ito, T. Nakayama, and K. Matsumoto, "Orientation-controlled molecule-by-molecule polymer wire growth by the carrier-gas-type organic chemical vapor deposition and the molecular layer deposition," *Appl. Phys. Lett.* **91**, 033103-1-3 (2007).

surface exhibit absorption peaks in wavenumber regions around 1200 to 1300 cm^{-1} and 850 cm^{-1}. The former and the latter peaks are attributed respectively to the vibration of in-plane direction (denoted by "in-plane") and the vibration of surface-normal direction (denoted by "surface-normal") in benzene rings. The absorbance ratio of in-plane to surface-normal ($I_{\text{In-Plane}}/I_{\text{Surface-Normal}}$) is larger by a factor of about 2 in the SAM surface than in the non-SAM surface. This suggests that polymer wires tend to grow along nearly vertical directions in the SAM surface as the model shown in Figure 3.21 while wires tend to grow along nearly horizontal directions in the non-SAM surface.

3.4.1.2   Organic CVD from SAM

In order to confirm the poly-AM wire growth direction described above, we carried out polymer wire growth by the carrier gas–type organic CVD on SAM surfaces and non-SAM surfaces [9]. Molecular cell temperature $T_{\text{Molecular Cell}}$ and carrier gas flow rate R_{Flow} were 50°C and 4 NL/min, respectively. The results are shown in Figure 3.24. In the case of the non-SAM surface, film thickness increases at the initial stage, followed by saturation. On the other hand, film thickness increases linearly with growth time in the case of the SAM surface. These results can be explained in terms of reaction site densities on surfaces.

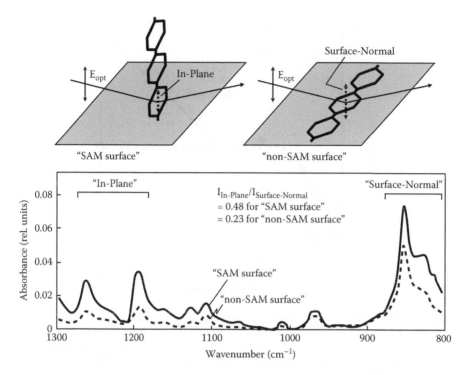

FIGURE 3.23 FTIR-RAS spectra of poly-AM grown by the carrier gas–type MLD (step m = 20). From T. Yoshimura, S. Ito, T. Nakayama, and K. Matsumoto, "Orientation-controlled molecule-by-molecule polymer wire growth by the carrier-gas-type organic chemical vapor deposition and the molecular layer deposition," *Appl. Phys. Lett.* **91**, 033103-1-3 (2007).

As shown in the model illustrated in Figure 3.25, TPA and PPDA molecules adsorb on an Au-coated substrate horizontally in the case of the non-SAM surface, resulting in horizontal poly-AM wire growth. Since the –CHO group and the –NH₂ group are hydrophilic, TPA and PPDA molecules are preferably adsorbed to –CHO and –NH₂ groups at the edge of poly-AM wires while they do not adsorb to the other parts of poly-AM wires that are hydrophobic. The density of hydrophilic sites that exist at the edge of the poly-AM wires decreases as poly-AM wires grow, and then the growth rate of poly-AM films decreases, resulting in the film thickness saturation. Here, although CH=N, which appears at each unit of body parts of the wires, has H-bond acceptor characteristics, the polarity of CH=N is smaller compared to that of NH₂ and CHO. So, the model shown in Figure 3.25 is still viable as the 0th approximation.

As Figure 3.26 shows, the contact angle of a water droplet is 44.6° on a SiO₂ surface. The angle increases after poly-AM is grown on the surface. The contact angles of water droplets on the surfaces increase with growth time followed by saturation. This indicates that the surface changes from hydrophilic states to hydrophobic states, confirming the reduction of the hydrophilic site density at the surface.

In the case of the SAM surface, on the other hand, TPA and PPDA molecules adsorb on the substrate vertically. So, poly-AM wires grow vertically, keeping the

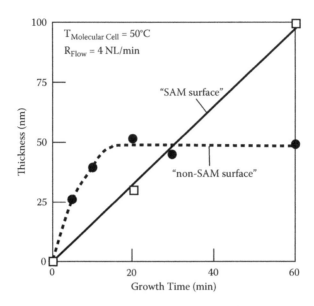

FIGURE 3.24 Growth time dependence of the poly-AM film thickness in the carrier gas–type organic CVD. From T. Yoshimura, S. Ito, T. Nakayama, and K. Matsumoto, "Orientation-controlled molecule-by-molecule polymer wire growth by the carrier-gas-type organic chemical vapor deposition and the molecular layer deposition," *Appl. Phys. Lett.* **91**, 033103-1-3 (2007).

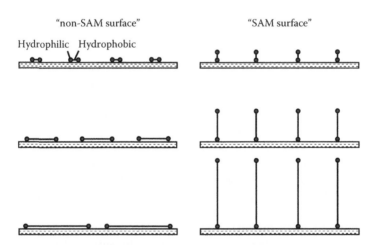

FIGURE 3.25 Model for poly-AM growth by the carrier gas–type organic CVD. From T. Yoshimura, S. Ito, T. Nakayama, and K. Matsumoto, "Orientation-controlled molecule-by-molecule polymer wire growth by the carrier-gas-type organic chemical vapor deposition and the molecular layer deposition," *Appl. Phys. Lett.* **91**, 033103-1-3 (2007).

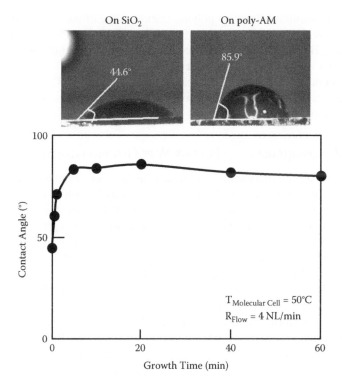

FIGURE 3.26 Growth time dependence of the contact angle of water droplets on the surfaces. From T. Yoshimura, S. Ito, T. Nakayama, and K. Matsumoto, "Orientation-controlled molecule-by-molecule polymer wire growth by the carrier-gas-type organic chemical vapor deposition and the molecular layer deposition," *Appl. Phys. Lett.* **91**, 033103-1-3 (2007).

density of the hydrophilic sites constant. Therefore, poly-AM film thickness does not exhibit saturation.

It is clearly indicated from these results that nearly vertical molecule-by-molecule growth of conjugated polymer wires can be achieved on substrates with SAM by the carrier gas–type MLD. This enables three-dimensional self-aligned growth by using seed cores, which is ultrasmall, for example, nm-scale Au blocks with SAM on the top surfaces and/or sidewalls. The seed cores are distributed with designed configurations on a substrate. Polymer wires are expected to grow vertically from the top surfaces and horizontally from the sidewalls to form polymer wire networks by providing molecular gases to the substrate.

It should be noted that, in the vacuum deposition polymerization, film thickness of poly-AM increases with growth time without saturation for the non-SAM surface. The reason why the growth saturation that occurs in the carrier gas–type organic CVD does not occur in vacuum deposition polymerization can be explained in terms of the previously described advantage (1) of the carrier gas–type organic CVD. During growth of poly-AM, H_2O molecules are generated as by-products.

In the case of vacuum deposition polymerization, some of the generated H_2O molecules may weakly adsorb on the surface. Then, TPA and PPDA molecules can adsorb to the H_2O molecules to initiate poly-AM growth from the adsorbed molecules. In other words, H_2O molecules act as growth cores. Therefore, growth saturation does not occur. In the case of the carrier gas–type organic CVD, the generated H_2O molecules on the surface are blown away by carrier gas molecules to prevent the growth initiation.

3.4.2 MONOMOLECULAR STEP POLYMER WIRE GROWTH FROM SEED CORES

Since required polymer/molecular wire lengths range from ~1 nm to several hundred nm depending on the application, it is important to clarify the polymer wire lengths that can be grown with perfect monomolecular steps by MLD. Using the carrier gas–type MLD, we investigated precise processes of monomolecular step growth of poly-AM wires from seed core molecules of amino-alkanethiol SAMs on Au film [10].

The MLD process is the same as that shown in Figure 3.21. A simulation based on the molecular orbital method (Fujitsu WinMOPACK) reveals that a poly-AM wire grows from the seed core molecule with a slight helix waving, as Figure 3.27 shows.

In the present experiment, the maximum step count was 18. The carrier gas–type MLD was performed in the following conditions: substrate temperature, 25°C; molecular cell temperatures, 25°C and 50°C for TPA and PPDA, respectively; blowing duration of molecular gases onto the surface, 1 min or 5 min; pumping duration for removing residual molecular gases, 5 min; and carrier gas flow rate, 4 NL/min.

In Figure 3.28, FTIR-RAS spectra of poly-AM wires grown by the carrier gas–type MLD with a molecular gas blowing duration of 5 min are shown for step 6.

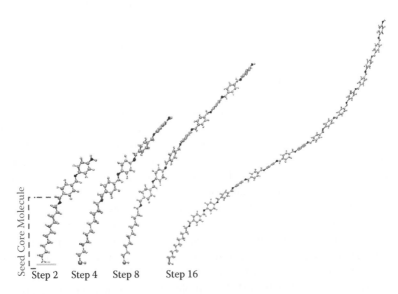

FIGURE 3.27   Simulation of poly-AM wire growth by the molecular orbital method.

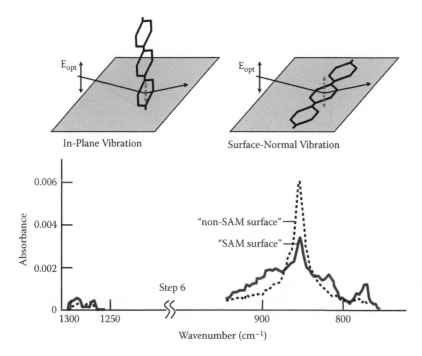

FIGURE 3.28 FTIR-RAS spectra of poly-AM wires grown by the carrier gas–type MLD with a molecular gas blowing duration of 5 min. From T. Yoshimura and Y. Kudo, "Monomolecular-step polymer wire growth from seed core molecules by the carrier-gas type molecular layer deposition (MLD)," *Appl. Phys. Express* **2**, 015502-1-3 (2009).

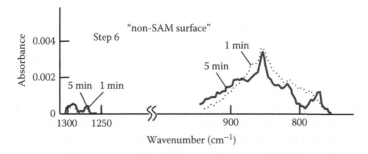

FIGURE 3.29 FTIR-RAS spectra of poly-AM wires grown by the carrier gas–type MLD with a molecular gas blowing duration of 1 min. From T. Yoshimura and Y. Kudo, "Monomolecular-step polymer wire growth from seed core molecules by the carrier-gas type molecular layer deposition (MLD)," *Appl. Phys. Express* **2**, 015502-1-3 (2009).

As described in Section 3.4.1, the absorbance ratio of $I_{\text{In-Plane}}/I_{\text{Surface-Normal}}$ is larger in the SAM surface than in the non-SAM surface, indicating that poly-AM wires tend to grow in upward directions in the SAM surface, while the wires tend to grow in horizontal directions in the non-SAM surface. It can be seen from Figure 3.29 that absorption spectra for a blowing duration of 1 min are almost the same as those for

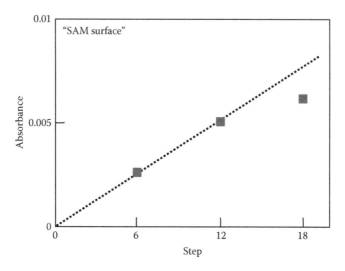

FIGURE 3.30   Step count dependence of peak absorbance in the region around 850 cm^{-1} for poly-AM wires of SAM surface. From T. Yoshimura and Y. Kudo, "Monomolecular-step polymer wire growth from seed core molecules by the carrier-gas type molecular layer deposition (MLD)," *Appl. Phys. Express* **2**, 015502-1-3 (2009).

a blowing duration of 5 min, suggesting that a molecular gas blowing duration of 1 min is enough to complete the self-limiting monomolecular growth.

Figure 3.30 shows the step count dependence of peak absorbance in the region around 850 cm^{-1} for poly-AM wires of the SAM surface. Until step 12, the absorbance linearly increases with step count. This indicates that perfect monomolecular step growth is achieved in a step count range from 1 to 12, which approximately corresponds to wire lengths of 0.5 to ~6 nm. In step 18, however, the absorbance deviates from the linear relationship, implying that the MLD process causes growth imperfection in long wires.

Although the reason for the deviation of the absorbance from the linear relationship is not clear at present, one possible reason is *double reactions* at the top surface during growth, as suggested by George [24]. The double reactions, in which two reactive groups of a molecule provided on the surface make chemical bonds with top reactive groups of two polymer wires simultaneously, may remove the active sites at the surface to reduce the MLD growth rate. Another possible reason is *tangling* of poly-AM wires, as indicated by Liskola [25]. The tangling that might be caused by the helix-shaped growth characteristics of poly-AM wires shown in Figure 3.27 may make the wire growth direction disordered.

In Figure 3.31, step count dependence of the absorbance ratio $I_{In\text{-}Plane}/I_{Surface\text{-}Normal}$ in poly-AM wires is shown. Until step 12, the absorbance ratio is larger in the SAM surface than in the non-SAM surface by a factor of 2~3. The result clearly indicates that, from step 1 to step 12, poly-AM wires tend to grow in upward directions in the SAM surface, while in the non-SAM surface, the wires tend to grow in horizontal

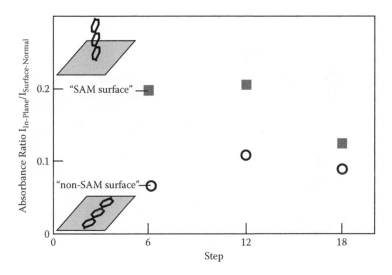

FIGURE 3.31 Step count dependence of absorbance ratio $I_{\text{In-Plane}}/I_{\text{Surface-Normal}}$ in poly-AM wires.

directions. In step 18, the absorbance ratio in the SAM surface decreases, getting close to the absorbance ratio in the non-SAM surface. This result suggests that the growth direction of the poly-AM wires of the SAM surface is tilted. Possible reasons for the tilt might be the previously described double reactions and tangling.

In order to achieve perfect monomolecular step upward growth from seed core molecules beyond the current limitation (~ step 12), precise optimization of the substrate temperature will be necessary. Iijima and Takahashi reported that 4,4'-diamino-diphenyl ether molecules adsorb on a surface vertically at a substrate temperature of 40°C while they adsorb horizontally at 10°C [26]. If TPA and PPDA molecules have the same tendency, namely, they adsorb on the surface vertically at high substrate temperature, it is expected that the double reactions can be suppressed by raising the substrate temperature because the probability that the source molecules react with two active sites at the surface simultaneously is reduced when the source molecules stand vertically. Gas flow rates, which are one of the important parameters in MLD, should also be optimized.

3.5 HIGH-RATE MLD

In conventional carrier gas–type MLD equipment, gas injection into a chamber and gas removal from the chamber are necessary for molecular gas switching, resulting in an increase in processing time. For example, in our conventional MLD equipment, it takes about 5 min for the gas injection and removal. In order to reduce the process time and increase the growth rate, the domain-isolated MLD was proposed [27].

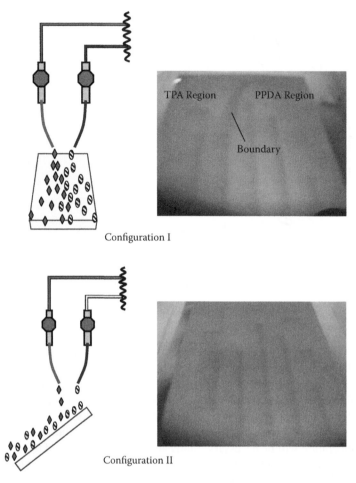

TPA Region   PPDA Region

Boundary

Configuration I

Configuration II

FIGURE 3.32 Influences of molecular gas flow on polymer film growth patterns in the carrier gas–type CVD. From K. Matsumoto and T. Yoshimura, "Electro-optic waveguides with conjugated polymer films fabricated by the carrier-gas-type organic CVD for chip-scale optical interconnects," *Proc. SPIE* **6899**, 98990E-1-9 (2008).

3.5.1 Influences of Molecular Gas Flow on Polymer Film Growth

It was found that carrier gas prevents outside gases from penetrating into the region in which the carrier gas flows. Figure 3.32 shows influences of molecular gas flow on polymer film growth patterns in the carrier gas–type CVD [28]. Two kinds of molecular gas flow configurations, I and II, were examined. In configuration I, where the substrate is tilted by 45° and TPA gas flow is parallel to PPDA gas flow, poly-AM grows only near the boundary between the TPA region and the PPDA region. The polymer is not deposited under the nozzles where only one kind of gas, that is, TPA or PPDA, exists. This indicates that source molecules cannot penetrate into the other regions, namely, the carrier gas blocks penetration of molecules from outside. Conversely, in configuration II, where TPA gas flow overlaps PPDA gas flow, poly-AM was formed in wide regions since the joining position of the two kinds of the molecular gases is spread over the substrate surface.

These results indicate a possibility that a plurality of molecular domains can coexist closely when the gas flow condition is adjusted.

3.5.2 DOMAIN-ISOLATED MLD

As described in Section 3.5.1, carrier gas blocks molecular penetration from outside when the gas flow configuration is appropriately adjusted. The unique characteristics enable us to develop the domain-isolated MLD, which enables high-rate MLD.

The *rotation type* domain-isolated MLD using two kinds of source molecules is schematically illustrated in Figure 3.33 (a). A nozzle for molecule A (denoted by A) and a nozzle for molecule B (denoted by B) are set on a substrate holder. Between the two nozzles, a nozzle for purge gas (denoted by P) is located. Substrates are put on the holder. By blowing gas containing molecules A, gas containing molecules B, and purge gas onto the substrates from these nozzles, a molecule A domain, a molecule B domain, and a purge gas curtain (PGC) are generated. the PGC prevents molecules in the two domains from mixing, and isolate the molecule A domain and the molecule B domain. By rotating the substrate holder, each substrate passes through the two domains alternately to achieve molecule-by-molecule growth of polymer wires by MLD.

Figure 3.33(b) is for the case using four kinds of source molecules. Nozzles for molecule A, molecule B, molecule C (denoted by C), and molecule D (denoted by D) are set around the center of the substrate holder. Between the four nozzles, nozzles for purge gas are located. By blowing gas containing molecule A, gas containing molecule B, gas containing molecule C, gas containing molecule D, and purge gas onto the substrates from these nozzles, a molecule A domain, a molecule B domain, a molecule C domain, a molecule D domain, and PGCs are generated. By rotating the substrate holder, each substrate passes through the four domains sequentially to achieve MLD.

Figure 3.34 shows *train-type* domain-isolated MLD using four kinds of source molecules. Nozzles for molecules A, B, C, and D are set in parallel to generate molecule A domains, molecule B domains, molecule C domains, and molecule D domains. Between these domains, PGCs are inserted. By moving substrates on a conveyor, each substrate sequentially passes through the domains to achieve MLD. When the gas blowing is switched by opening and closing the nozzle valves, arbitrary sequences of molecules are available. It is possible to bring the substrates back to the starting point to repeat MLD, which saves the number of required nozzles.

In order to examine the blocking ability of PGCs against molecular mixing, TPA/N_2 gas and $PPDA/N_2$ gas were simultaneously introduced onto a glass substrate in *rotation-type* MLD shown in Figure 3.33(a) without rotation. With increasing flow rates of N_2 for the PGC from 2.0 NL/min to 8.0 NL/min, light absorption due to poly-AM films deposited by TPA/PPDA mixing was drastically

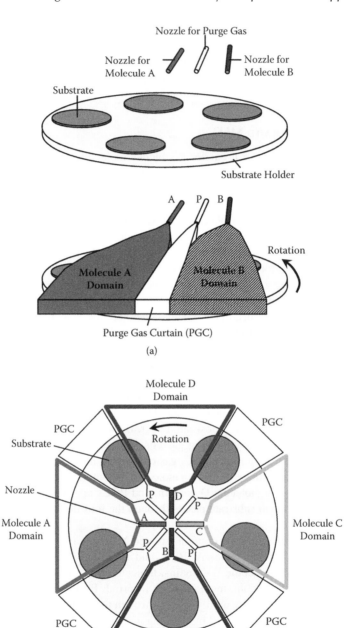

FIGURE 3.33 Domain-isolated MLD (Rotation Type). (a) For two kinds of source molecules and (b) for four kinds of source molecules.

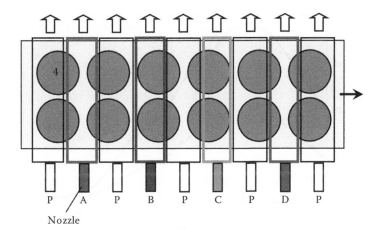

FIGURE 3.34 Domain-isolated MLD (Train Type).

suppressed, indicating that the PGC has enough blocking ability to isolate the molecular domains.

Using a PGC with 8.0 NL/min, domain-isolated MLD of poly-AM with rotation counts of 9 and 20 (step counts of 18 and 40) were attempted. The rotation speed of the spin stage was 0.25 rpm. As can be seen in Figure 3.35(a), the absorption peak of poly-AM is about 2 times larger for a step count of 40 than for a step count of 18, suggesting that the thickness of poly-AM films is approximately proportional to the step counts. The time to complete the 40-step growth is about 80 min, while it is about 200 min in the conventional MLD equipment. Therefore, the growth rate increases by 2.5 times in the domain-isolated MLD, as compared with the conventional MLD.

The domain-isolated MLD using three kinds of source molecules, TPA, PPDA, and ODH, were also attempted. Figure 3.35(b) shows absorption spectra of films grown with step counts of 27 and 40. The growth time was fixed at 60 min for both cases, implying that the rotation speed is 1.5 times higher for a step count of 40 than for a step count of 27. It is found that the absorption peak height of the film, that is, the film thickness, increases by 1.3 times when the step count increases from 27 to 40. The film thickness ratio of 1.3 is fairly close to the step count ratio of 40/27 = 1.5, suggesting the feasibility of the domain-isolated MLD for three kinds of source molecules.

3.6 SELECTIVE WIRE GROWTH

As described in Section 3.4, by using seed cores, precisely controlled three-dimensional growth of polymer wires and molecular wires is possible with designated locations and orientations by MLD. MLD enables similar controlled polymer/molecular wire growth with the assistance of selective wire growth. In Figure 3.36, the selective wire growth methods with MLD are summarized.

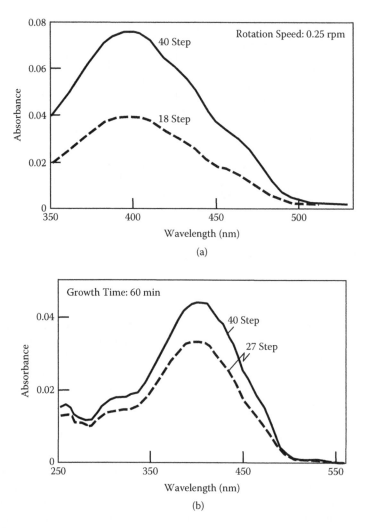

FIGURE 3.35 Absorption spectra of poly-AM grown by domain-isolated MLD (a) using TPA and PPDA with steps 18 and 40, and (b) using TPA, PPDA, and ODH with steps 27 and 40.

In the present section, the selective wire growth methods, such as selective growth and selectively-aligned growth, which utilize surface treatment of substrates and electric fields, are presented.

3.6.1 SELECTIVE GROWTH ON SURFACES WITH PATTERNED TREATMENT

The concept of selective growth of polymers using molecules A and B is shown in Figure 3.37. Molecules A and B are connected by chemical reactions to grow polymer wires. Surface treatment is applied to a substrate surface with designed patterns. In the example shown in Figure 3.37, the adsorption strength of molecules to the surface is stronger in the surface-treated region than in the untreated region.

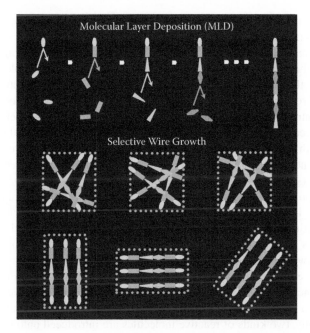

FIGURE 3.36 Selective wire growth methods. From T. Yoshimura, Trends in Thin Solid Films Research (ed. A. R. Jost) Ch. 5 "Self-Organized Growth of Polymer Wire Networks with Designed Molecular Sequences for Wavefunction-Controlled Nano Systems." Nova Science Publishers, New York (2007).

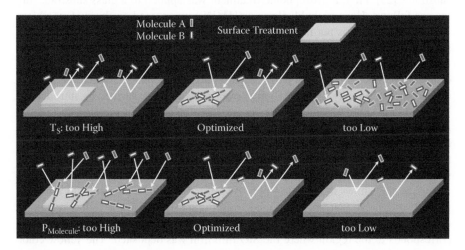

FIGURE 3.37 Selective growth. From T. Yoshimura, Trends in Thin Solid Films Research (ed. A. R. Jost) Ch. 5 "Self-Organized Growth of Polymer Wire Networks with Designed Molecular Sequences for Wavefunction-Controlled Nano Systems." Nova Science Publishers, New York (2007).

In order to realize selective growth, optimization of substrate temperature T_S and molecular gas pressure $P_{Molecule}$ is required. When T_S is too high, the average time τ_S for molecules to remain at the surface becomes short, resulting in re-evaporation of molecules in all the surface regions before being combined with other molecules. When T_S is too low, the molecules are aggregated all over the surface. At optimized T_S, in the untreated region, molecules re-evaporate. In the treated region, τ_S becomes longer than in the untreated region due to stronger adsorption strength, enhancing collisions between molecules A and B. Then, molecules A and B can be combined to make new molecules with larger molecular weight, preventing the molecule from re-evaporation, and consequently, polymer wires grow. Thus, selective growth can be realized.

When $P_{Molecule}$ is too high, molecules are combined in all the surface regions, resulting in polymer wire growth in both the treated and untreated regions. When $P_{Molecule}$ is too low, molecules re-evaporate before collisions happen with other molecules. At optimized $P_{Molecule}$, molecules re-evaporate in the untreated region while polymer wires grow in the surface-treated region, where τ_S is long enough to induce collisions between molecules A and B, enabling selective growth.

A process example of selective growth is shown in Figure 3.38. First, a hydrophobic treatment is applied to a substrate surface. Then, a hydrophilic treatment is applied with a designed pattern on the surface, forming a hydrophilic/hydrophobic pattern. Finally, two kinds of reactive molecules are introduced onto the surface to grow a polymer thin film using chemical reactions between the molecules. When the hydrophilic surface is more favorable for molecule adsorption than the hydrophobic surface, reactions between the molecules occur only on the hydrophilic region, resulting in selective growth of a polymer thin film.

An experimental demonstration of selective growth was performed for poly-AM [TPA/PPDA] using patterned hydrophobic/hydrophilic surface treatment of glass substrates [29]. First, a hydrophobic treatment was applied to a glass substrate surface as follows. Hexamethyldisilazane (HMDS) was coated on the surface. After 1 h, the substrate was put into dichloromethane to remove excess HMDS from the surface to obtain a hydrophobic surface on the glass substrate. Next, a SiO thin film, 100 nm thick, was deposited with a designed pattern on the surface using vacuum evaporation. Since SiO is hydrophilic, a hydrophilic/hydrophobic pattern was formed on the surface. Finally, poly-AM wires were grown on the surface by introducing TPA and PPDA into a vacuum chamber for 150 min with a gas pressure around 0.1 Pa at $T_S = 25$°C. Since the hydrophilic SiO surface is more favorable for TPA and PPDA molecule adsorption than the hydrophobic HMDS surface, reactions between the molecules are expected to occur only on the hydrophilic region, resulting in selective growth of poly-AM wires.

Figure 3.39 shows the result of poly-AM growth on the substrate with a hydrophilic/hydrophobic pattern. It is found from Figure 3.39(a) that a yellow rectangle (a dark rectangle in the monochromatic photograph) exists. The rectangular part corresponds to the region where the SiO thin film was deposited. The yellow color arises from 11-nm-thick poly-AM thin film that was confirmed by absorption spectra described later. In the surrounding region, where HMDS exists, the substrate exhibits no color, indicating little sign of poly-AM thin film growth. These results demonstrate that a poly-AM thin film selectively grows on the SiO region. The enlarged

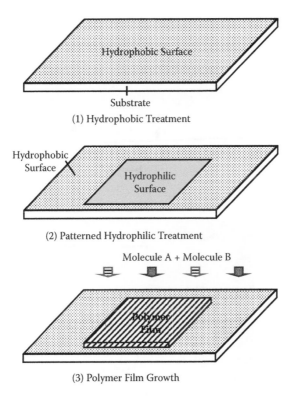

(1) Hydrophobic Treatment

(2) Patterned Hydrophilic Treatment

(3) Polymer Film Growth

FIGURE 3.38 Process example of selective growth. From T. Yoshimura, N. Terasawa, H. Kazama, Y. Naito, Y. Suzuki, and K. Asama, "Selective growth of conjugated polymer thin films by the vapor deposition polymerization," *Thin Solid Films* **497**, 182–184 (2006).

photograph in Figure 3.39(b) reveals that the boundary between the SiO region and the HMDS region has a sharp contrast, suggesting the potentiality of a high-resolution patterning.

Figure 3.40 shows absorption spectra of the SiO region and the HMDS region. An absorption band of poly-AM with a peak at ~410 nm appears for the SiO region while there is no absorption for the HMDS region. Therefore, the result confirms that poly-AM thin film is selectively grown on the SiO region, in other words, on the hydrophilic region.

The same selective growth of poly-AM can be realized by using triphenyldiamine (TPD) instead of HMDS as shown in Figure 3.41 [28]. When poly-AM film was grown by the carrier gas–type organic CVD on a glass substrate with patterned TPD thin films deposited by vacuum evaporation through a metal mask, poly-AM did not grow on TPD, on which PPDA and TPA molecules weakly adsorb since TPD is hydrophobic. Poly-AM selectively grew on the hydrophilic glass surface, on which PPDA and TPA molecules strongly adsorb.

As future challenges, selective growth of poly-AM on nano-scale patterns is interesting for photonic crystals and three-dimensional nano-scale optical circuits in optoelectronic large-scale integrated circuits (LSIs).

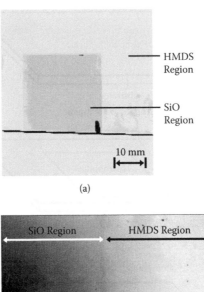

(a)

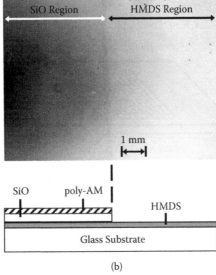

(b)

FIGURE 3.39 (a) Observation of selective growth for poly-AM, and (b) magnified photograph and schematic diagram near the boundary between the hydrophilic and the hydrophobic regions. From T. Yoshimura, N. Terasawa, H. Kazama, Y. Naito, Y. Suzuki, and K. Asama, "Selective growth of conjugated polymer thin films by the vapor deposition polymerization," *Thin Solid Films* **497**, 182–184 (2006).

3.6.2 Selectively-Aligned Growth on Atomic-Scale Anisotropic Structures

In polymer functional devices like those described in Section 3.8, it is important to control polymer wire alignment. Selectively-aligned polymer wire growth techniques will provide a wide range of device structures. For example, if polymer wires are aligned selectively within a stripe with a width of submicron or micron scale, a channel optical waveguide can be constructed without etching because

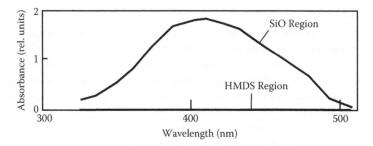

FIGURE 3.40 Absorption spectra of the SiO region and the HMDS region. From T. Yoshimura, N. Terasawa, H. Kazama, Y. Naito, Y. Suzuki, and K. Asama, "Selective growth of conjugated polymer thin films by the vapor deposition polymerization," *Thin Solid Films* **497**, 182–184 (2006).

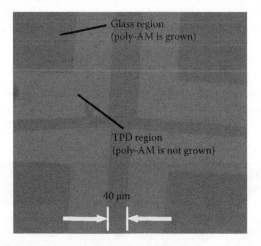

FIGURE 3.41 Selective growth of poly-AM using patterned TPD thin films. From K. Matsumoto and T. Yoshimura, "Electro-optic waveguides with conjugated polymer films fabricated by the carrier-gas-type organic cvd for chip-scale optical interconnects," *Proc. SPIE* **6899**, 98990E-1-9 (2008).

such selective alignment induces a refractive index difference between the stripe and surrounding regions. The polymer wire alignment is favorable for enhancing second-order or third-order optical nonlinearities of the optical waveguides. Carrier mobility in polymer thin-film transistors (TFTs) might also be increased by the alignment.

Kanetake et al. reported that aligned polydiacetylene film is grown by vacuum evaporation on rubbing polydiacetylene films [30]. Patel et al. showed that epitaxial growth induces selective alignment of spin-coated polydiacetylene films by patterning the rubbing polymer underlayer [31]. To develop an all-dry process, we attempted the selectively-aligned growth of poly-AM and polyimide wires on surfaces with anisotropic structures [32].

3.6.2.1   Concept

The concept of polymer wire alignment process by selectively-aligned growth utilizing patterned surface treatment with anisotropic structures is illustrated in Figure 3.42. First, an obliquely-evaporated dielectric film is deposited with designed patterns on a substrate, which is tilted along the y-axis by θ°. Here, θ is a tilting angle of the substrate surface normal direction from the evaporant beam direction. The dielectric film has ultra-fine anisotropic structures along the y-axis. Next, molecules A and B are introduced onto the surface to grow polymer wires. Wires grown on the obliquely-evaporated dielectric film are oriented in a direction parallel (or perpendicular) to the y-axis, while wires grown directly on the substrate surface are oriented in random directions. Thus, selectively-aligned growth is achieved.

The driving force for the polymer wire alignment might be interaction between the anisotropic structures of the dielectric film and molecules. The molecules tend to be placed on the dielectric film in a configuration at potential energy minimum. It should be noted that any surfaces having anisotropic potential energy distributions can be used for selectively-aligned growth instead of the obliquely-evaporated dielectric films.

3.6.2.2   Growth

To demonstrate the proof of concept of selectively-aligned growth, poly-AM was grown on an obliquely-evaporated SiO_2 film. In Figure 3.43, a schematic illustration of a cross-section of poly-AM wires grown on the obliquely-evaporated SiO_2 film

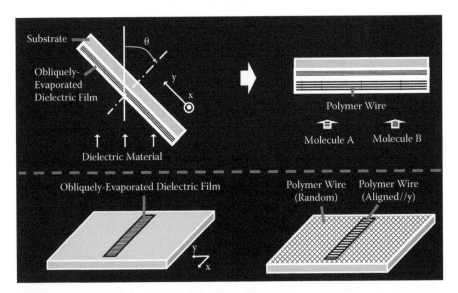

FIGURE 3.42   The polymer wire alignment process by selectively-aligned growth utilizing patterned surface treatment with anisotropic structures. From T. Yoshimura, Trends in Thin Solid Films Research (ed. A. R. Jost) Ch. 5 "Self-Organized Growth of Polymer Wire Networks with Designed Molecular Sequences for Wavefunction-Controlled Nano Systems." Nova Science Publishers, New York (2007).

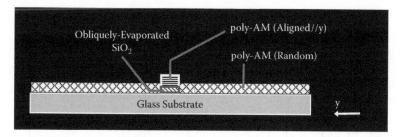

FIGURE 3.43 Schematic illustration of a cross-section of poly-AM wires grown on obliquely evaporated SiO$_2$ film. From T. Yoshimura, Trends in Thin Solid Films Research (ed. A. R. Jost) Ch. 5 "Self-Organized Growth of Polymer Wire Networks with Designed Molecular Sequences for Wavefunction-Controlled Nano Systems." Nova Science Publishers, New York (2007).

by selectively-aligned growth is shown. First, a SiO$_2$ underlayer was deposited by the oblique evaporation on a quartz substrate at 10^{-3} Pa. The substrate was tilted by θ° along the y-axis. Film thickness ranged from 50 to 200 nm. Next, poly-AM films of 160 to 500 nm were deposited by introducing TPA and PPDA onto the surface. The molecular gases were introduced into a vacuum chamber from K cells heated at 100–130°C. The gas pressures were 1 to 8 × 10^{-1} Pa, controlled by the cell temperatures. The substrate temperature was about 25°C. The obtained film was insoluble in methanol and acetone, confirming polymer formation.

3.6.2.3 Optical Characterization for Selective Alignment of Polymer Wires

Figure 3.44 shows absorption spectra of the poly-AM for light polarization parallel and perpendicular to the y-axis. For θ = 0°, that is, a sample with an underlayer normally evaporated, the film shows no dichroism. For θ = 45°, on the other hand, absorption for E$_{Opt}$//y is about ten times larger than for E$_{Opt}\perp$ y. The conjugated polymer has a large absorption for light polarization in the polymer wire direction. Therefore, the result indicates that the polymer wire is aligned along the y-axis in the poly-AM film on the SiO$_2$ underlayer deposited by the oblique evaporation with tilting along the y-axis.

By using previously described phenomenon, selective alignment of the poly-AM wires was attempted for a 10-μm-wide stripe pattern of SiO$_2$ via the following process (Figure 3.45(a)).

1. A photo-resist pattern window was formed on a quartz substrate for the 10-μm-wide stripe.
2. An 80-nm-thick SiO$_2$ film was deposited on it by oblique evaporation with θ = 45° or θ = 0°. The substrate tilting direction, that is, the y-axis, was either perpendicular or parallel to the stripe.
3. The photo-resist was removed by the lift-off process, forming the 10-μm-wide SiO$_2$ stripe.
4. A 160-nm polymer film was deposited over the entire substrate by vacuum deposition polymerization using TPA and PPDA.

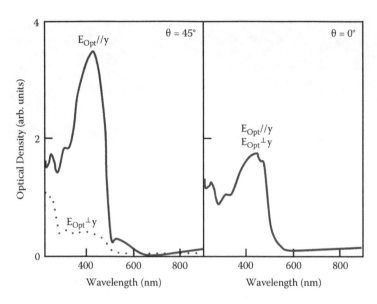

FIGURE 3.44 Absorption spectra of the poly-AM grown by selectively aligned growth with $\theta = 45°$ and $0°$ for light polarization parallel and perpendicular to the y-axis. From T. Yoshimura, K. Motoyoshi, S. Tatsuura, W. Sotoyama, A. Matsuura, and T. Hayano, "Selectively aligned polymer film growth on obliquely-evaporated SiO$_2$ pattern by chemical vapor deposition," *Jpn. J. Appl. Phys.* **31**, L980–L982 (1992).

Figure 3.45(b) shows microscopic observation of the poly-AM with polarized light. In the case where $\theta = 45°$, the obliquely-evaporated SiO$_2$ region is darker than the surrounding region for light polarization parallel to the y-axis ($E_{Opt}//y$), while the obliquely-evaporated SiO$_2$ region is brighter than the surrounding region for light polarization perpendicular to the y-axis ($E_{Opt}\perp y$). The results indicate that the wires on the obliquely-evaporated SiO$_2$ are aligned along the y-axis, and wires on the glass substrate surface are randomly oriented as shown in Figure 3.43. In the case where $\theta = 0°$, on the other hand, contrast between the SiO$_2$ region and the surrounding region does not depend on light polarization, indicating that the poly-AM wires are random over the entire surface. These results confirm that selective alignment of polymer wires is realized by SiO$_2$ underlayers deposited by oblique evaporation. Selective alignment was also observed in thick poly-AM films with thicknesses of 500 nm or more. By increasing θ from $0°$ to $60°$, the degree of alignment was increased.

When, as shown in Figure 3.46(a), a 100-nm-thick SiO$_2$ underlayer was obliquely evaporated by tilting the substrate along the y-axis on a SiO$_2$ layer obliquely evaporated by tilting the substrate along the x-axis, the contrast between the stripe region and the surrounding region was completely inverted with rotating polarization from E//x to E//y (Figure 3.46(b)). This indicates that the wire direction in the stripe region is at right angles to that in the surrounding region. That is, the wires in the stripe region are aligned along the y-axis, and the wires in the surround region are aligned along the x-axis, as illustrated in Figure 3.46. This result means that the polymer wire directions are determined only by the SiO$_2$ layer

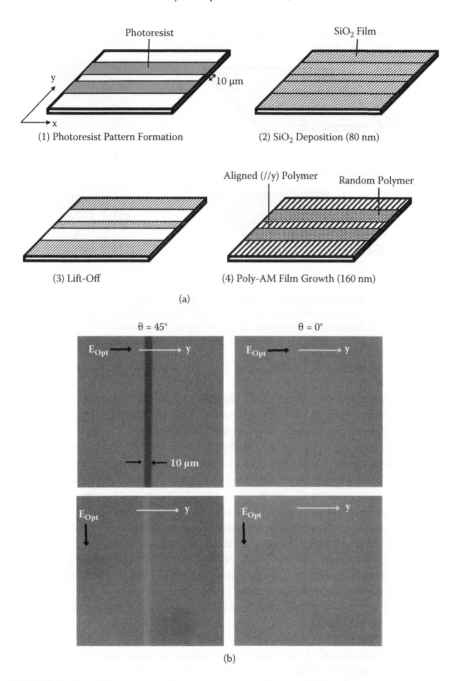

FIGURE 3.45 (a) Process for selective alignment of poly-AM wires, and (b) microscopic image observation of the poly-AM with polarized light. From T. Yoshimura, K. Motoyoshi, S. Tatsuura, W. Sotoyama, A. Matsuura, and T. Hayano, "Selectively aligned polymer film growth on obliquely evaporated SiO₂ pattern by chemical vapor deposition," *Jpn. J. Appl. Phys.* **31**, L980–L982 (1992).

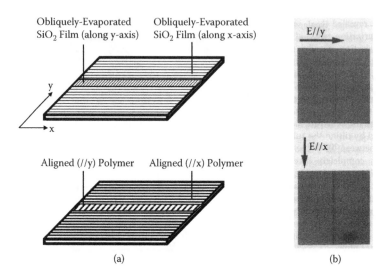

FIGURE 3.46   Control of polymer wire directions by stacked obliquely evaporated SiO$_2$ films. (a) Process and (b) microscopic image observation with polarized light. From T. Yoshimura, K. Motoyoshi, S. Tatsuura, W. Sotoyama, A. Matsuura, and T. Hayano, "Selectively aligned polymer film growth on obliquely evaporated SiO$_2$ pattern by chemical vapor deposition," *Jpn. J. Appl. Phys.* **31**, L980–L982 (1992).

coming in contact with the polymer wires. Therefore, it is suggested that the polymer wire directions can freely be controlled with designated patterns on a substrate by stacking SiO$_2$ underlayers obliquely evaporated with different tilting directions.

To examine the birefringence induced by the aligned polymer wires, microscopic image observation under the cross-nicol condition was carried out using the setup shown in Figure 3.47. As Figure 3.48 shows, in the poly-AM film on the stripe pattern of SiO$_2$ obliquely evaporated with $\theta = 45°$, a bright 10-μm-wide stripe is observed, indicating that birefringence occurs selectively on the obliquely-evaporated SiO$_2$ film. In the case of $\theta = 0°$, no birefringence is observed. Estimation based on absorption spectra fringes revealed that the refractive index is about 1.65 for polarization perpendicular to the poly-AM wires and 1.95 for polarization parallel to the poly-AM wires. Using the anisotropicity of the refractive index, optical waveguides can be constructed without etching, and various other functional photonic devices will be realized. An investigation on a possibility of TFTs of feature sizes of 10 nm is also attractive.

Figure 3.49 shows microscopic images under the cross-nicol condition for polyimide grown with $\theta = 45°$ from PMDA and DDE by the same process as described previously for poly-AM. When the y-direction of the sample is perpendicular to the polarization direction of the incident light, the whole area of the sample is dark. When the sample is rotated and the y-direction of the sample deviates from the direction perpendicular to the polarization direction, similar to the case of poly-AM, a bright 10-μm-wide stripe is observed. This indicates that birefringence occurs

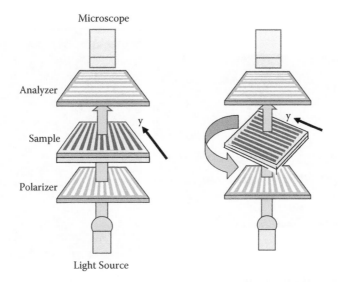

FIGURE 3.47 A setup for microscopic image observation under the cross-nicol condition. From K. Matsumoto and T. Yoshimura, "Electro-optic waveguides with conjugated polymer films fabricated by the carrier-gas-type organic CVD for chip-scale optical interconnects," *Proc. SPIE* **6899**, 98990E-1-9 (2008).

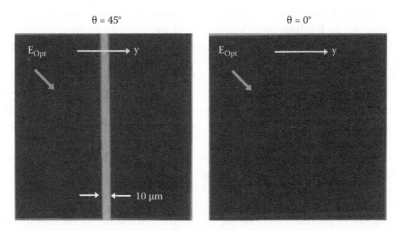

FIGURE 3.48 Microscopic images under the cross-nicol condition for poly-AM grown by selectively-aligned growth with $\theta = 45°$ and $0°$.

selectively on the obliquely-evaporated SiO_2 region, demonstrating that selectively-aligned growth is applicable to polyimide systems.

Figure 3.50 shows microscopic images for birefringence measurement in poly-AM grown by the carrier gas–type organic CVD on a 200-nm-thick SiO film that is obliquely evaporated by tilting along the y-axis [28]. Growth time was 20 min. During the film growth, molecular gas pressure was kept around 100 Pa. For

$\theta = 45°$

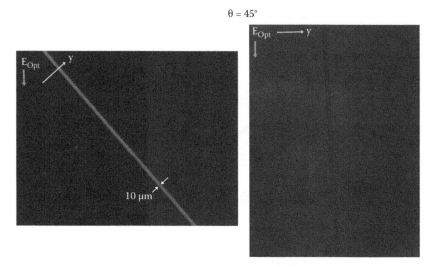

FIGURE 3.49 Microscopic images under the cross-nicol condition for polyimide grown by selectively-aligned growth with $\theta = 45°$.

light polarization parallel to the y-axis, the microscope image is dark. When the substrate is rotated by 45° from the initial configuration, the stripe region, where the SiO underlayer is deposited by oblique evaporation, is brighter than the surrounding region due to birefringence. The results of the microscope image observation revealed that poly-AM wires are selectively aligned along the y-axis on the SiO underlayer pattern, similar to the case of SiO_2.

Observation of obliquely-evaporated SiO films for underlayers and poly-AM films was carried out by the atomic force microscope (AFM) [33]. As can be seen from Figure 3.51(a), in the obliquely-evaporated SiO film, a grain morphology along the y-axis is observed, revealing that the film has an anisotropic structure. In the poly-AM film on the obliquely-evaporated SiO film, as shown in Figure 3.51(b), line-shaped patterns along the y-axis are observed. This result suggests that polymer wires are aligned along the y-axis. In the glass substrate region, as shown in Figure 3.51(c), structures with definite orientation are not observed.

As described above, selectively-aligned growth is expected to be applicable to various kinds of polymers and can use various dielectric films for the underlayers.

3.6.3 ELECTRIC-FIELD-ASSISTED GROWTH

The concept of selectively-aligned growth with a strong electric field, which is called *electric-field-assisted growth*, using molecules A and B is shown in Figure 3.52. Here, at least one of the two kinds of molecules has electric dipoles. In the present case, molecule B has electric dipoles. Molecules A and B are provided onto a substrate surface, on which slit-type electrodes are formed. When molecule B comes near the surface, it tends to be aligned in the direction of the electric field generated between the electrodes. Since the molecular alignment is induced before the

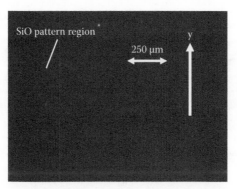

(a) Polarized direction//y-axis

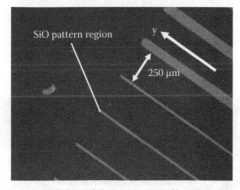

(b) Polarized direction : rotated by 45°

FIGURE 3.50 Microscopic images under the cross-nicol condition for poly-AM grown by the carrier gas–type organic CVD on obliquely-evaporated SiO film. From K. Matsumoto and T. Yoshimura, "Electro-optic waveguides with conjugated polymer films fabricated by the carrier-gas-type organic CVD for chip-scale optical interconnects," *Proc. SPIE* **6899**, 98990E-1-9 (2008).

molecules are tightly inserted into polymer wires, the resistance against molecular rotation is small and molecules can be aligned without raising temperatures above glass transition temperature. It is expected that polymer wires, in which molecule B tends to have unique orientation, grow in a direction parallel to the electric field, achieving electric-field-assisted growth. In addition, in this growth method, little contamination is involved because no solvent is used and impurities in sources and by-products are easily removed in vacuum.

Experimental demonstration of electric-field-assisted growth was performed for polyamic acid [PMDA/DNB] by using PMDA (molecule A) and DNB (molecule B) [34]. The reaction between PMDA and DNB is described in Section 3.3.1 (Figure 3.8). These molecules were evaporated separately from temperature-controlled K cells in a vacuum chamber to provide the molecules onto a glass substrate surface. The background pressure was 5×10^{-6} Pa. A slit-type electrode with a 10-μm gap was

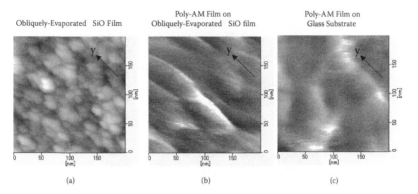

Obliquely-Evaporated SiO Film	Poly-AM Film on Obliquely-Evaporated SiO film	Poly-AM Film on Glass Substrate
(a)	(b)	(c)

FIGURE 3.51   Atomic force microscope (AFM) images of obliquely-evaporated SiO film and poly-AM films. From Y. Suzuki, H. Kazama, N. Terasawa, Y. Naito, T. Yoshimura, Y. Arai, and K. Asama, "Selective growth of conjugated polymer thin film with nano scale controlling by chemical vapor depositions (CVD) toward 'Nanonics,'" *Proc. SPIE* **5938**, 59310G-1-8 (2005).

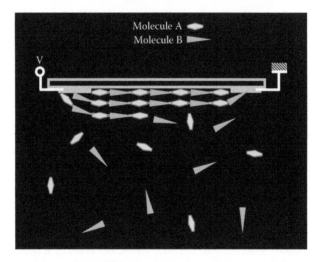

FIGURE 3.52   Selectively-aligned growth with a strong electric field—electric-field-assisted growth. From T. Yoshimura, Trends in Thin Solid Films Research (ed. A. R. Jost) Ch. 5 "Self-Organized Growth of Polymer Wire Networks with Designed Molecular Sequences for Wavefunction-Controlled Nano Systems." Nova Science Publishers, New York (2007).

formed on the substrate. An electric field of 0.78 MV/cm for aligning the molecules was applied to the gap during film deposition.

Table 3.1 shows the substrate temperature dependence of deposition rate and film appearance. Infrared (IR) measurements revealed that films show many characteristic polyamic acid absorption lines in the wavenumber region of 1000–2000 cm^{-1}. We also found that the films become insoluble in methanol after being annealed at 220°C for 3 h. These results confirm that the as-deposited films are polyamic acid, which turns to polyimide by annealing. At $T_s = 25°C$, the films are not clear but at $T_s = 65°C$ and $110°C$ they are clear.

Figure 3.53 shows the effect of co-evaporation on the film deposition rate. At $T_s = 25°C$, for supplying PMDA only, DNB only, or both, considerable deposition rates are detected. Residual molecules having no connection with polymer wires

TABLE 3.1

Substrate Temperature Dependence of Deposition Rate and Film Appearance

T_s (°C)	Gas Pressure (Pa)	Deposition Rate (nm/min)	Film Appearance
25	6×10^{-4}	25	Frosted
65	3×10^{-3}	12	Clear
110	3×10^{-3}	5	Clear

Source: T. Yoshimura, S. Tatsuura, and W. Sotoyama, "Chemical vapor deposition of polyamic acid thin films stacking non-linear optical molecules aligned with a strong electric field," *Thin Solid Films* **207**, 9–11 (1992).

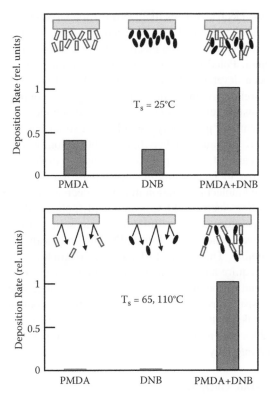

FIGURE 3.53 Effect of co-evaporation on the film deposition rate. From T. Yoshimura, S. Tatsuura, and W. Sotoyama, "Chemical vapor deposition of polyamic acid thin films stacking non-linear optical molecules aligned with a strong electric field," *Thin Solid Films* **207**, 9–11 (1992).

form island-like aggregates, resulting in a frosted film. For $T_s = 65$ and $110°C$, the deposition rates are very low, owing to re-evaporation from the substrate when only PMDA or DNB is supplied. When co-evaporating PMDA and DNB together, the reaction on the substrate reduces the total vapor pressure, and the deposition rate is therefore increased by mass action. This indicates that residual molecules are removed automatically, forming a stoichiometric clear polymer film.

Figure 3.54(a) shows the transmission spectra for the polyamic acid film deposited at $T_s = 65°C$ and for methanol solutions of PMDA and DNB. The film shows an absorption band near a band in the spectrum of DNB, confirming that DNB molecules are inserted into the polymer wire. The splitting of the band might be attributed to a change in the electronic state in DNB in the wires. Figure 3.54(b) shows the absorption spectra of the film deposited at $T_s = 65°C$ in the region between electrodes. The film is 160 nm thick. The absorption for polarization in the direction of the electric field applied during film deposition is slightly larger than that for polarization perpendicular to the electric field direction, suggesting that DNB molecules can be aligned by electric-field-assisted growth. This would make subsequent poling with heating unnecessary. It should be noted that direct polyimide film formations with molecules having electric dipoles would be possible by growth at a high substrate temperature.

Molecules having electric dipoles like DNB exhibit large second-order nonlinear optical properties. In the case of DNB, the second-order optical nonlinearity arises from the amino group acting as the electron donor and the nitro group acting as the electron acceptor. Poled polymer thin films containing such nonlinear optical molecules have been developed for electro-optic (EO) devices, but they have two serious problems—relaxation in the field-induced alignment of the nonlinear optical molecules and the insufficient EO coefficient of poled polymers.

The EO coefficient r of poled polymers obtained so far has been comparable to that of $LiNbO_3$, and not much higher than $LiNbO_3$. Nonlinear optical molecules with a large nonlinearity, for example, long conjugated molecules, are required to enhance r. To improve molecular alignment, we must reduce the resistance against electric-field-induced molecular rotation in poling, decrease the poling temperature, and increase the poling electric field. Conventionally poling is done in a rubber condition at a high temperature after the film is formed by spin coating. The surrounding polymer matrix prevents molecules, especially long ones with a large nonlinearity, from rotating smoothly. Thermal disturbance also degrades molecular alignment. High T_g polymers require a high temperature, which may destroy molecules. Spin-coated films also tend to contain contamination, which reduces the electric field.

Electric-field-assisted growth would resolve all these problems of the conventional method for poled polymer fabrication since it is a solvent/contamination-free process and it enables easy molecular alignment at low temperatures before the molecules are inserted into polymer wires.

3.6.4 HEAD-TO-TAIL GROWTH

The concept of the head-to-tail growth is shown in Figure 3.55. Here, a source molecule has two kinds of reactive groups at the head and tail positions. Chemical

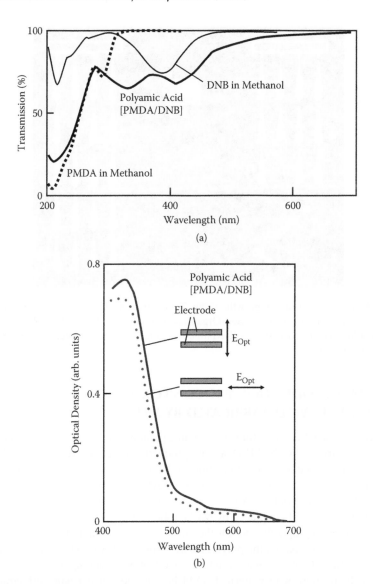

FIGURE 3.54 (a) Transmission spectra of PMDA, DNB, and polyamic acid [PMDA/DNB] deposited at $T_S = 65°C$, and (b) absorption spectra of the polyamic acid film for polarized light in the region between electrodes. From T. Yoshimura, S. Tatsuura, and W. Sotoyama, "Chemical vapor deposition of polyamic acid thin films stacking non-linear optical molecules aligned with a strong electric field," *Thin Solid Films* **207**, 9–11 (1992).

reactions selectively occur between the two kinds of groups. By providing the molecules onto a substrate surface, it is expected that polymer wires grow in a unique head-to-tail configuration. This method might be effective to enhance nonlinear optical properties in polymer wires. For head-to-tail growth, special molecules are necessary to carry out experimental demonstrations.

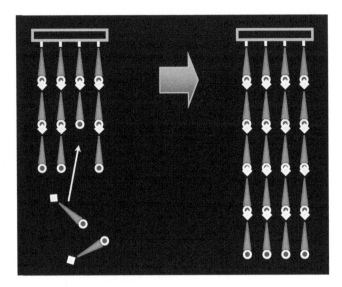

FIGURE 3.55 Head-to-tail growth. From T. Yoshimura, Trends in Thin Solid Films Research (ed. A. R. Jost) Ch. 5 "Self-Organized Growth of Polymer Wire Networks with Designed Molecular Sequences for Wavefunction-Controlled Nano Systems." Nova Science Publishers, New York (2007).

3.7 MASS PRODUCTION PROCESS FOR NANO-SCALE DEVICES FABRICATED BY MLD

In order to reduce the cost of thin-film organic photonic/electronic devices consisting of polymer wires grown by MLD, a mass production process is required. To do this, we proposed molecular nano duplication (MND), as shown in Figure 3.56 [23]. On a growth substrate, surface treatment such as SAM formation is applied in designed nano-scale patterns. On the SAM, removable molecules that have removable bonds connected to the SAM are placed by MLD, followed by polymer wire growth. After attaching a picking-up substrate on the surface, the removable bonds are cut by a chemical agent to separate the polymer wires from the SAM. Thus, nano-scale patterns consisting of polymer wires are transferred onto the picking-up substrate. Since the growth substrate with patterned SAM is reusable, further nano lithography is not required for nano-scale pattern duplication. By combining MND with selectively occupied repeated transfer (SORT) [35], which is described in Section 7.5, a material-saving mass production process will be provided.

3.8 EXAMPLES OF GOALS ACHIEVED BY MLD

By combining MLD with seed-core-assisted growth or the selective wire growth, various kinds of thin-film organic photonic/electronic devices and molecular nano systems will be constructed [2,3]. In the present section, some examples of the goals of MLD are briefly described.

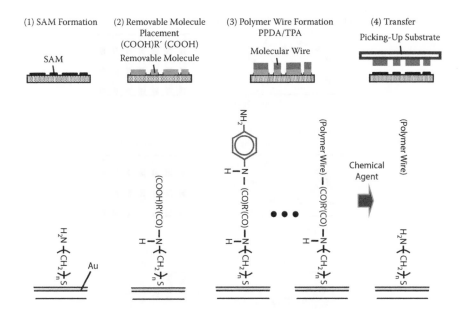

FIGURE 3.56 Molecular nano duplication. From T. Yoshimura, Y. Suzuki, N. Shimoda, T. Kofudo, K. Okada, Y. Arai, and K. Asama, "Three-dimensional chip-scale optical interconnects and switches with self-organized wiring based on device-embedded waveguide films and molecular nanotechnologies," *Proc. SPIE* **6126**, 612609-1-15 (2006).

3.8.1 FUNCTIONAL ORGANIC DEVICES

Figure 3.57 shows examples of various functional organic devices consisting of conjugated polymer wires grown by MLD with seed-core-assisted growth or the selective wire growth. For comparison, conventional devices consisting of random polymer wires fabricated by spin coating are illustrated. In electro-optic devices, polymer wires with controlled molecular sequences are grown in a definite direction parallel to the driving electric fields, which enhance the optical nonlinearity of EO waveguides. In photo-refractive device, in which photoconductive agents and carrier traps are added to the EO material, the same effect is expected. In electroluminescent (EL) devices and photovoltaic devices, polymer wires, in which a p-type region and an n-type region are formed in each polymer wire, are aligned from the lower electrode to the upper electrode. This structure improves carrier mobility in the devices. In electrochromic devices, p-type and n-type electrochromic molecules, and electron-blocking molecules are arranged in a polymer wire in this sequence. The configuration is suitable to generate color centers with trapped electrons arising from electric-field-induced polarization. In TFTs, the polymer wires are aligned from source to drain electrodes. This configuration improves carrier mobility in the polymer wires. When feature sizes less than 10 nm are required, seed cores made by nano lithography, such as electron beam lithography and scanning tunneling microscopy (STM), can be applied.

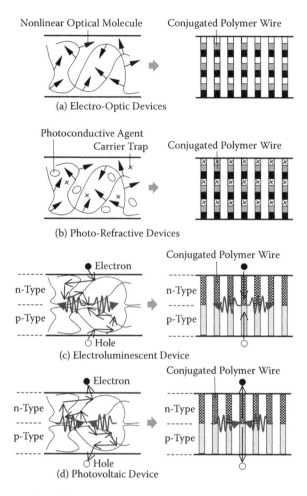

FIGURE 3.57 Examples of various functional organic devices consisting of conjugated polymer wires and improvement of the device performance by MLD with seed-core-assisted growth or selective wire growth.

3.8.2 INTEGRATED NANO-SCALE OPTICAL CIRCUITS

MLD with combining MND is preferable for constructing integrated nano-scale optical circuits consisting of high-index contrast (HIC) waveguides and photonic crystals. Low-cost mass production for heterogeneous integration of nano-scale optical circuits might be realized by using photolithographic packaging with SORT (PL-Pack with SORT) [35], whose details are described in Section 7.5. SORT is a resource-saving heterogeneous integration process with multistep device transfers using all-photolithographic methods without conventional flip-chip assembly packaging. Figures 3.58 and 3.59 show proposals of the nano-scale optoelectronic (OE) packaging [23]. In Figure 3.58, HIC waveguides are grown on a substrate with patterned SAM by MLD. The waveguides are picked up onto a picking-up substrate by MND. The waveguides

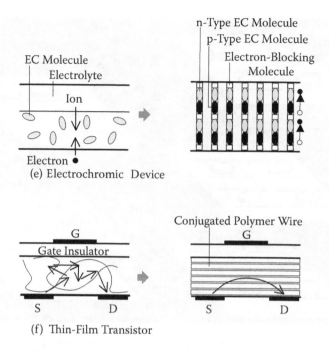

EC Molecule
Electrolyte

Ion

Electron ●
(e) Electrochromic Device

n-Type EC Molecule
p-Type EC Molecule
Electron-Blocking
Molecule

G
Gate Insulator

S   D

(f) Thin-Film Transistor

Conjugated Polymer Wire
G

S   D

FIGURE 3.57   (CONTINUED) Examples of various functional organic devices consisting of conjugated polymer wires and improvement of the device performance by MLD with seed-core-assisted growth or selective wire growth.

are transferred to a final substrate by SORT via a supporting substrate. In Figure 3.59, HIC waveguides and photonic crystals are integrated into an optical waveguide film.

When the nano-scale optical circuits are used for chip-scale optical interconnects, the circuits are distributed with a pitch of ~100 μm corresponding to pad pitch of LSI chips. In this case, since the sizes of the nano-scale optical circuits are in the order of 10 μm, the optical circuits exist with very low density in the interconnect systems. SORT, in which the nano-scale optical circuits fabricated in dense two-dimensional arrays are distributed to necessary sites, enables us to save materials and process costs for the nano-scale optical circuit integration.

3.8.3 MOLECULAR CIRCUITS

Conjugated polymer wires can be regarded as a future fundamental material after the carbon nanotube. The polymer wires are expected to be suitable to construct the molecular circuits schematically illustrated in Figure 3.60 [3,4,23] for the following reasons.

- High carrier mobility characteristics of their π-conjugated systems
- Small wire diameter, which is smaller than that of carbon nanotubes.
- Molecular sequence controllability in the wire with monomolecular scale, that is, precise wavefunction shape controllability, which is not available in carbon nanotubes.

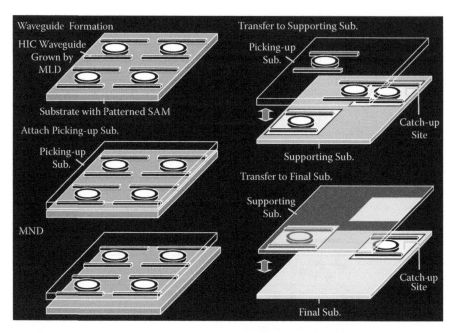

FIGURE 3.58 SORT process with MND for transfers of nano-scale optical circuits. From T. Yoshimura, Y. Suzuki, N. Shimoda, T. Kofudo, K. Okada, Y. Arai, and K. Asama, "Three-dimensional chip-scale optical interconnects and switches with self-organized wiring based on device-embedded waveguide films and molecular nanotechnologies," *Proc. SPIE* **6126**, 612609-1-15 (2006).

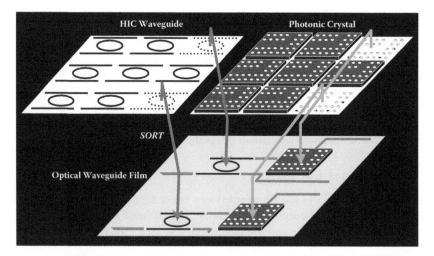

FIGURE 3.59 Heterogeneous integration of nano-scale optical circuits using SORT. From T. Yoshimura, Y. Suzuki, N. Shimoda, T. Kofudo, K. Okada, Y. Arai, and K. Asama, "Three-dimensional chip-scale optical interconnects and switches with self-organized wiring based on device-embedded waveguide films and molecular nanotechnologies," *Proc. SPIE* **6126**, 612609-1-15 (2006).

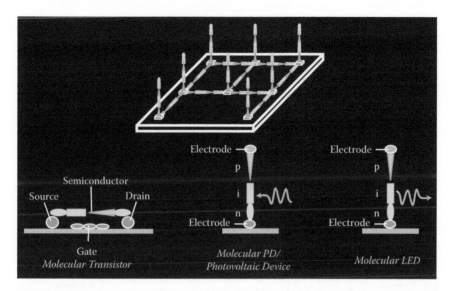

FIGURE 3.60 Concept of molecular circuits consisting of molecular transistors, PDs, photovoltaic devices, and LEDs. From T. Yoshimura, Trends in Thin Solid Films Research (ed. A. R. Jost) Ch. 5 "Self-Organized Growth of Polymer Wire Networks with Designed Molecular Sequences for Wavefunction-Controlled Nano Systems." Nova Science Publishers, New York (2007).

The molecular circuits consist of molecular lines/electrodes, molecular transistors, molecular photodiodes (PDs), molecular photovoltaic devices, and molecular light-emitting diodes (LEDs). In order to construct the molecular circuits, the self-organized growth of the polymer/molecular wire network described in Section 3.4 is effective.

In molecular transistors, a semiconductor conjugated polymer wire is aligned from a source electrode to a drain electrode, passing above a gate electrode. Molecular PDs, photovoltaic devices, and LEDs are made of conjugated polymer wires with pin junctions that are sandwiched by a pair of conjugated polymer wires for electrodes.

Figure 3.61 shows tentative examples of conjugated polymer wires with pin junctions. For the donor/acceptor substitution-type pin junctions, donors and acceptors are respectively substituted in the p-type region and the n-type region. For the backbone-type pin junction, an n-type backbone and a p-type backbone are connected. In the tentative examples, poly-AM and poly-OXD are assumed for the backbones to make a junction. Although p- and n-type characterizations of poly-AM and poly-OXD are uncertain, if poly-AM and poly-OXD are formed within a polymer wire with molecular sequence arrangements like ABABABABADCDCDCDCD using the reactions shown in Figure 3.12 with molecule A: OA, molecule B: ODH, molecule C: TPA, and molecule D: PPDA, some kind of *pin-like* junctions might be formed in the poly-OXD–poly-AM interface region. For LED operation, electrons and holes are combined in the intrinsic, i, region under a forward bias condition to emit light beams. For PD operation, electrons and holes are generated by absorbing

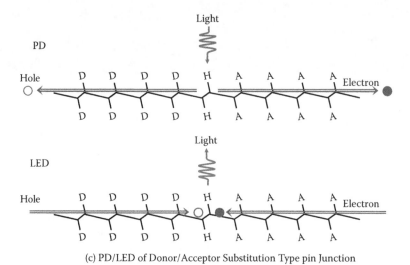

(a) Donor/Acceptor Substitution Type pin Junction

(b) Backbone Type pin Junction

(c) PD/LED of Donor/Acceptor Substitution Type pin Junction

FIGURE 3.61 Tentative examples of conjugated polymer wires with pin junctions. From T. Yoshimura, Trends in Thin Solid Films Research (ed. A. R. Jost) Ch. 5 "Self-Organized Growth of Polymer Wire Networks with Designed Molecular Sequences for Wavefunction-Controlled Nano Systems." Nova Science Publishers, New York (2007).

light in the i region under a backward bias condition to induce currents and voltage. Photovoltaic operation is also available in the same structure as the PD.

REFERENCES

1. T. Yoshimura, S. Tatsuura, and W. Sotoyama, "Polymer films formed with monolayer growth steps by molecular layer deposition," *Appl. Phys. Lett.* **59**, 482–484 (1991).
2. T. Yoshimura, E. Yano, S. Tatsuura, and W. Sotoyama, "Organic functional optical thin film, fabrication and use thereof," U.S. Patent 5,444,811 (1995).
3. T. Yoshimura, "Self-organized growth of polymer wire networks with designed molecular sequences for wavefunction-controlled nano systems," in *Trends in thin solid films research*, edited by A. R. Jost. Nova Science Publishers, New York (2007).

4. T. Yoshimura, *Molecular nano systems: Applications to optoelectronic computers and solar energy conversion*, Corona Publishing Co., Ltd., Tokyo (2007) [in Japanese].

5. T. Yoshimura, "Liquid phase deposition," Japanese Patent, Tokukai Hei 3-60487 (1991) [in Japanese].

6. T. Yoshimura, "Growth methods of polymer wires and thin films," Japanese Patent Tokukai 2008-216947 (2008) [in Japanese].

7. T. Suntola, "Atomic layer epitaxy," *Material Science Reports* **4**(7) (1989).

8. M. Pessa, R. Makela, and T. Suntola, "Characterization of surface exchange reactions used to grow compound films," *Appl. Phys. Lett.* **38**, 131–133 (1981).

9. T. Yoshimura, S. Ito, T. Nakayama, and K. Matsumoto, "Orientation-controlled molecule-by-molecule polymer wire growth by the carrier-gas-type organic chemical vapor deposition and the molecular layer deposition," *Appl. Phys. Lett.* **91**, 033103-1–3 (2007).

10. T. Yoshimura and Y. Kudo, "Monomolecular-step polymer wire growth from seed core molecules by the carrier-gas type molecular layer deposition (MLD)," *Appl. Phys. Express* **2**, 015502-1–3 (2009).

11. Y. Takahashi, M. Iijima, K. Inagawa, and A. Itoh, "Synthesis of aromatic polyimide film by vacuum deposition polymerization," *J. Vac. Sci. Technol.* **A 5**, 2253–2256 (1987).

12. M. Iijima, Y. Takahashi, and E. Fukuda, "Vacuum deposition polymerization," *Nikkei New Materials*, December 11, 93–100 (1989) [in Japanese].

13. M. Putkonen, J. Harjuoja, T. Sajavaara, and L. Niinisto, "Atomic layer deposition of polyimide thin films," *J. Mater. Chem.*, DOI:10.1039/b612823h Paper (2007).

14. L. Salmi, M. Vehkamaki, E. Puukilainen, and M. Ritala, "ALD of inorganic-organic nanolaminates," AVS, 8th International Conference on Atomic Layer Deposition, WedM1b-2, Bruges, Belgium (2008).

15. S. M. George, B. Yoon, and A. A. Dameron, "Surface chemistry for molecular layer deposition of organic and hybrid organic-inorganic polymers," *Acc. Chem. Res.* **42**, 498–508 (2009).

16. X. Liang and A. Weimer, "Photoactivity passivation of TiO$_2$ nanoparticles using molecular layer deposition (MLD) polymer films," *J. Nanopart. Res.*, DOI 10.1007/s11051-009-9587-0 (2009).

17. P. W. Loscutoff, H. Zhou, and S. F. Bent, "Molecular layer deposition of multicomponent organic films for nanoelectronics," AVS, 9th International Conference on Atomic Layer Deposition, Monterey, California, 148 (2009).

18. H. Murata, S. Ukishima, H. Hirano, and T. Yamanaka, "A novel fabrication technique and new conjugated polymers for multilayer polymer light-emitting diodes," *Polymer for Advanced Technologies* **8**, 459–464 (1996).

19. T. Yoshimura, S. Tatsuura, W. Sotoyama, A. Matsuura, and T. Hayano, "Quantum wire and dot formation by chemical vapor deposition and molecular layer deposition of one-dimensional conjugated polymer," *Appl. Phys. Lett.* **60**, 268–270 (1992).

20. K. Kiyota, T. Yoshimura, and M. Tanaka, "Electrophotographic behavior of ZnO sensitized by two dyes," *Photogr. Sci. Eng.* **25**, 76–79 (1981).

21. T. Yoshimura, K. Kiyota, H. Ueda, and M. Tanaka, "Contact potential difference of ZnO layer adsorbing p-type dye and n-type dye," *Jpn. J. Appl. Phys.* **18**, 2315–2316 (1979).

22. H. Kokado, T. Nakayama, and E. Inoue, "A new model for spectral sensitization of photoconduction in zinc oxide powder," *J. Phys. Chem. Solids* **34**, 1–8 (1973).

23. T. Yoshimura, Y. Suzuki, N. Shimoda, T. Kofudo, K. Okada, Y. Arai, and K. Asama, "Three-dimensional chip-scale optical interconnects and switches with self-organized wiring based on device-embedded waveguide films and molecular nanotechnologies," *Proc. SPIE* **6126**, from Photonics West 2006, San Jose, California, 612609- 1-15 (2006).

24. D. Seghete, B. Yoon, A. S. Cavanagh, and S. M. George, "New approaches to molecular layer deposition using ring-opening and heterobifunctional reactants," AVS, 8th International Conference on Atomic Layer Deposition, WedM1b-3, Bruges, Belgium (2008).

25. Private comment from Dr. E. Liskola in AVS, 8th International Conference on Atomic Layer Deposition, Bruges, Belgium (2008).

26. M. Iijima and Y. Takahashi, "Techniques for vapor deposition polymerization: Preparation of organic functional thin films," *OYO BUTURI* **66**, 1084–1088 (1997) [in Japanese].

27. D. Kim and T. Yoshimura, "The domain-isolated molecular layer deposition (DI-MLD) for fast polymer wire growth," AVS, 9th International Conference on Atomic Layer Deposition, Monterey, California, 243 (2009).

28. K. Matsumoto and T. Yoshimura, "Electro-optic waveguides with conjugated polymer films fabricated by the carrier-gas-type organic CVD for chip-scale optical interconnects," *Proc. SPIE* **6899**, from Photonics West 2008, San Jose, California, 98990E- 1-9 (2008).

29. T. Yoshimura, N. Terasawa, H. Kazama, Y. Naito, Y. Suzuki, and K. Asama, "Selective growth of conjugated polymer thin films by the vapor deposition polymerization," *Thin Solid Films* **497**, 182–184 (2006).

30. T. Kanetake, K. Ishikawa, T. Koda, Y. Tokura, and K. Takeda, "Highly oriented polydiacetylene films by vacuum deposition," *Appl. Phys. Lett.* **51**, 1957–1959 (1987).

31. J. S. Patel, S. D. Lee, G. L. Baker, and J. A. Shelburne, "Epitaxial growth of aligned polydiacetylene films on anisotropic orienting polymers," *Appl. Phys. Lett.* **56**, 131–133 (1990).

32. T. Yoshimura, K. Motoyoshi, S. Tatsuura, W. Sotoyama, A. Matsuura, and T. Hayano, "Selectively aligned polymer film growth on obliquely evaporated SiO₂ pattern by chemical vapor deposition," *Jpn. J. Appl. Phys.* **31**, L980–L982 (1992).

33. Y. Suzuki, H. Kazama, N. Terasawa, Y. Naito, T. Yoshimura, Y. Arai, and K. Asama, "Selective growth of conjugated polymer thin film with nano scale controlling by chemical vapor depositions (CVD) toward 'Nanonics,'" *Proc. SPIE* **5938**, from Optics & Photonics 2005, San Diego, California, 59310G- 1–8 (2005).

34. T. Yoshimura, S. Tatsuura, and W. Sotoyama, "Chemical vapor deposition of polyamic acid thin films stacking non-linear optical molecules aligned with a strong electric field," *Thin Solid Films* **207**, 9–11 (1992).

35. T. Yoshimura, M. Ojima, Y. Arai, and K. Asama, "Three-dimensional self-organized micro optoelectronic systems for board-level reconfigurable optical interconnects: Performance modeling and simulation," *IEEE J. Select. Topics in Quantum Electron.* **9**, 492–511 (2003).

4 Fabrication of Multiple-Quantum Dots (MQDs) by MLD

Multiple quantum dot (MQD) in polymer wires is one of the distinguished structures accomplished by molecular layer deposition (MLD).

In the present chapter, experimental demonstrations of the MQD construction are presented.

4.1 FUNDAMENTALS OF QUANTUM DOTS

Control of dimensionality is an important issue to improve various kinds of optical and electronic properties of materials. Figure 4.1 shows a schematic illustration of three-, two-, one-, and zero-dimensional systems, which respectively correspond to bulks, quantum wells, quantum wires, and quantum dots. The dimensionality of materials affects energy dependence of the density of states [1], which gives rise to variations of optical and electronic properties.

Figure 4.2 shows shell shapes for given energy of a free electron in three-, two-, and one-dimensional systems. Here, k represents the wavenumber. In the three-dimensional system, the shell is a spherical surface with a radius of k and a thickness of dk. In the two-dimensional system, the shell is a circle with a radius of k and a thickness of dk, and in the one-dimensional system, the shell is two short lines with a length of dk.

From these shell shapes, density of states can be calculated based on the free-electron model, where wavefunctions and energy of electrons are expressed as

$$\psi(\mathbf{r}) = \left(\frac{1}{L^3}\right)^{1/2} e^{i\mathbf{k}\bullet\mathbf{r}},\tag{4.1}$$

$$E_{\mathbf{k}} = \frac{\hbar^2}{2m}k^2.\tag{4.2}$$

Here, $\mathbf{k}$ is the wavenumber vector, m is the mass of an electron, and L^3 is the volume where electrons exist. When L is sufficiently large and k can be regarded as a continuous quantity,

$$k = \left(\frac{2m}{\hbar^2}E\right)^{1/2},\tag{4.3}$$

$$kdk = \frac{m}{\hbar^2}dE\tag{4.4}$$

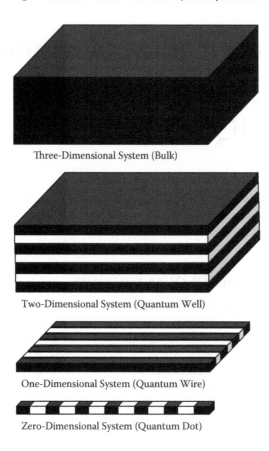

Three-Dimensional System (Bulk)

Two-Dimensional System (Quantum Well)

One-Dimensional System (Quantum Wire)

Zero-Dimensional System (Quantum Dot)

FIGURE 4.1 Three-dimensional system (bulk), two-dimensional system (quantum well), one-dimensional system (quantum wire), and zero-dimensional system (quantum dot).

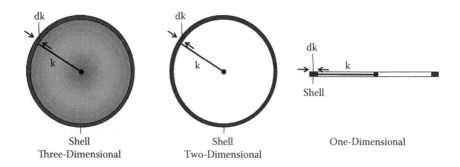

FIGURE 4.2 Shell shapes for given energy of a free electron in three-, two-, and one-dimensional systems.

Equations (4.3) and (4.4) are derived from Equation (4.2).

For the three-dimensional system, the number of states N_{3D} contained in the shell is given by the following expression because two states with up and down spin states exist in a volume of $(2\pi/L)^3$ in the k-space.

$$N_{3D} = \left[2/(2\pi/L)^3\right]\int_{shell} dk = \left[2/(2\pi/L)^3\right]4\pi k^2 dk \tag{4.5}$$

By substituting Equations (4.3) and (4.4) into Equation (4.5),

$$N_{3D} = D(E)dE, \tag{4.6}$$

$$D(E) = (L^3/2\pi^2)(2m/\hbar^2)^{3/2}\sqrt{E} \tag{4.7}$$

are obtained. Here, $D(E)$ represents the density of states at energy E.

For two-dimensional systems, the number of states N_{2D} contained in the shell is given by the following expression because two states exist in an area of $(2\pi/L)^2$.

$$N_{2D} = \left[2/(2\pi/L)^2\right]\int_{shell} dk = \left[2/(2\pi/L)^2\right]2\pi k dk \tag{4.8}$$

By substituting Equations (4.3) and (4.4) into Equation (4.8)

$$D(E) = (L^2/\pi)(m/\hbar^2) \tag{4.9}$$

is obtained.

For one-dimensional systems, the number of states N_{1D} contained in the shell is given by the following expression because two states exist in a length of $2\pi/L$.

$$N_{1D} = \left[2/(2\pi/L)\right]\int_{shell} dk = \left[2/(2\pi/L)\right]2dk \tag{4.10}$$

By substituting Equations (4.3) and (4.4) into Equation (4.10)

$$D(E) = (L/\pi)\sqrt{2m/\hbar^2}/\sqrt{E} \tag{4.11}$$

is obtained.

Figure 4.3 shows schematic diagrams of the density of states for the three-, two-, and one-dimensional systems. In three-dimensional systems, the density of states monotonically increases with energy. In two-dimensional systems, the density of states does not depend on energy. In one-dimensional systems, the density of states monotonically decreases with energy.

In Figure 4.4, dependence of electron confinement and density-of-state profiles on dimensionality is schematically summarized. In three-dimensional systems, electrons freely move in three directions, which causes the density of states to be small near the band edges and to increase with distance from the band edges. As a result, the energy gap between the conduction bands and the valence bands is diffused, giving rise to a broad absorption band. In two-dimensional systems, the translational motion freedom of electrons is reduced from three to two directions, which suppresses the increase in density of states with distance from the edges. In

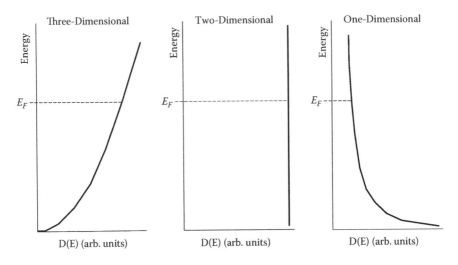

FIGURE 4.3 Schematic diagrams of the density of states for three-, two-, and one-dimensional systems.

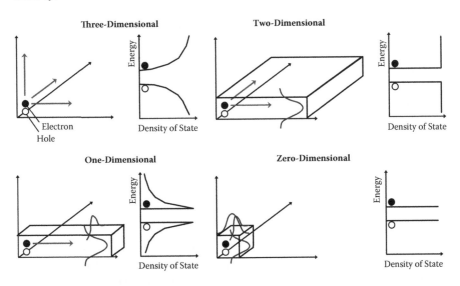

FIGURE 4.4 Dependence of electron confinement and density-of-state profiles on dimensionality.

one-dimensional systems, the electron translational motion is allowed only in one direction, which causes the density of states to be the largest at the band edges and decrease with distance from the edges. As a result, the energy gap is defined clearly, giving rise to a sharp absorption band near the energy gap. In zero-dimensional systems, which provide a situation similar to that of atoms or small molecules, electron translational motion is not allowed, which causes the δ-function-like density-of-states profile, resulting in a sharp absorption band near the energy gap.

As described in Chapter 1, quantum dots correspond to dimensionality between zero and one. The 0.5-dimensional systems correspond to long quantum dots, and the zero-dimensional systems correspond to short quantum dots. Dimensionality control between zero and one, that is, the sizes of quantum dots, can be achieved by inserting molecular units that disconnect the spread of π-electrons in conjugated polymer wires.

4.2   QUANTUM DOT CONSTRUCTION IN CONJUGATED POLYMERS BY MLD

4.2.1   MQD FABRICATION BY ARRANGING TWO KINDS OF MOLECULES

Molecular sequence control in polymer wires is interesting from the perspective of controlling optical and electronic properties of polymer wires to provide various kinds of photonic/electronic devices such as nonlinear optical circuits, photovoltaic devices, electroluminescent (EL) devices, biophotonic devices, molecular devices, and so on. As described in Chapter 5, it was predicted by the molecular orbital (MO) calculations that electro-optic (EO) effects in conjugated polymer wires are greatly enhanced by controlling the wavefunction shapes in the wire to optimize the wavefunction separation and overlap between the excited state and the ground state [2]. For wavefunction shape optimization, dimensionality control somewhere between zero and one (i.e., quantum dot length control) is important. The dimensionality control is also effective to improve photovoltaic devices with polymer MQDs, which means polymer wires with MQD, as described in Section 9.1.

In order to realize such wavefunction control and quantum dot length control in polymer wires, molecular sequence control is essential. A promising approach to fabricate practical films of these polymers is connecting molecules one by one in order of preference. This monomolecular-step polymer wire growth process can be achieved by MLD.

For experimental demonstrations of dimensionality control between zero and one in conjugated polymer wires, we constructed quantum dots in the wires using source molecules as shown in Figure 4.5 [3]. The reaction between –CHO and –NH$_2$ is the same as described in Section 3.3.1 for poly-AM formation. Molecules A and B are connected by double bonds through –CHO and –NH$_2$ groups to form polymer wires. For molecule A, terephthalaldehyde (TPA) was used. For molecule B, p-phenylenediamine (PPDA), diaminodiphenyl ether (DDE), 4,4'-diaminodiphenyl sulfide (DDS), and 4,4'-diaminodiphenylmethane (DDM) were used. We fabricated quantum wire structures with long π-conjugated systems and quantum dot structures with confined π-conjugated systems in one-dimensional polymer wires. Since only two kinds of molecules, that is, molecules A and B are used in this experiment, the same polymer structures are obtained either by MLD or by vacuum deposition polymerization.

Molecules were introduced into a vacuum chamber in gas phase from temperature-controlled K cells to perform vacuum deposition polymerization. Molecular gas pressures were $1–8 \times 10^{-1}$ Pa, which was controlled by cell temperatures ranging from 120 to 150°C. The thickness of deposited films was measured, assuming that the film density is 1, using a quartz oscillator thickness monitor contacting the

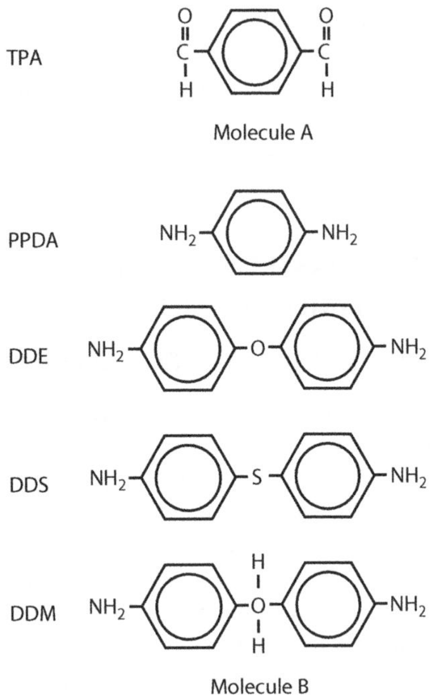

FIGURE 4.5 Source molecules for making quantum dot structures based on poly-AM. From T. Yoshimura, S. Tatsuura, W. Sotoyama, A. Matsuura, and T. Hayano, "Quantum wire and dot formation by chemical vapor deposition and molecular layer deposition of one-dimensional conjugated polymer," *Appl. Phys. Lett.* **60**, 268–270 (1992).

substrate holder. The substrate temperature T_S was measured using a thermocouple on the holder.

Figures 4.6(a) and (b) show the absorption spectra of poly-AM [TPA/PPDA] film grown using TPA and PPDA. The film growth rate was adjusted by cell temperatures, that is, molecular gas pressures. The growth rate increases with gas pressure. The spectra were measured at room temperature with incident light angles of 0° and 60°. Absorption bands appear in 400 to 500 nm with a sharp absorption peak of exciton observed near 500 nm, while methanol solutions of TPA and PPDA exhibit little absorption in the visible region. This indicates that long conjugated systems, that is, quantum wires, grow. The half width of the exciton absorption line for the long-wavelength side is about 50 meV. Assuming that the exciton is similar in size to that in polydiacetylene, the conjugated length is estimated to be more than the exciton length, namely, 5 nm [4]. When the film is deposited at a rate of 10 nm/min, the exciton peak is stronger for 0° than for 60°. In film deposited at 3 nm/min, the exciton peak is more dominant for 60° than for 0°. In general, exciton absorption is strong for light polarization in the wire direction. These results suggest that polymer wires stretch to the in-plane direction for 10 nm/min, while for 3 nm/min, wires tend to stretch out of the in-plane direction. Scanning electron microscope (SEM) observations with magnification of ×25,000 reveal that a beltlike texture appears on

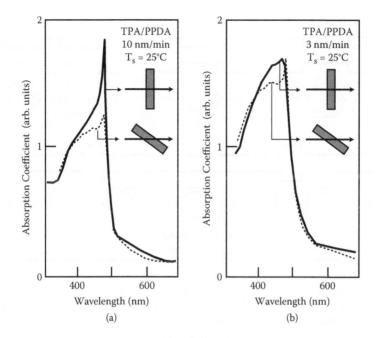

FIGURE 4.6   Absorption spectra of poly-AM [TPA/PPDA] film measured at room temperature with incident light angles of 0° and 60°. From T. Yoshimura, S. Tatsuura, W. Sotoyama, A. Matsuura, and T. Hayano, "Quantum wire and dot formation by chemical vapor deposition and molecular layer deposition of one-dimensional conjugated polymer," *Appl. Phys. Lett.* **60**, 268–270 (1992).

the surface of the film deposited with 10 nm/min, while no such surface structure appears for 3 nm/min. The difference in surface morphology is attributable to the wire orientation.

Note that exciton absorption becomes strong with an increased deposition rate. No exciton peak is observed for a slow deposition rate of 0.05 nm/min. We also found that the exciton absorption is enhanced by maintaining the substrate temperature to −10°C during deposition. These results indicate that, to obtain long conjugated systems, deposition conditions must be carefully optimized. It is speculated that thermal disturbance would induce a disorder in the conjugated wires.

Figure 4.7 diagrams molecules and polymer wires with their corresponding electronic potential energy curves. Benzene rings in TPA, PPDA, and DDE are regarded as a short quantum dot of ~0.5 nm in length. Poly-AM [TPA/DDE] is a polymer MQD formed by combining TPA and DDE. In the polymer MQD, the π-conjugated part is regarded as a long quantum dot with barriers of −O−. The quantum dot contains three benzene rings and the length is ~2 nm. In poly-AM [TPA/PPDA] formed by combining TPA and PPDA, π-conjugation is spread all along the wire.

Figure 4.8 shows the absorption spectra of TPA, PPDA, DDE, poly-AM [TPA/DDE], and poly-AM [TPA/PPDA]. In poly-AM [TPA/PPDA], a sharp peak of excitons is observed near 500 nm in wavelength. In poly-AM [TPA/DDE] the absorption

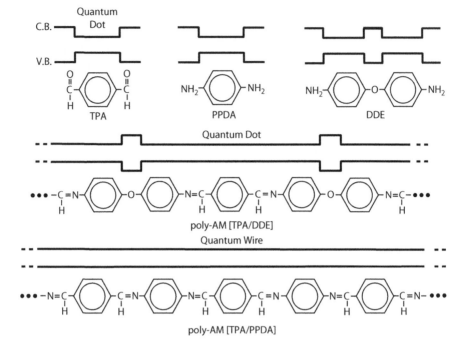

FIGURE 4.7 Molecules and polymer wires, where quantum dots and quantum wires are formed, with their corresponding electronic potential energy curves. From T. Yoshimura, S. Tatsuura, W. Sotoyama, A. Matsuura, and T. Hayano, "Quantum wire and dot formation by chemical vapor deposition and molecular layer deposition of one-dimensional conjugated polymer," *Appl. Phys. Lett.* **60**, 268–270 (1992).

band appears in the shorter-wavelength (higher-energy) side of that in poly-AM [TPA/PPDA] due to electron confinement in the quantum dot structure of ~2 nm in length. In TPA, PPDA, and DDE, the absorption band exhibits further higher-energy shift, indicating strong electron confinement in the short quantum dots.

By using DDS or DDM as molecule B, we fabricated poly-AM [TPA/DDS] and poly-AM [TPA/DDM]. In the former, –S– makes a barrier, and in the latter, –CH$_2$– makes a barrier as –O– does in poly-AM [TPA/DDE]. Figure 4.9 shows the influence of barrier bonds on the absorption peak energy. In the order of –S–, –O–, and –CH$_2$–, the absorption peak shifts to the higher-energy side, indicating that the barrier height increases to produce a strong electron confinement.

Energy of electrons confined in a quantum dot with infinite potential barrier height is expressed as follows in the free electron model.

$$E_n = \hbar^2\pi^2 n^2/(2m*m_0 L^2).$$ (4.12)

Here, $n = 1, 2, ..., m_0$ is electron mass, $m*$ is effective mass in units of m_0, and L is dot length. Assuming that the effective mass is 1 and the absorption peak energy for $L = \infty$ equals that of poly-AM [TPA/PPDA], the excitation energy for the lowest allowed transition is calculated as a function of L as shown in Figure 4.10.

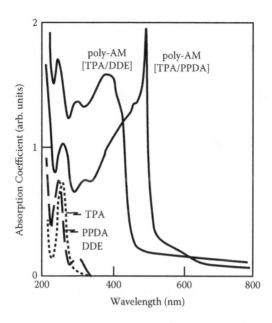

FIGURE 4.8   Absorption spectra of TPA, PPDA, DDE, poly-AM [TPA/DDE], and poly-AM [TPA/PPDA]. From T. Yoshimura, S. Tatsuura, W. Sotoyama, A. Matsuura, and T. Hayano, "Quantum wire and dot formation by chemical vapor deposition and molecular layer deposition of one-dimensional conjugated polymer," *Appl. Phys. Lett.* **60**, 268–270 (1992).

Experimental and calculated results agree fairly well, confirming the electron confinement in the quantum dots.

These results demonstrate the possibility of controlling dimensionality to enhance the optical nonlinearity in conjugated polymer wires and to optimize electron confinement effects for photovoltaic devices.

4.2.2   MQDs FABRICATED BY ARRANGING THREE KINDS OF MOLECULES

As described in Section 4.2.1, MQD structures can be constructed in polymer wires by arrangements of two kinds of molecules. In this case, however, the MQD structures obtained by MLD are not different from those obtained by vacuum deposition polymerization because the two molecules are automatically connected alternately even in vacuum deposition polymerization.

On the contrary, when three kinds of molecules are used for polymer wire growth, the available polymer wire structures are completely different between MLD and vacuum deposition polymerization. In the case of vacuum deposition polymerization, the three kinds of molecules are contained in the polymer wires with random sequences. In the case of MLD, on the other hand, the three kinds of molecules are contained in the polymer wires with designated arrangements. Therefore, the essential advantage of MLD appears when more than three kinds of molecules are used for polymer wire growth.

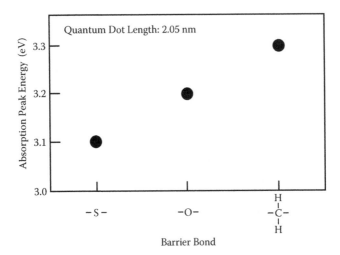

FIGURE 4.9 Influence of barrier bonds on the absorption peak energy. From T. Yoshimura, S. Tatsuura, W. Sotoyama, A. Matsuura, and T. Hayano, "Quantum wire and dot formation by chemical vapor deposition and molecular layer deposition of one-dimensional conjugated polymer," *Appl. Phys. Lett.* **60**, 268–270 (1992).

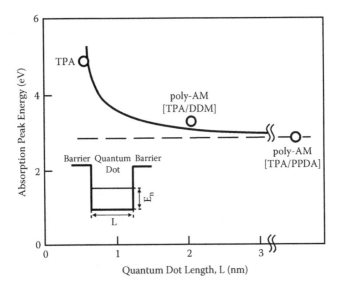

FIGURE 4.10 Dependence of absorption peak energy on quantum dot lengths. From T. Yoshimura, S. Tatsuura, W. Sotoyama, A. Matsuura, and T. Hayano, "Quantum wire and dot formation by chemical vapor deposition and molecular layer deposition of one-dimensional conjugated polymer," *Appl. Phys. Lett.* **60**, 268–270 (1992).

Considering the previously described background, we attempted to construct MQDs in polymer wires by MLD using three kinds of molecules, which is known as *three-molecule MLD* [5,6]. Figure 4.11 shows three kinds of molecules, that is, TPA, PPDA, and oxalic dihydrazide (ODH), and reactions between them for three-molecule MLD. TPA and PPDA are connected with a double bond, allowing the wavefunction of π-electrons to be spread over both molecules. The connection of TPA and ODH gives a series of single bonds, which cut the wavefunction at the bond. By using these bond characteristics, quantum dots can be formed in polymer wires.

Figure 4.12 shows a three-molecule MLD process for constructing polymer MQD of OTPT, where molecules are connected in a sequence of ---ODH-TPA-PPDA-TPA-ODH---. The region between two ODHs is regarded as a quantum dot since the wavefunction of π-electrons is spread through the region. The quantum dot length is around 2 nm. In Figure 4.13, structures of poly-AM quantum wire, and polymer MQDs of OTPTPT, OTPT, and OT are summarized. For OTPTPT, molecules are connected in a sequence of ---ODH-TPA-PPDA-TPA-PPDA-TPA-ODH---, resulting in a long quantum dot length of ~3 nm. For OT, ODH and TPA are alternately connected, resulting in a very short quantum dot length of less than 1 nm. For poly-AM, TPA and PPDA are alternately connected, so the wavefunction of π-electrons are spread through the polymer wire, creating a quantum wire.

In the present experiment, no seed cores are formed on the glass substrates. Therefore, two-sided growth MLD (shown in Figure 3.1(b)) occurs to provide symmetric growth of polymer wires. For example, in the case of OTPT, polymer wires grow by MLD as shown in Figure 4.14.

Poly-AM, OTPTPT, OTPT, and OT were experimentally fabricated by three-molecule MLD. The MLD equipment is the carrier gas type. Carrier gas of nitrogen (N_2) is employed to introduce molecular gases of TPA, PPDA, and ODH in temperature-controlled molecular cells onto a glass substrate surface. MLD operation can be obtained by sequentially switching the three valves for molecular gas injection. Excess gases are removed by a rotary pump. The molecular cell temperature is 50°C for PPDA and ODH, and 25°C for TPA. The substrate surface is exposed to

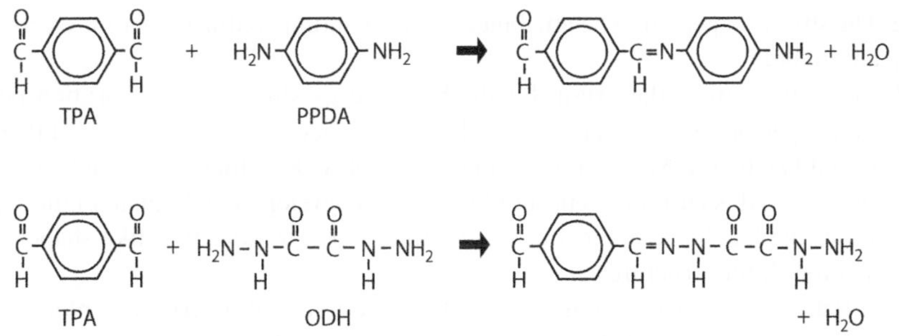

FIGURE 4.11 Molecules and reactions for three-molecule MLD. From T. Yoshimura, A. Oshima, D. Kim, and Y. Morita, "Quantum dot formation in polymer wires by three-molecule molecular layer deposition (MLD) and applications to electro-optic/photovoltaic devices," *ECS Transactions* **25** (4), "Atomic layer deposition applications 5," 15–25 (2009).

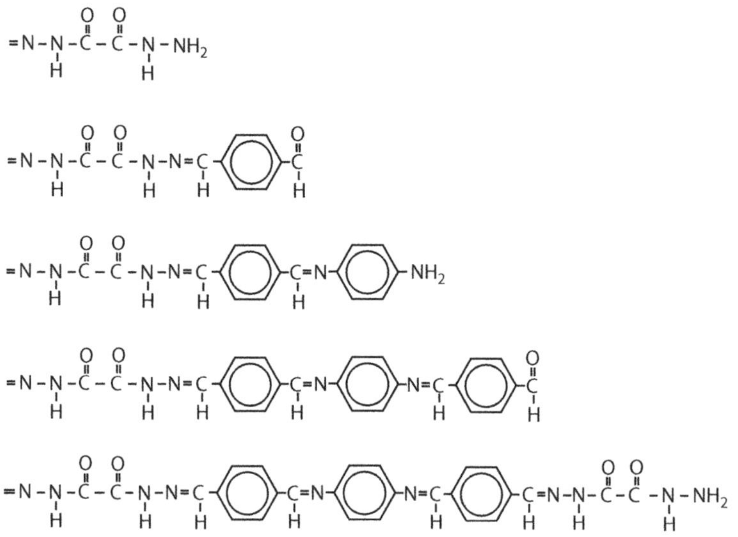

FIGURE 4.12 Three-molecule MLD process for constructing a polymer MQD of OTPT. From T. Yoshimura, A. Oshima, D. Kim, and Y. Morita, "Quantum dot formation in polymer wires by three-molecule molecular layer deposition (MLD) and applications to electro-optic/photovoltaic devices," *ECS Transactions* **25** (4), "Atomic layer deposition applications 5," 15–25 (2009).

molecular gas for 5 min in each step. For switching molecular gases, after removing previous molecular gas by closing the valve for the gas and pumping for 5 min, the valve for the next molecular gas is opened. Switching step count is around 20. This implies that six quantum dots are contained in a polymer wire for OTPTPT, ten quantum dots for OTPT, and twenty quantum dots for OT.

Figure 4.15 shows absorption spectra of poly-AM quantum wire, and polymer MQDs of OTPTPT, OTPT, and OT fabricated by three-molecule MLD. In the order of poly-AM, OTPTPT, OTPT, and OT, namely, with decreasing the quantum dot lengths, the absorption peak energy shifts to the shorter-wavelength (higher-energy) side. The shift is expected to be attributed to the quantum confinement of π-electrons in quantum dots.

In order to confirm the origin for the high energy shift of the absorption peak energy, we performed calculations of absorption spectra for poly-AM, OTPTPT, OTPT, and OT by the MO method using WinMOPACK, which is the MO calculation software and is currently a part of SCIGRESS developed by Fujitsu. Figure 4.16 shows their molecular structures after structural optimization. Wirelike shapes are obtained for all the structures.

In Figure 4.17, experimental results of absorption peak energy are plotted by rectangles as a function of the quantum dot length. In the figure, absorption peak energy calculated using the MO method is also plotted by circles. Since there is a 0.3-eV offset between the calculated value and the experimental value in the absorption peak energy of the poly-AM quantum wire, the calculated absorption peak energy in Figure 4.17 is plotted by taking the experimental absorption peak

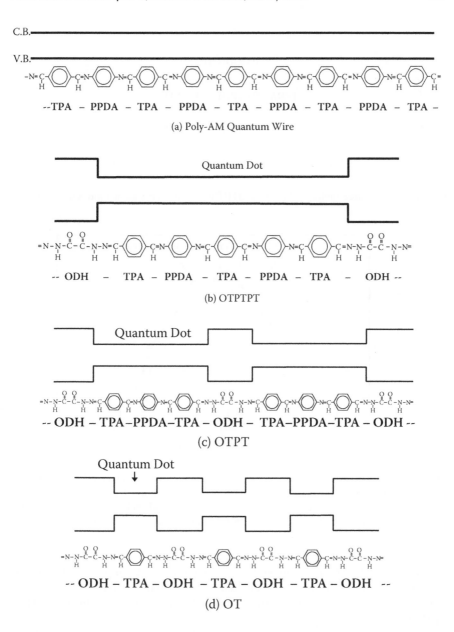

FIGURE 4.13 Structures of a poly-AM quantum wire and polymer MQDs of OTPTPT, OTPT, and OT formed by three-molecule MLD. From T. Yoshimura, A. Oshima, D. Kim, and Y. Morita, "Quantum dot formation in polymer wires by three-molecule molecular layer deposition (MLD) and applications to electro-optic/photovoltaic devices," *ECS Transactions* **25** (4), "Atomic layer deposition applications 5," 15–25 (2009).

ODH

TPA-ODH-TPA

PPDA-TPA-ODH-TPA-PPDA

TPA-PPDA-TPA-ODH-TPA-PPDA-TPA

ODH-TPA-PPDA-TPA-ODH-TPA-PPDA-TPA-ODH

FIGURE 4.14   Three-molecule MLD for OTPT on a glass substrate without seed cores.

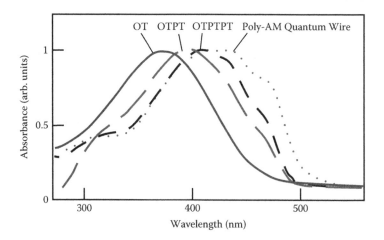

FIGURE 4.15   Absorption spectra of a poly-AM quantum wire, OTPTPT, OTPT, and OT. From T. Yoshimura, A. Oshima, D. Kim, and Y. Morita, "Quantum dot formation in polymer wires by three-molecule molecular layer deposition (MLD) and applications to electro-optic/photovoltaic devices," *ECS Transactions* **25** (4), "Atomic layer deposition applications 5," 15–25 (2009).

energy of the poly-AM quantum wire as the standard. It is found from the figure that the experimental results and the calculated results are fairly coincident, suggesting that the absorption peak energy shift is attributed to the quantum confinement effect in MQD.

In addition, a result derived from the free-electron model in a quantum dot is presented in Figure 4.17. Energy of electrons confined in a quantum dot with infinite potential barrier height is expressed by Equation (4.12) in Section 4.2.1. Assuming that the absorption peak energy for L = ∞ equals that of the poly-AM quantum wire, the excitation energy for the lowest allowed transition is plotted. It is found that the experimental results and the calculated results based on the free-electron model agree well. This suggests that the π-electrons in the polymer wires can be regarded as free electrons in quantum dots.

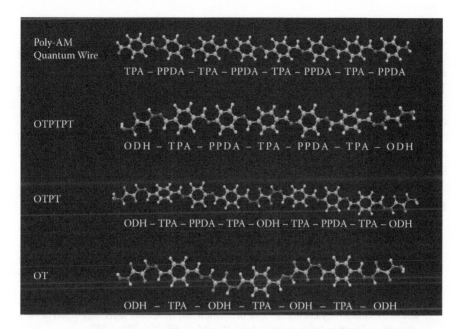

FIGURE 4.16 Molecular structures of poly-AM quantum wire, OTPTPT, OTPT, and OT after structural optimization. From T. Yoshimura, A. Oshima, D. Kim, and Y. Morita, "Quantum dot formation in polymer wires by three-molecule molecular layer deposition (MLD) and applications to electro-optic/photovoltaic devices," *ECS Transactions* **25** (4), "Atomic layer deposition applications 5," 15–25 (2009).

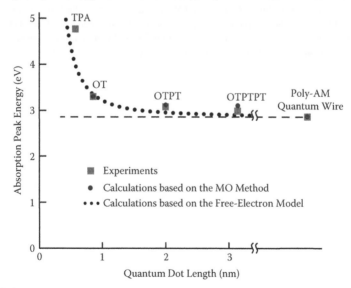

FIGURE 4.17 Dependence of absorption peak energy on quantum dot lengths. From T. Yoshimura, A. Oshima, D. Kim, and Y. Morita, "Quantum dot formation in polymer wires by three-molecule molecular layer deposition (MLD) and applications to electro-optic/photovoltaic devices," *ECS Transactions* **25** (4), "Atomic layer deposition applications 5," 15–25 (2009).

As described previously, we succeeded in controlling molecular arrangements in conjugated polymer wires with designated sequences of three kinds of molecules by three-molecule MLD and fabricating polymer MQDs. Peak energy shift of the absorption spectra of the MQDs due to the quantum confinement effect was observed. The method of controlling the arrangements of molecules of more than three kinds will enable us to control the wavefunction shapes in the polymer wires with significant freedom, providing high-performance nonlinear optical circuits, photovoltaic devices, and molecular devices. By introducing donors and acceptors into the wires, new conjugated polymer structures for use in high-performance photonic/electronic devices will be fabricated.

REFERENCES

1. T. Yoshimura, *Molecular nano systems: Applications to optoelectronic computers and solar energy conversion*, Corona Publishing Co., Ltd., Tokyo (2007) [in Japanese].
2. T. Yoshimura, "Enhancing second-order nonlinear optical properties by controlling the wave function in one-dimensional conjugated molecules," *Phys. Rev.* **B40**, 6292–6298 (1989).
3. T. Yoshimura, S. Tatsuura, W. Sotoyama, A. Matsuura, and T. Hayano, "Quantum wire and dot formation by chemical vapor deposition and molecular layer deposition of one-dimensional conjugated polymer," *Appl. Phys. Lett.* **60**, 268–270 (1992).
4. B. I. Greene, J. Orenstein, R. R. Millard, and L. R. Williams, "Nonlinear optical response of excitons confined to one dimension," *Phys. Rev. Lett.* **58**, 2750–2753 (1987).
5. A. Oshima and T. Yoshimura, "Controlling sequences of three molecules and quantum dot lengths in conjugated polymer wires by molecular layer deposition," AVS, 9th International Conference on Atomic Layer Deposition, Monterey, California, 149 (2009).
6. T. Yoshimura, A. Oshima, D. Kim, and Y. Morita, "Quantum dot formation in polymer wires by three-molecule molecular layer deposition (MLD) and applications to electro-optic/photovoltaic devices," *ECS Transactions* **25** (4) "Atomic layer deposition applications 5," from the 216th ECS Meeting, Vienna, Austria, 15–25 (2009).

5 Theoretical Predictions of Electro-Optic (EO) Effects in Polymer Wires

One of the important applications of molecular layer deposition (MLD) is nonlinear optical polymers that are the key materials for photonic devices such as light modulators, optical switches, tunable filters, wavelength converters, and so on.

In the present chapter, nonlinear optical effects, especially EO effects in polymer wires with designated molecular sequences, are predicted theoretically [1–3].

5.1 MOLECULAR ORBITAL METHOD

In order to design EO polymer wires, the molecular orbital (MO) method is used. In the present section, the MO method is briefly reviewed.

Figure 5.1 shows models for Hamiltonian in the MO calculations. In this example, the system consists of N electrons. The Hamiltonian of the whole system is written as follows.

$$H = \sum_{i=1} \left[-\frac{\hbar^2}{2m} \Delta_i + V(\mathbf{q}_i) \right] + \sum_{i>j} \frac{e^2}{|\mathbf{q}_i - \mathbf{q}_j|} \tag{5.1}$$

Here, $i(i = 1, 2, \ldots, N)$ and $j(j = 1, 2, \ldots, N)$ represent i^{th} electron and j^{th} electron, respectively. $\mathbf{q}_i$ is the position of the ith electron, $\mathbf{q}_j$ is the position of the jth electron, Δ_i is the Laplacian for the ith electron, $V(\mathbf{q}_i)$ is the interaction energy between the ith electron and nuclei, m is the mass of an electron, and e is the charge of an electron. The second term represents interaction energy arising from the Coulomb force between electrons.

A one-electron Hamiltonian for the ith electron is expressed as follows:

$$H_i = -\frac{\hbar^2}{2m} \Delta_i + V(\mathbf{q}_i) + \sum_{i \neq j} \int_{-\infty}^{\infty} \phi_j^*(\mathbf{q}_j) \frac{e^2}{|\mathbf{q}_i - \mathbf{q}_j|} \phi_j(\mathbf{q}_j) d\mathbf{q}_j \tag{5.2}$$

Here, $\phi_j(\mathbf{q}_j)$ is the wavefunction of the jth electron. $\phi_j(\mathbf{q}_j)$ is called the molecular orbital. The second term represents the interaction energy between the ith electron and the jth electron, where the jth electron is regarded as an electron cloud. The interaction energy is calculated as an expected value determined by using the charge density distribution of the jth electron, $e\phi_j^*(\mathbf{q}_j) \phi_j(\mathbf{q}_j)$.

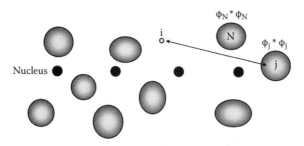

(a) Model for Hamiltonian of Whole System

(b) Model for Hamiltonian of the ith Electron

FIGURE 5.1   Models for Hamiltonian in the molecular orbital calculations.

Schrödinger equations for the whole system Hamiltonian and the one-electron Hamiltonian are respectively given as follows:

$$i\hbar \frac{\partial \Psi}{\partial t} = H\Psi \tag{5.3}$$

$$i\hbar \frac{\partial \phi_i}{\partial t} = H_i \phi_i \tag{5.4}$$

Here, Ψ is a wavefunction of the whole system.

The molecular orbitals can be expressed as linear combinations of wavefunctions in atoms (atomic orbitals) χ_r.

$$\phi_i(\mathbf{q}_i) = \sum_r c_{ir}\chi_r \quad i = 1, 2, \cdots, N \tag{5.5}$$

An example of the relationship between ϕ_i and χ_r is schematically shown in Figure 5.2 for a case of C_2H_2. ϕ_i is built up by χ_r with amplitude of c_{ir}. c_{ir}s are determined by the variation principle, namely, by finding a set of c_{ir}s that minimize ε_i in the following formula.

$$\int_{-\infty}^{\infty} \phi_i^*(\mathbf{q}_i) H_i \phi_i(\mathbf{q}_i) d\mathbf{q}_i = \varepsilon_i \int_{-\infty}^{\infty} \phi_i^*(\mathbf{q}_i) \phi_i(\mathbf{q}_i) d\mathbf{q}_i \tag{5.6}$$

The minimized ε_i gives the energy eigen value for the molecular orbital of $\phi_i = \Sigma_r c_{ir}\chi_r$. In a system with N electrons, N molecular orbitals are derived. However, in real systems,

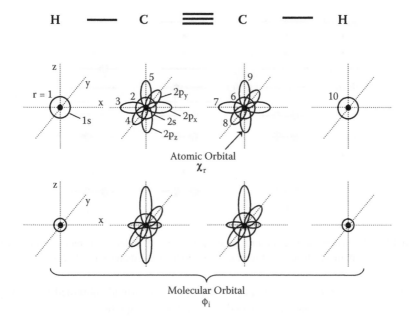

FIGURE 5.2   Schematic illustration of the relationship between atomic orbitals and molecular orbitals.

two spin states (α, β) are involved for one molecular orbital, resulting in $2N$ molecular orbitals. So, we should renumber the molecular orbitals including spin states as ϕ_i ($i = 1, 2, \cdots, 2N$).

Figure 5.3 shows schematic illustrations for energy levels of molecular orbitals and electron occupation in many-electron wavefunctions built up from the molecular orbitals. The many-electron wavefunctions represent the actual electronic state of the whole system. The wavefunction for the ground state is expressed as follows,

$$\Psi_g = \frac{1}{\sqrt{N!}} \sum_p (-1)^p \hat{P} \phi_1(\mathbf{q}_1) \phi_2(\mathbf{q}_2) \cdots \phi_N(\mathbf{q}_N), \qquad (5.7)$$

where, $\hat{P}$ is a permutation operator and p is the number of permutations. Electrons occupy molecular orbitals $\phi_1, \phi_2, \cdots, and\ \phi_N$.

In the 0th approximation, many-electron wavefunctions for the excited states are written as follows

$$\Phi_{i \to j} = \frac{1}{\sqrt{N!}} \sum_p (-1)^p \hat{P} \phi_1(\mathbf{q}_1) \phi_2(\mathbf{q}_2) \cdots \phi_{i-1}(\mathbf{q}_{i-1}) \phi_{i+1}(\mathbf{q}_{i+1}) \cdots \phi_N(\mathbf{q}_N) \phi_j(\mathbf{q}_i) \quad (5.8)$$

where, $i \to j$ represents that an excited state, where an electron is excited from ϕ_i to ϕ_j. $\Phi_{i \to j}$ is called the *configuration function*.

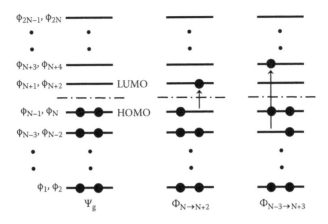

FIGURE 5.3 Schematic illustrations for energy levels of molecular orbitals and electron occupation in many-electron wavefunctions built up from the molecular orbitals.

In order to improve the accuracy of the molecular orbitals obtained previously, the variation principle is again applied to the following equation.

$$\int_{-\infty}^{\infty} \Psi_g^* H \Psi_g \, d\mathbf{q}_1 d\mathbf{q}_2 \cdots d\mathbf{q}_N = E \int_{-\infty}^{\infty} \Psi_g^* \Psi_g \, d\mathbf{q}_1 d\mathbf{q}_2 \cdots d\mathbf{q}_N \tag{5.9}$$

Then, a set of improved c_is that minimize E are obtained. The minimized E gives the ground state energy E_g for Ψ_g.

In general, the excited states are expressed as linear combinations of $\Phi_{i \to j}$ as follows.

$$\Psi_n = \sum_{i,j} C_{n,i \to j} \Phi_{i \to j} \tag{5.10}$$

$C_{n,\,i \to j}$s are determined by the variation principle, namely, by finding a set of $C_{n,\,i \to j}$s that minimize E_n in the following formula.

$$\int_{-\infty}^{\infty} \Psi_n^* H \Psi_n d\mathbf{q}_1 d\mathbf{q}_2 \cdots d\mathbf{q}_N = E_n \int_{-\infty}^{\infty} \Psi_n^* \Psi_n d\mathbf{q}_1 d\mathbf{q}_2 \cdots d\mathbf{q}_N \tag{5.11}$$

The minimized E_n gives the energy eigen values for the excited states of $\Psi_n = \Sigma_{i,j} C_{n,i \to j} \Phi_{i \to j}$.

Thus, wavefunctions and energy eigen states are obtained.

5.2 NONLINEAR OPTICAL EFFECTS

Nonlinear optical phenomena can be understood using the spring analogy shown in Figure 5.4. In a spring, displacement x is proportional to external force F for small F. With increasing F, the spring stretches rapidly with a nonlinear relationship between

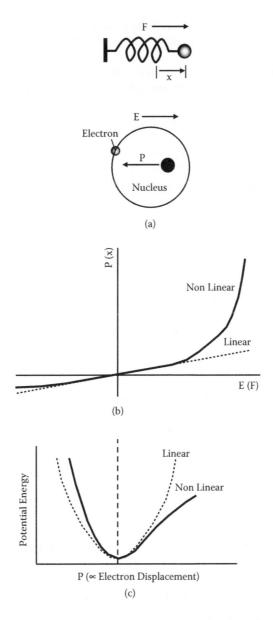

FIGURE 5.4   Model for the nonlinear optical phenomena using the spring analogy.

x and F. Nonlinear optical phenomena can be similarly described by replacing x with polarization P and F with electric field E as follows.

$$P = \varepsilon_0 \chi^{(1)} E + \varepsilon_0 \chi^{(2)} E^2 + \varepsilon_0 \chi^{(3)} E^3 + \cdots \tag{5.12}$$

Here, ε_0 is the dielectric constant in vacuum, $\chi^{(1)}$ is linear susceptibility, and $\chi^{(2)}$ and $\chi^{(3)}$ are nonlinear susceptibilities. When E is small, P is proportional to E.

With increasing E, the contribution of the E^2 or E^3 term becomes dominant. The E^2 and E^3 terms induce second-order and third-order nonlinear optical effects, respectively.

E can be expressed by a sum of electric fields of light with an angular frequency of ω, $E_{opt} = Ee^{-j\omega t}$, and an externally applied electric field, E_{DC}, as follows:

$$E = E_{opt} + E_{DC} \tag{5.13}$$

The term E^2 gives $E_{opt}^2, E_{DC}E_{opt}, E_{DC}^2$ which are origins of the second-order nonlinear optical effects, and the term E^3 gives $E_{opt}^3, E_{opt}^2 E_{DC}, E_{opt}E_{DC}^2, E_{DC}^3$ which are origins of the third-order nonlinear optical effects. When we focus on the term E_{opt}^2, Equation (5.12) can be written as

$$P = \varepsilon_0 \chi^{(1)} E_0 e^{-j\omega t} + \varepsilon_0 \chi^{(2)} E_0^2 e^{-j2\omega t}. \tag{5.14}$$

Polarization oscillating at 2ω generates light with angular frequency of 2ω, that is, the second harmonic generation (SHG) arises. When we focus on the term E_{opt}^3, Equation (5.12) can be written as

$$P = \varepsilon_0 \chi^{(1)} E_0 e^{-j\omega t} + \varepsilon_0 \chi^{(3)} E_0^3 e^{-j3\omega t}. \tag{5.15}$$

Polarization contains a 3ω component, generating light with angular frequency of 3ω, that is, the third harmonic generation (THG) arises.

When we focus on the term $E_{DC}E_{opt}$, Equation (5.12) can be written as

$$P = \varepsilon_0 \chi^{(1)} E_{opt} + \varepsilon_0 \chi^{(2)} E_{opt} E_{DC} = \left(\varepsilon_0 \chi^{(1)} + \varepsilon_0 \chi^{(2)} E_{DC} \right) E_{opt}. \tag{5.16}$$

Since the refractive index, n, is a function of $\varepsilon_0 \chi^{(1)} + \varepsilon_0 \chi^{(2)} E_{DC}$, n can be expressed as follows.

$$n = f\left(\varepsilon_0 \chi^{(1)} + \varepsilon_0 \chi^{(2)} E_{DC} \right) \tag{5.17}$$

Considering the relationship of $\varepsilon_0 \chi^{(1)} \gg \varepsilon_0 \chi^{(2)} E_{DC}$, n is rewritten using a Taylor expansion as

$$n \approx f\left(\varepsilon_0 \chi^{(1)} \right) + f'\left(\varepsilon_0 \chi^{(1)} \right) \varepsilon_0 \chi^{(2)} E_{DC}. \tag{5.18}$$

By putting $n_0 = f\left(\varepsilon_0 \chi^{(1)} \right)$ and $r \propto f'\left(\varepsilon_0 \chi^{(1)} \right) \varepsilon_0 \chi^{(2)}$

$$n = n_0 + const\ r\ E_{DC} \tag{5.19}$$

is obtained. This formula implies that the refractive index changes linearly with an applied electric field, which is the Pockels effect. r is the Pockels coefficient.

When we focus on the term $E_{DC}{}^2 E_{opt}$, Equation (5.12) can be written as

$$P = (\varepsilon_0 \chi^{(1)} + \varepsilon_0 \chi^{(3)} E_{DC}{}^2) E_{opt}. \tag{5.20}$$

As in the Pockels effect,

$$n = n_0 + const \ R \ E_{DC}{}^2 \tag{5.21}$$

is derived. This implies that the refractive index changes in proportion with E^2, which is the Kerr effect. R is the Kerr coefficient. The Pockels effect and the Kerr effect are used for various kinds of EO devices such as light modulators, optical switches, tunable filters, and so on.

The relationship between E and P can be translated into the electronic potential energy vs. polarization curve as Figures 5.4(b) and (c) show. When E and P are linearly related, the potential energy curve is parabolic. This corresponds to the fact that the potential energy of a spring is proportional to x^2. When nonlinear terms predominate, the potential energy curve deviates from the parabola.

In Figure 5.5, various kinds of nonlinear optical effects are drawn visually. SHG can be understood by the illustration shown in Figure 5.5(a). Due to the term E^2 in Equation (5.12), the potential energy vs. polarization curve becomes asymmetric. Therefore, when light with angular frequency of ω is introduced, asymmetric responses of P are induced, resulting in a P component with 2ω. The P oscillating at 2ω emits light with angular frequency of 2ω. For THG, due to the term E^3 in Equation (5.12), the potential energy vs. polarization curve deviates from the parabola symmetrically as shown in Figure 5.5(b). Then, when light with an angular frequency of ω is introduced, the P component with 3ω is induced, emitting light with an angular frequency of 3ω.

For the Pockels effect, in which the refractive index changes linearly with an applied electric field E_{DC}, the potential energy vs. polarization curve deviates from the parabola asymmetrically as shown in Figure 5.5(c). When light is introduced, in the model shown in Figure 5.5(c), a larger P disturbance, ΔP, is induced for $E_{DC} > 0$ than for $E_{DC} < 0$ since the slope of the curve is smaller for $E_{DC} > 0$ than for $E_{DC} < 0$. Large ΔP tends to give rise to a large refractive index. Thus, with an increase of the applied electric field, the refractive index increases.

Measures of second-order and third-order optical nonlinearities for a molecule are respectively given by molecular second-order and third-order nonlinear susceptibilities β and γ. For aggregates of many molecules, macroscopic measures of second-order and third-order optical nonlinearities can be given by $\chi^{(2)}$ originating from βs and $\chi^{(3)}$ originating from γs. EO coefficient r, which is proportional to $\chi^{(2)}$, is used as the measure for the Pockels effect.

In general, the potential curve deviation becomes large in highly polarizable systems. Figure 5.6 schematically illustrates electron-cloud polarizations induced by an electric field. In dielectric materials such as lithium niobate (LN), electronic

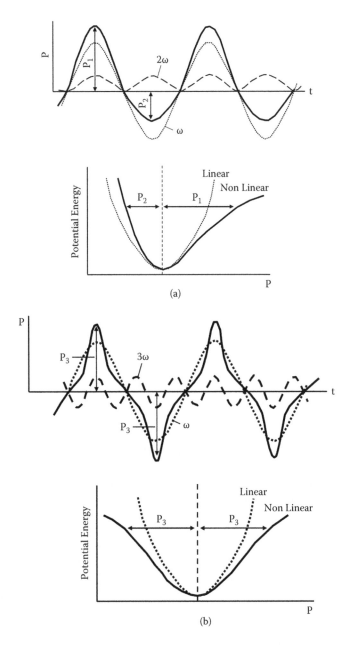

FIGURE 5.5 Visual explanation for (a) second harmonic generation, (b) third harmonic generation, and (c) the Pockels effect.

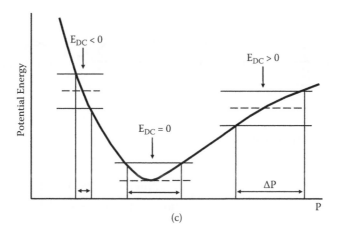

FIGURE 5.5   (CONTINUED) Visual explanation for (a) second harmonic generation, (b) third harmonic generation, and (c) the Pockels effect.

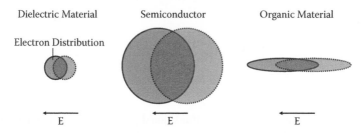

FIGURE 5.6   The reason why organic materials are promising.

polarization is not large because electrons are localized near atoms, regarded as a strong spring. In semiconductors, electron clouds are widely spread, regarded as a weak spring. Therefore, compared to dielectric materials, large electronic polarization is attainable in semiconductors. Organic materials, especially polymers with long conjugated wires, have electrons delocalized one dimensionally, that is, they are regarded as natural quantum wires. The one-dimensional characteristics are favorable for efficient electronic polarization induced by an electric field in the wire direction. This is why organic materials seem promising for nonlinear optical devices.

5.3   PROCEDURE FOR EVALUATION OF THE EO EFFECTS BY THE MOLECULAR ORBITAL METHOD

Molecular second-order nonlinear susceptibility, or hyperpolarizability in the wire direction, is shown in Equation (5.22) based on Ward's expression [4,5]

$$\beta = -\frac{e^3}{2\hbar^2}\left[\sum_{\substack{g \neq n' \\ n \neq g \\ n \neq n'}} r_{gn'}r_{n'n}r_{ng}\frac{3\omega_{n'g}\omega_{ng}+\omega^2}{(\omega_{n'g}^2-\omega^2)(\omega_{ng}^2-\omega^2)} + \sum_n r_{gn}^2\Delta r_n\frac{3\omega_{ng}^2+\omega^2}{(\omega_{ng}^2-\omega^2)^2}\right]$$

(5.22)

for the Pockels effect, and

$$\beta = -\frac{e^3}{2\hbar^2}\left[\begin{array}{l}\sum_{\substack{g \neq n' \\ n \neq g \\ n \neq n'}} r_{gn'}r_{n'n}r_{ng}\frac{2(\omega_{n'g}\omega_{ng}+2\omega^2)}{(\omega_{n'g}^2-4\omega^2)(\omega_{ng}^2-\omega^2)} \\[2ex] + \sum_{\substack{g \neq n' \\ n \neq g \\ n \neq n'}} r_{gn'}r_{n'n}r_{ng}\frac{\omega_{n'g}\omega_{ng}-\omega^2}{(\omega_{n'g}^2-\omega^2)(\omega_{ng}^2-\omega^2)} \\[2ex] + \sum_n r_{gn}^2\Delta r_n\frac{3\omega_{ng}^2}{(\omega_{ng}^2-\omega^2)(\omega_{ng}^2-4\omega^2)}\end{array}\right]$$

(5.23)

for SHG.

In Equations (5.22) and (5.23), r_{gn} is the transition dipole moment between the ground and excited states and $r_{n'n}$ is that between two excited states. Δr_n is the difference in the dipole moment between the excited and ground states. $\hbar\omega_{ng}$ is the excitation energy from the ground state to the excited state. $\hbar\omega$ is the energy of incident photons.

r_{gn}, $r_{n'n}$, and Δr_n are expressed by the following equations,

$$r_{gn} = \int \Psi_g{}^* r\Psi_n d^3x,$$

(5.24)

$$r_{n'n} = \int \Psi_{n'}{}^* r\Psi_n d^3x,$$

(5.25)

$$\Delta r_n = \int \Psi_n{}^* r\Psi_n d^3x - \int \Psi_g{}^* r\Psi_g d^3x,$$

(5.26)

where, Ψ_g is the wavefunction of the ground state and Ψ_n is the wavefunction of the excited state. These wavefunctions are expressed by Equations (5.7) and (5.10), and are constructed by molecular orbitals ϕ_i. Then, r_{gn}, $r_{n'n}$, and Δr_n are expressed as follows [5].

$$r_{gn} = \int \Psi_g {}^* r \Psi_n d^3 x = \sqrt{2} \sum_{i,j} C_{n,i \to j} m_{ij} \qquad (5.27)$$

$$r_{n'n} = \int \Psi_{n'} {}^* r \Psi_n d^3 x = \sum_{i,j,k,l} {}' C_{n',i \to j} C_{n,k \to l} \left(\delta_{ik} m_{jl} - \delta_{jl} m_{ik} \right) \qquad (5.28)$$

$$\Delta r_n = \int \Psi_n {}^* r \Psi_n d^3 x - \int \Psi_g {}^* r \Psi_g d^3 x$$

$$= \sum_{i,j,k,l} {}' C_{n,i \to j} C_{n,k \to l} \left(\delta_{ik} m_{jl} - \delta_{jl} m_{ik} \right) + \sum_{i,j} C_{n,i \to j}{}^2 \left(m_{jj} - m_{ii} \right) \qquad (5.29)$$

$$m_{ij} = \int \phi_i {}^* r \phi_j d^3 x \qquad (5.30)$$

Here, the primed summations indicate to exclude terms, in which $i = k$ and $j = 1$ simultaneously. r_{gn}, $r_{n'n}$, and Δr_n can be calculated from Equations (5.27) through (5.30) once a set of molecular orbitals are obtained by the MO method. Using the r_{gn}, $r_{n'n}$, and Δr_n, β can be calculated from Equations (5.22) and (5.23).

In the present simulation [1–3], the Austin Model 1 (AM1) method, a semi-empirical molecular orbital method was used for calculations of molecular orbitals. AM1 was developed by modifying the core repulsion function in the modified neglect of the diatomic overlap (MNDO) method. Using these molecular orbitals, the ground state wavefunction Ψ_g, a product of the occupied orbitals, and configuration functions, $\Phi_{i \to j}$, a product of the orbitals with one-electron excitation from occupied orbital ϕ_i, to unoccupied orbital ϕ_j were made. Excited states Ψ_n, expressed by linear combinations of configuration functions as Equation (5.10), were determined by single-excitation configuration interaction (CIS). CI calculation involved eight unoccupied orbitals above the lowest unoccupied molecular orbital (LUMO) and six occupied orbitals below the highest occupied molecular orbital (HOMO). 2-methyl-4-nitroaniline (MNA) was used as the standard molecule. From the calculation, β of 13×10^{-30} esu for SHG with a fundamental wavelength of 1.06 μm was obtained. This agrees fairly well with the 12×10^{-30} esu reported by Garito et al. [6].

5.4 QUALITATIVE GUIDELINES FOR IMPROVING OPTICAL NONLINEARITIES

Figure 5.7 shows the microscopic origins of second-order and third-order optical non-linearities. As mentioned in Section 5.2, the potential energy curve in linear systems is parabolic. In nonlinear systems, however, it is not. This potential curve deviation from the parabola induces a variety of nonlinear phenomena. Second-order optical nonlinearity arises from the asymmetric potential curve deviation (see Figure 5.7(a)). Third-order optical nonlinearity arises from the symmetric potential curve deviation

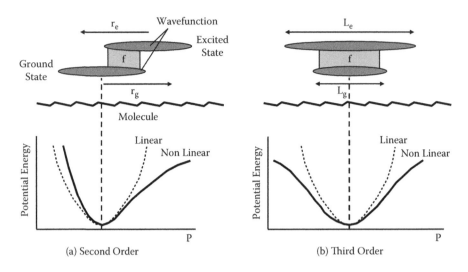

FIGURE 5.7 Microscopic origins for (a) second-order and (b) third-order optical nonlinearities.

(see Figure 5.7(b)). The larger the potential curve deviation becomes, the larger the induced nonlinear optical effect becomes.

The driving force for this potential curve deformation is explained in terms of the shape of the wavefunction as shown in Figures 5.7(a) and (b). Consider the ground state g and one excited state e, that is, a two-level model, for simplicity. When perturbation is induced in a molecule by incident light or electric field, mixing of the excited state wavefunction into the ground state wavefunction occurs, deforming the potential curve. Therefore, the first approach to improving optical nonlinearity lies in promoting the mixing rate of ground and excited states, that is, promoting the oscillator strength, f, between the ground and excited states. Here, $f \propto r_{ge}^2$, which can be increased by promoting the wavefunction overlap between the ground and excited states.

However, increasing only the wavefunction mixing rate is not in itself sufficient to increase the potential curve deformation. For example, when the ground and excited state wavefunctions have a similar shape, mixing the wavefunctions results in only a small change in electron distribution. Consequently, there is only a small potential curve deformation. Thus, another approach to improving optical nonlinearity is promoting the wavefunction difference between the ground and excited states. From the arguments mentioned above, the qualitative guidelines described in subsections 5.4.1 and 5.4.2 are derived.

5.4.1 FOR SECOND-ORDER OPTICAL NONLINEARITY

1. Promote wavefunction overlap between the ground and excited states, increasing oscillator strength $f \propto r_{ge}^2$.
2. Promote wavefunction separation between the ground and excited states, increasing the dipole moment difference $\Delta r = r_e - r_g$. Here, $r_e = \int \Psi_e^* r \Psi_e d^3x$ and $r_g = \int \Psi_g^* r \Psi_g d^3x$.

3. Consequently, second-order optical nonlinearity is proportional to the product of f and Δr, that is,

$$\beta \propto r_{ge}^2 (r_e - r_g)$$

The result is consistent with the two-level model expression for molecular second-order nonlinear susceptibility in Equation (5.22),

$$\beta = -\frac{e^3}{2\hbar^2} r_{ge}^2 \Delta r_e \frac{3\omega_{eg}^2 + \omega^2}{(\omega_{eg}^2 - \omega^2)^2} \propto r_{ge}^2 \Delta r_e = r_{ge}^2 (r_e - r_g) \qquad (5.31)$$

5.4.2 FOR THIRD-ORDER OPTICAL NONLINEARITY

1. Promote wavefunction overlap between the ground and excited states, increasing the oscillator strength $f \propto r_{ge}^2$.
2. Promote differences in the spread of wavefunctions between the ground and excited states, increasing $\Delta L = L_e - L_g$. Here $L_e = \int \Psi_e * r^2 \Psi_e d^3 x = (r^2)_e$ and $L_g = \int \Psi_g * r^2 \Psi_g d^3 x = (r^2)_g$ are measures of wavefunction spread.
3. Consequently, third-order optical nonlinearity is proportional to the product of f and ΔL, that is,

$$\gamma \propto r_{ge}^2 \left((r^2)_e - (r^2)_g \right). \qquad (5.32)$$

For both second-order and third-order optical nonlinearity, note that r_{ge}^2 and $r_e - r_g$, or r_{ge}^2 and $(r^2)_e - (r^2)_g$ are not independent, but are in a trade-off relationship. Therefore, it is concluded that the balance of the wavefunction overlap and wavefunction separation, or the balance of the wavefunction overlap and wavefunction spread difference, must be considered in optimizing nonlinear optical effects.

5.5 ENHANCEMENT OF SECOND-ORDER OPTICAL NONLINEARITY BY CONTROLLING WAVEFUNCTIONS

The effect of wavefunction shapes of the excited state (n) and the ground state (g) on r_{gn}, $\Delta r = r_n - r_g$, and β is schematically illustrated in Figure 5.8. From left to right, wavefunction separation between the excited and the ground states increases while wavefunction overlap between the two states decreases, that is, Δr increases and r_{gn} decreases in a trade-off relationship. According to Equation (5.31), in the two-level model, β is expressed by $\beta \propto r_{gn}^2 \Delta r$. So, β becomes the maximum in the middle wavefunction shape. Both the wavefunction separation and wavefunction overlap must simultaneously be considered to optimize β.

Furthermore, as described in Section 5.5.2, Δr and r_{gn} depend on the dimensionality of the conjugated systems. Therefore, it is concluded that β is expected to have its

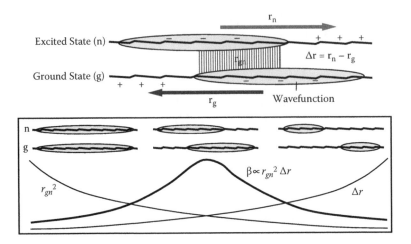

FIGURE 5.8 Effect of wavefunction shapes on r_{gn}, Δr, and β. From T. Yoshimura, "Enhancing second-order nonlinear optical properties by controlling the wave function in one-dimensional conjugated molecules," *Phys. Rev.* **B40**, 6292–6298 (1989).

maximum in a wavefunction shape of intermediate wavefunction separation with an appropriate conjugated system dimensionality existing somewhere between zero and one, that is, between the quantum dot condition and the quantum wire condition.

5.5.1 EFFECTS OF WAVEFUNCTION SHAPES

In order to examine different wavefunction shapes, four types of conjugated polymer wires shown in Figure 5.9 were considered. The wires have backbones of polydiacetylene (PDA) structures and substituted donor (D) and acceptor (A) groups. DA, DAAD, DADA, and DDAA show the types of donor and acceptor substitution, and the numbers following them are the number of carbon sites (N_c) in the wire. As Figure 5.10 shows, the donor and the acceptor have a tendency to push electrons away and pull electrons closer, respectively, when electrons are excited. So, wavefunction shapes can be optimized by adjusting donor/acceptor substitution sites in the wires using push and pull effects as well as by adjusting wire lengths. In the present model, the donor is NH_2 and the acceptor is NO_2. β was calculated by varying N_c from 10 to 34.

Figure 5.11 shows molecular orbitals near the Fermi surface for DA34, DAAD34, DADA34, and DDAA30. Here, GM ($M = 1, 2, 3$) denotes the occupied molecular orbitals and EN ($N = 1, 2, 3$) unoccupied molecular orbitals. G1 is HOMO and E1 is LUMO. In DA34, DAAD34, and DADA34, the charge separation in the wire direction appears mainly in the HOMO and LUMO. In other molecular orbitals, electrons in the wires tend to be delocalized. In DDAA30, however, considerable charge separation appears in E2 and E3 orbitals as in LUMO.

Figure 5.12 shows $C^2_{n, GM \rightarrow EN}$ as a function of GM ($M = 1,2,. .., 6$) and EN ($N = 1,2, ..., 8$) for the first excited states ($n = 1$) in DA10, DA34, DAAD10, DAAD34, DADA10, DADA34, DDAA14, and DDAA30. $C^2_{n, GM \rightarrow EN}$ is the fraction of the configuration function $\Phi_{i \rightarrow j}$ involved in Equation (5.10). The height of the pyramid

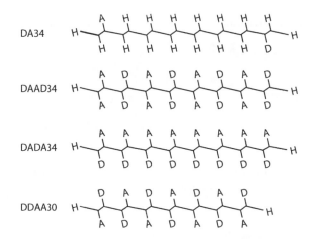

FIGURE 5.9   Four types of conjugated polymer wires with polydiacetylene backbones. From T. Yoshimura, "Enhancing second-order nonlinear optical properties by controlling the wave function in one-dimensional conjugated molecules," *Phys. Rev.* **B40**, 6292–6298 (1989).

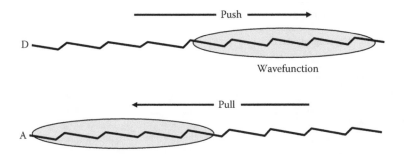

FIGURE 5.10   Push–pull effects on wavefunction induced by donor–acceptor groups.

indicates $C^2_{n,GM \to EN}$. For the short polymer wires with $N_c = 10$ or 14, the first excited states consist mainly of the configuration functions with small GM and EN, which means that the configuration interaction can be calculated sufficiently using only a few occupied and unoccupied molecular orbitals. For the long polymer wires with $N_c = 30$ or 34, the pyramids are distributed more widely than for the short molecules. The heights of the pyramids tend to decrease rapidly with increasing GM and EN. This suggests that six occupied orbitals and eight unoccupied orbitals are enough to calculate the configuration interaction for the polymer wires shown in Figure 5.9.

Combining these results of $C^2_{n,\ GM \to EN}$ shown in Figure 5.12 with the shape of the molecular orbitals shown in Figure 5.11, the wavefunction shapes are clarified. Figure 5.13 shows structures and schematic wavefunction shapes for DA34, DAAD34, and DDAA30. In DA34, the Gl -> El component, corresponding to the HOMO-to-LUMO transition, is extremely small. The wavefunction separation then becomes small, which corresponds to the wavefunction condition on the left side in Figure 5.8. In DAAD34, finite Gl -> *EN* and *GM* -> El components appear, resulting

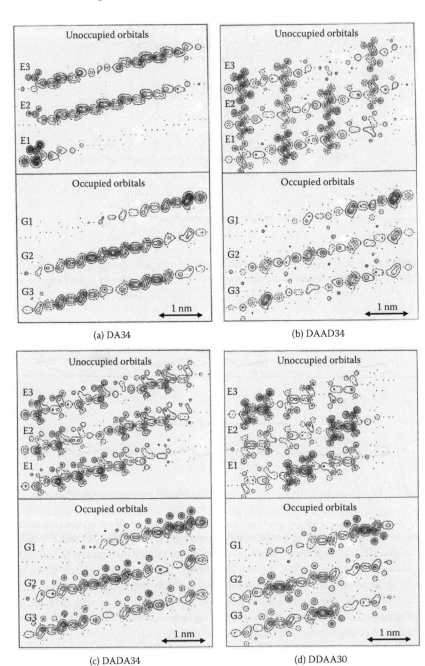

FIGURE 5.11 Contour diagrams of molecular orbitals near the Fermi surface in plane $z = 0.06$ nm for (a) DA34, (b) DAAD34, (c) DADA34, and (d) DDAA30. From T. Yoshimura, "Enhancing second-order nonlinear optical properties by controlling the wave function in one-dimensional conjugated molecules," *Phys. Rev.* **B40**, 6292–6298 (1989).

in a considerable wavefunction separation, which corresponds to the condition in the center in Figure 5.8. In DDAA30, Gl -> EN components predominate, and furthermore, the EN orbitals exhibit considerable charge separation, inducing the large wavefunction separation like on the right in Figure 5.8. Thus, it is found that wavefunction separations increase in the order of DA34, DAAD34, and DDAA30.

Figure 5.14 shows the calculated molecular second-order nonlinear susceptibility for DA34, DAAD34, DADA34, and DDAA30 at a detuning energy of 0.2 eV from the first excited states. Here, ρ_β represents β normalized by molecular wire lengths. Although calculations were carried out by using Equation (5.22) involving 48 excited states, the major contribution to ρ_β was from the first excited state. Thus, the overall tendencies in Figure 5.14 can be explained in terms of the wavefunction of the first excited state. Δr increases and r_{gn} decreases in the order of DA34, DAAD34, DADA34, and DDAA30. Consequently, ρ_β becomes small in DA34 and DDAA30, and becomes largest between them, in DAAD34, which exhibits a medium wavefunction separation. This parallels the tendency of the qualitative guideline shown in Figure 5.8 and indicates that a balance between r_{gn} and Δr is important to improve β.

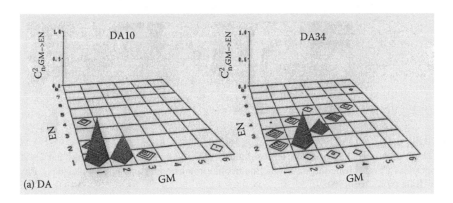

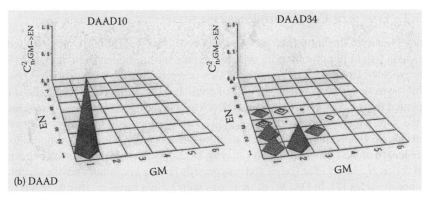

FIGURE 5.12 $C^2_{n,\,GM\to EN}$ for the first excited states in (a) DA, (b) DAAD, (c) DADA, and (d) DDAA. From T. Yoshimura, "Enhancing second-order nonlinear optical properties by controlling the wave function in one-dimensional conjugated molecules," *Phys. Rev.* **B40**, 6292–6298 (1989).

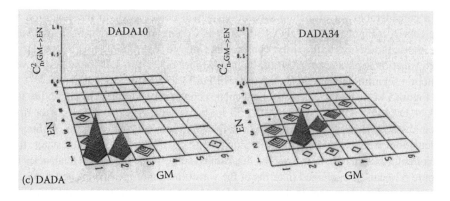

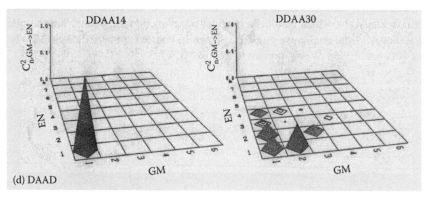

FIGURE 5.12   (CONTINUED) $C^2_{n,\,GM\text{-}>EN}$ for the first excited states in (a) DA, (b) DAAD, (c) DADA, and (d) DDAA. From T. Yoshimura, "Enhancing second-order nonlinear optical properties by controlling the wave function in one-dimensional conjugated molecules," *Phys. Rev.* **B40**, 6292–6298 (1989).

5.5.2 Effects of Conjugated Wire Lengths

Morley reported that molecular second-order nonlinear susceptibility increases with the chain length in polyenes with donor–acceptor substitution at the chain ends [7]. It is expected from the result that the PDA wires substituted by the donor–acceptor exhibit wire-length dependence of optical nonlinearity. In Figure 5.15, calculated ρ_β is plotted for DA, DAAD, DADA, and DDAA as a function of N_c, which corresponds to the wire length. N_c dependence of ρ_β is greatly affected by the donor–acceptor distribution. In DA and DAAD, ρ_β is maximized at $N_c = 18$, which corresponds to a wire length of ~2.2 nm, and in DDAA, at $N_c = 22$. Morley reported similar behavior in β versus N_c curves in polyenes and polyphenyls having one donor and one acceptor at opposite sides [7]. In DADA, ρ_β increases with N_c in a range of $10 \leq N_c \leq 34$.

ρ_β is smallest in DDAA. DAAD has the largest ρ_β of about 3000×10^{-30} esu/nm at $N_c = 18$. Assuming PDA wire density of 1.3×10^{14} 1/cm², the expected maximum EO coefficient r of DAAD is about 3000 pm/V that is 100 times larger than the EO coefficient r_{33} of LN.

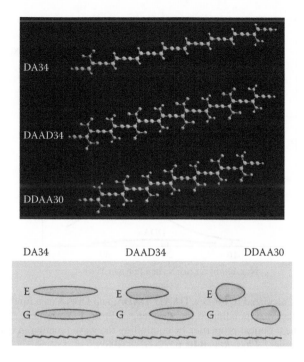

FIGURE 5.13 Schematic wavefunction shapes for DA34, DAAD34, and DDAA30 speculated from the results shown in Figures 5.11 and 5.12.

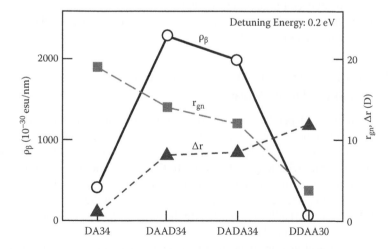

FIGURE 5.14 Calculated r_{gn}, Δr, and ρ_β for DA34, DAAD34, DADA34, and DDAA30. ρ_β
is β per 1 nm of wire length. From T. Yoshimura, "Enhancing second-order nonlinear optical
properties by controlling the wave function in one-dimensional conjugated molecules," *Phys.
Rev.* **B40**, 6292–6298 (1989).

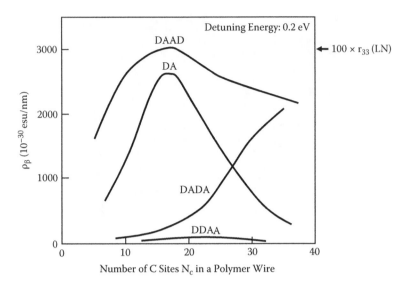

FIGURE 5.15 Calculated ρ_β for DA, DAAD, DADA, and DDAA as a function of the number of carbon sites corresponding to the wire length. From T. Yoshimura, "Enhancing second-order nonlinear optical properties by controlling the wave function in one-dimensional conjugated molecules," *Phys. Rev.* **B40**, 6292–6298 (1989).

Thus, it is expected that a large EO effect will be obtained by controlling wavefunction dimensionality as well as shapes, which can be realized by adjusting conjugated system lengths and donor–acceptor substitution sites in polymer wires. The maximum optical nonlinearity can be obtained in a dimensionality between zero and one, that is, between a quantum dot condition and a quantum wire condition. For example, in the case of DAAD, maximum optical nonlinearity is obtained in a quantum dot with a length of ~2.2 nm.

5.5.3 RELATIONSHIP BETWEEN WAVEFUNCTIONS AND TRANSITION DIPOLE MOMENTS

To investigate the N_c dependence precisely, the transition dipole moments r_{gn} and Δr are plotted as a function of N_c in Figure 5.16. In DA, r_{gn} increases and Δr decreases with increasing N_c. Consequently, as Figure 5.15 shows, ρ_β becomes maximum at an intermediate polymer wire length of $N_c = 18$. In DAAD, r_{gn} and Δr exhibit N_c dependence similar to DA, with a peak at $N_c = 18$ in the ρ_β versus N_c curve in Figure 5.15. In DADA, Δr and r_{gn} increases with N_c. This is reflected in the increase in ρ_β with N_c. In DDAA, r_{gn} is markedly smaller than in the other three types of polymer wires, with ρ_β reduced as shown in Figure 5.15.

Because the wavefunction shape affects β via dipole moments, in order to determine the relationship between wavefunctions and β, we must clarify the relationship between the wavefunctions and the dipole moments. As mentioned above, charge separation mainly appears in HOMO and LUMO. Therefore, the G1 -> *EN* and *GM* -> E1 composition in Ψ_n ($n = 1$) mainly contribute to Δr.

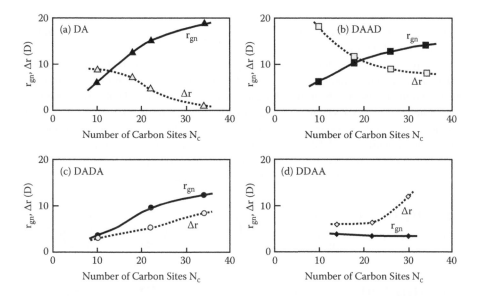

FIGURE 5.16 Dependence of r_{gn} and Δr on the number of carbon sites. From T. Yoshimura, "Enhancing second-order nonlinear optical properties by controlling the wave function in one-dimensional conjugated molecules," *Phys. Rev.* **B40**, 6292–6298 (1989).

In DA, as shown in Figure 5.12(a), the contribution of G1 -> *EN* and *GM* -> E1 composition to Ψ_n is reduced drastically with increasing N_c. This reduces the charge separations, then reduces Δr with increasing N_c. At the same time, the wavefunction overlap between the ground and excited states increases with N_c, enhancing r_{gn}. These results are consistent with Figure 5.16(a). For the DAAD, as Figure 5.12(b) shows, the wavefunction changes similarly to DA with increasing N_c, although not so drastically. This is consistent with Figure 5.16(b). For DA, G1 -> *EN* and *GM* -> E1 composition are less than in DAAD, which may account for the fact that in DA, Δr is smaller and r_{gn} is larger than in DAAD (Figures 16(a) and (b)). This is because DA has only one donor and one acceptor, so the electronic pull and push power would be smaller than in DAAD, which has many donors and acceptors.

For DADA, the contribution of the G1 -> *EN* composition to Ψ_n remains large when N_c increases from 10 to 34. This enhances Δr (Figure 5.16(c)). For DDAA30, the main composition is the G1 -> *EN* series and all three orbitals, E1, E2, and E3, exhibit considerable charge separation (Figure 5.11(d)). This increases Δr but reduces r_{gn} because the overlaps of G1/E1, G1/E2, and G1/E3 are small. Thus, r_{gn} in DDAA30 decreases while Δr increases (Figure 5.16(d)).

5.5.4 OPTICAL NONLINEARITY IN CONJUGATED WIRES WITH POLY-AM BACKBONES

So far, enhancement of optical nonlinearities in conjugated wire models with PDA backbones has been described. The same enhancement of optical nonlinearities is found by Matsuura et al. in conjugated wire models with poly-AM backbones [8], which are shown in Figure 5.17 together with calculated βs of the models for 633 nm in wavelength. DDDDAAAA exhibits the largest β, which corresponds to an EO coefficient of 1260 pm/V.

The reason why DDDDAAAA has larger β than AAAADDDD is that N at the edge of the wire has donor-like characteristics. In DDDDAAAA, pull–push strength is increased because the N at the edge and the donor region of DDDD are located on the same side. In AAAADDDD, on the other hand, pull–push strength is suppressed because the N at the edge and the donor region of DDDD are located on the opposite side. Consequently, β in DDDDAAAA becomes larger than β in AAAADDDD.

From of the perspective of polymer wire fabrication by MLD, the poly-AM backbone is useful since MLD of the poly-AM wires has already been demonstrated

HHHHHHHH
$b=15x10^{-30}$ esu

AAHHHHDD
$b=142x10^{-30}$ esu

DDHHHHAA
$36x10^{-30}$ esu

AADDAADD
$243x10^{-30}$ esu

DDAADDAA
$18x10^{-30}$ esu

AAAADDDD
$108x10^{-30}$ esu

DDDDAAAA
$453x10^{-30}$ esu

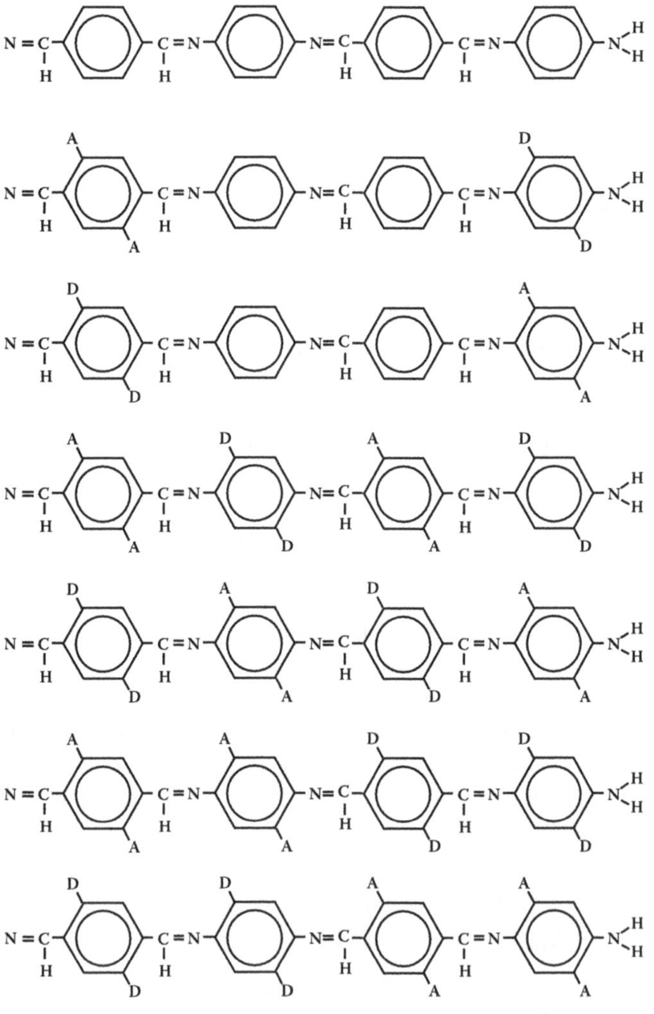

FIGURE 5.17 Models and βs of electro-optic conjugated wires with poly-AM backbones.

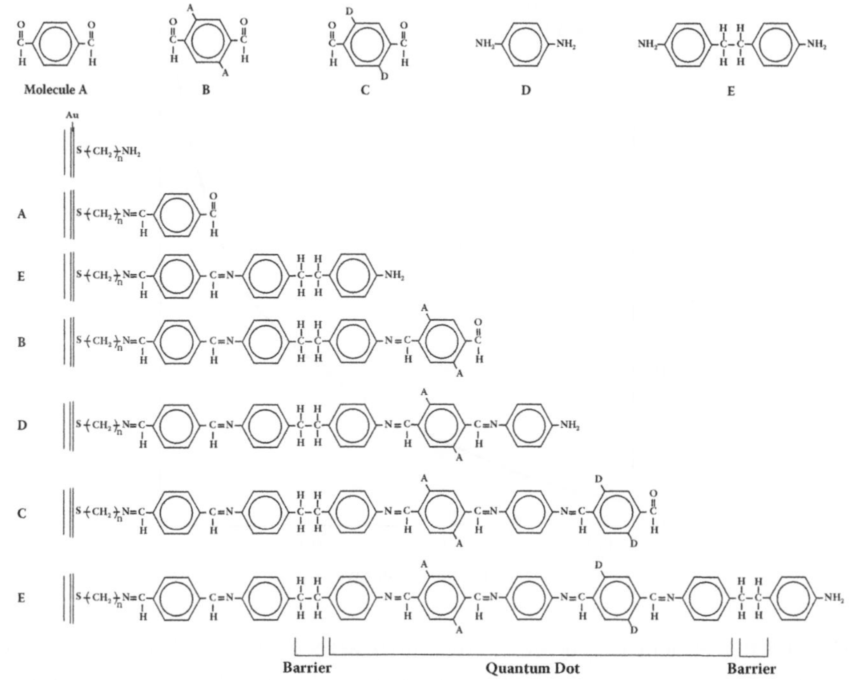

FIGURE 5.18 Possible example of MLD process for poly-AM wires with donors and acceptors.

experimentally as mentioned in Chapters 3 and 4. Possible examples of the MLD process for poly-AM wires with donors and acceptors is shown in Figure 5.18. Five kinds of molecules, molecules A, B, C, D, and E are used. Molecule A is terephthalaldehyde (TPA), and molecule D is p-phenylenediamine (PPDA). Molecule B has two –CHO groups and acceptors. Molecule C has two –CHO groups and donors. Molecule E is for terminating conjugated systems with three single bonds in the molecule. By connecting molecules by MLD in a sequence of molecules A, E, B, D, C, E, …, multiple quantum dot (MQD) structures with donors and acceptors, which is similar to the DDDDAAAA or AAAADDDD structures shown in Figure 5.17, will be constructed.

5.5.5 ENHANCEMENT OF OPTICAL NONLINEARITY BY SHARPENING ABSORPTION BANDS

Figure 5.19 shows the resonant enhancement of ρ_β near the first excited state for DAAD18. The absorption band width of polydiacetylene is about 0.1 eV, so a detuning energy of about 0.2 eV is necessary for low-loss operation. In this case, the EO coefficient is about 2 orders of magnitude larger than r_{33} in LiNbO$_3$. If the band width could be reduced 1 order of magnitude by sharpening the absorption band, the detuning energy required would only be ~0.02 eV, and the EO coefficient would be

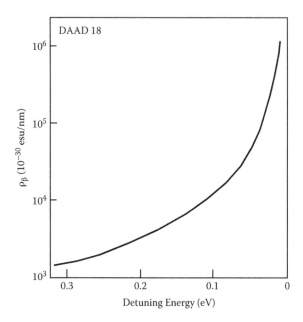

FIGURE 5.19   Photon energy dependence of ρ_β in DAAD18. From T. Yoshimura, "Enhancing second-order nonlinear optical properties by controlling the wave function in one-dimensional conjugated molecules," *Phys. Rev.* **B40**, 6292–6298 (1989).

about 10^4 times larger than r_{33} in LiNbO$_3$. This leads us to believe that to improve the nonlinear optical properties in organic materials, sharpening the absorption band [9] is important as well as controlling the wavefunction.

The uniformity of the conjugated length in materials is essential in sharpening. Reducing the exciton–phonon (or electron–phonon) coupling strength is also important. One way to do this might be to use rigid polymer structures. Bound excitons or exciton confinement in quantum dots in one-dimensional conjugated systems may be another way to sharpen absorption bands.

5.6   ENHANCEMENT OF THIRD-ORDER OPTICAL NONLINEARITY BY CONTROLLING WAVEFUNCTIONS

To improve the Kerr effect, which is one of the third-order nonlinear optical effects, the shape of the wavefunction was controlled by constructing quantum dot structures in one-dimensional conjugated systems [3] according to the qualitative guidelines mentioned in Section 5.4. Figures 5.20(a) and (b) give an example of a quantum dot structure and corresponding wavefunctions. Varying the dot length in the conduction band (ML) and that in the valence band (W) adjusts the wavefunction spread difference between the ground and excited states.

D is the dot width. Reducing D changes the system from two-dimensional to one-dimensional. γ per unit length for the light polarization with the well length direction was calculated from Equation (5.33). For simplicity, it was assumed that the ground state wavefunction has even parity, the excited state wavefunction odd parity, and wavefunctions are square waves. The results are shown in Figure 5.20(c). With reduced D, both $\left(r^2\right)_e - \left(r^2\right)_g$ and r_{ge}^2 increase, then γ increases rapidly. This clearly indicates that the one-dimensional system is better than a two-dimensional system for inducing large optical nonlinearity, as mentioned in Section 5.2. That is, a wavefunction extending perpendicular to the direction of light polarization contributes little to optical nonlinearity, simply diluting the wavefunction density and reducing optical nonlinearity.

With increasing W, that is, increasing the overlap and decreasing the difference in wavefunction spread between the ground and excited states, r_{ge}^2 increases and $\left(r^2\right)_e - \left(r^2\right)_g$ decreases, with γ peaking at an intermediate region in W.

Figures 5.21(a) and (b) show a structure and wavefunctions for a double quantum dot (DQD) with subdots at both sides. The wavefunction spread difference is controlled by the difference in dot depth between the conduction band and the valence band. With increasing dot depth, the wavefunction tends to localize on both sides. The calculated values of γ, $\left(r^2\right)_e - \left(r^2\right)_g$, and r_{ge}^2 are shown in Figure 5.21(c) as a function of R = C/D. R is a measure of the wavefunction extent in the valence band as defined in Figure 5.21(b). With increasing R, $\left(r^2\right)_e - \left(r^2\right)_g$ increases and r_{ge}^2 decreases. γ peaks at an intermediate region in R of about 1 when the balance

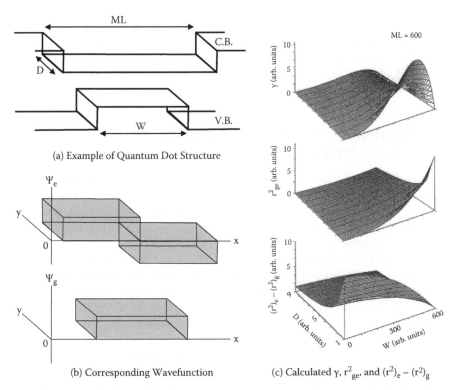

(a) Example of Quantum Dot Structure

(b) Corresponding Wavefunction

(c) Calculated γ, r^2_{ge}, and $(r^2)_e - (r^2)_g$

FIGURE 5.20 Enhancement of γ by controlling wavefunction shapes with quantum dot structures.

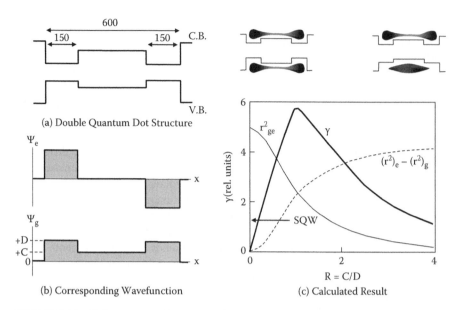

(a) Double Quantum Dot Structure

(b) Corresponding Wavefunction

(c) Calculated Result

FIGURE 5.21 Enhancement of γ by controlling wavefunction shape in double quantum dot structures.

between the wavefunction overlap and the wavefunction spread difference is optimum. In this case, γ is improved about five times over that for a single quantum dot (SQD). It was also found that further confinement of electrons in three or more subdots would make it possible to enhance γ more than ten times.

5.7 MULTIPLE QUANTUM DOTS (MQDS) IN CONJUGATED POLYMER WIRES

To apply the tailored materials described in Sections 5.5 and 5.6 to practical devices, it is necessary to fabricate the materials in ordered structures. This involves constructing MQD structures in one-dimensional conjugated polymer wires. One way to do this is to insert molecules for cutting the π-conjugation in polymer wires as described in Chapter 4.

The author found another way to form quantum dots [3]. It was simulated by the MO calculation that the energy gap of conjugated polymer wires with PDA backbones is reduced when donors and acceptors are substituted in the DAAD configuration. Figure 5.22 shows one of the examples. The energy gap is decreased from 3.3 eV of the intrinsic PDA wires to 1.6 eV by donor–acceptor substitutions. The phenomena can be used for MQD construction.

The mechanism of the energy gap narrowing in conjugated polymer wires by donor and acceptor substitution is as follows. From the viewpoint of charge distributions, it is found from Figure 5.23 that electrons are localized around donor sites in HOMO while they are localized around acceptor sites in LUMO. The electron distribution in HOMO corresponds to hole distributions in HOMO when the

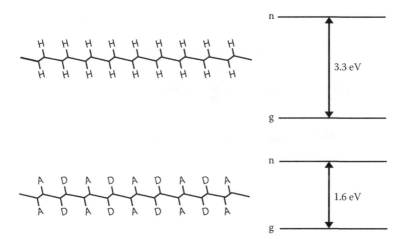

FIGURE 5.22   Energy gap narrowing in conjugated polymer wires by donor and acceptor substitution.

electron is excited. So, with decreasing the distance between donors and acceptors, electron–hole distance becomes short, increasing electric fields in the area, resulting in a large potential energy slope as shown in Figure 5.24. Then, energy gaps become narrow.

In Figure 5.25, DAAD18 and PDA with an inserted DAAD18 structure are compared. Although wavefunctions slightly penetrate into the barrier regions by tunneling in the conjugated polymer wire with an inserted DAAD18 structure, the molecular orbitals and β are almost the same for both. Therefore, using this effect, it will be possible to insert many DAAD structures into the conjugated polymer wire, constructing a polymer MQD of a polymer superlattice, as shown in Figure 5.26. This structure enables us to align wavefunctions of DAAD structures in the wire direction perfectly, which is favorable to attaining a large Pockels effect.

As mentioned above, using energy gap narrowing, it should be possible to form quantum dots in one-dimensional conjugated wires by modulating the energy gap with selective substitutions of donors and acceptors into the wires. A DQD structure with two 1.4-nm long dots is shown in Figure 5.27 together with an SQD structure. Quantum dots are formed in a PDA wire with 38 carbon sites. Areas with donor and acceptor substitution are quantum dots, and areas with hydrogen substitution are barriers. The energy gap in the quantum dot region is about 1.6 eV, about one half of that in the barrier region. Molecular orbitals show that electrons are spread throughout the conjugated wires in the SQD. In the DQD, electrons tend to be confined to both sides of the wire. Here, note that although electrons are localized on both sides in all E1 , E2, and E3 unoccupied orbitals, they are in the barrier region in the G3 occupied orbital. This suggests that the wavefunction tends to be delocalized throughout the wire more in the valence band than in the conduction band, reproducing the condition in Figure 5.21.

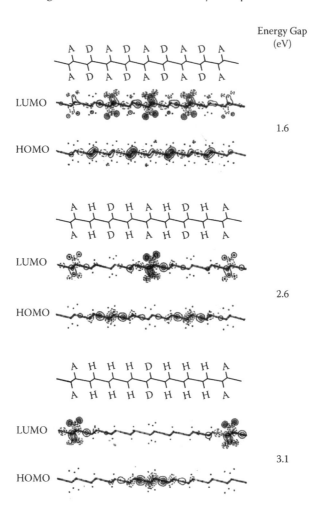

FIGURE 5.23   HOMO and LUMO in conjugated polymer wires with donor and acceptor substitution.

γ of DQD and SQD were calculated by the AM1 MO calculation. As expected, γ in the DQD is more than twice that in the SQD. This confirms that it may be possible to improve third-order optical nonlinearity by adjusting the well structures.

Research on artificial materials, like PDA, poly-AM or other conjugated systems with MQDs, has far-reaching consequences. If a way is developed to connect a molecule on a molecule one by one with direction control, the donor and acceptor substitution locations and the conjugated length could be controlled. MLD is the most promising method for making artificial materials. Figure 5.28 shows possible configurations of EO waveguides consisting of artificial materials fabricated by MLD.

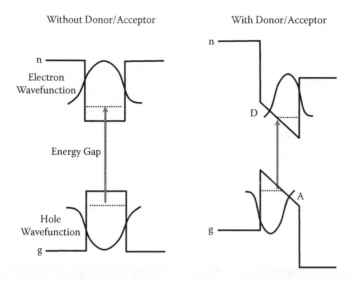

FIGURE 5.24 Mechanism by which the energy gap narrows in conjugated polymer wires by donor and acceptor substitution.

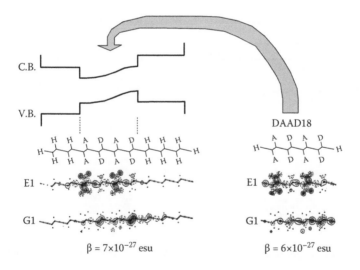

FIGURE 5.25 Insertion of DAAD18 structure into the conjugated polymer wire backbone.

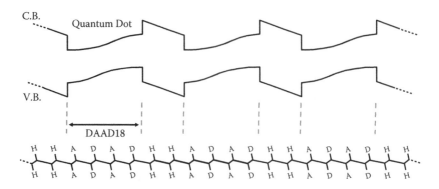

FIGURE 5.26 MQD structure of a polymer superlattice with corresponding electronic potential energy curves.

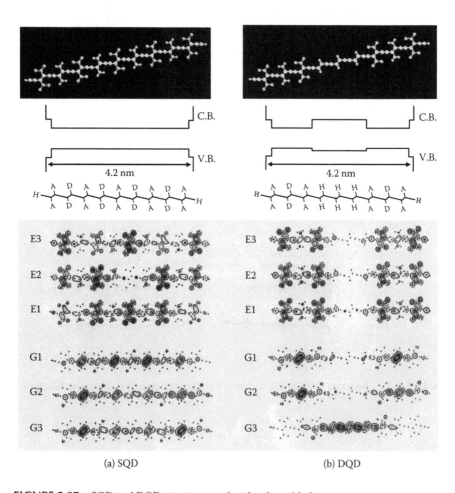

(a) SQD

(b) DQD

FIGURE 5.27 SQD and DQD structures and molecular orbitals.

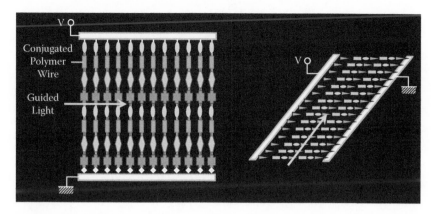

FIGURE 5.28 Possible configurations of EO waveguides consisting of artificial materials fabricated by MLD.

REFERENCES

1. T. Yoshimura, "Enhancing second-order nonlinear optical properties by controlling the wave function in one-dimensional conjugated molecules," *Phys. Rev.* **B40**, 6292–6298 (1989).
2. T. Yoshimura, "Theoretically predicted influence of donors and acceptors on quadratic hyperpolarizabilities in conjugated long-chain molecules," *Appl. Phys. Lett.* **55**, 534–536 (1989).
3. T. Yoshimura, "Design of organic nonlinear optical materials for electro-optic and all-optical devices by computer simulation," *FUJITSU Sc. Tech. J.* **27**, 115–131 (1991).
4. J. Ward, "Calculation of nonlinear optical susceptibilities using diagrammatic perturbation theory," *Rev. Mod. Phys.* **37**, 1–18 (1965).
5. S. J. Lalama and A. F. Garito, "Origin of the non-linear second-order optical susceptibilities of organic systems," *Phys. Rev.* **A 20**, 1179–1194 (1979).
6. A. F. Garito, C. C. Teng, K. W. Wong, and O. Zammani-Khamiri, "Molecular optics: Nonlinear optical processes in organic and polymer crystals," *Mol. Cryst. Liq. Cryst.* **106**, 219–258 (1984).
7. J. O. Morley, "Theoretical study of the electronic structure and hyperpolarizabilities of donor-acceptor comulenes and a comparison with the corresponding polyenes and polyynes," *J. Phys. Chem.* **99**, 10166–10174 (1995).
8. A. Matsuura and T. Hayano, "Theoretical prediction of the donor/acceptor site-dependence on first hyperpolarizabilities in conjugated systems containing azomethine bonds," *Mat. Res. Soc. Symp. Proc.* **291**, 503–508 (1993).
9. T. Yoshimura, "Estimation of enhancement in third-order nonlinear susceptibility induced by a sharpening absorption band" *Opt. Commun.* **70**, 535–537 (1989).

6 Design of Integrated Optical Switches

High-speed, thin and small light modulators and optical switches are the key devices in optical interconnects and optical switching systems; they improve the computer processing speed and the communication capacity. The polymer multiple quantum dot (MQD) fabricated by molecular layer deposition (MLD) is a promising future material for integrated light modulators and optical switches.

Mach-Zehnder interferometer and directional coupler type light modulators and optical switches utilizing the electro-optic (EO) effect are well known. However, these devices are largely affected by temperature and processing accuracy for device fabrication since they are phase-sensitive devices; consequently, their large-scale integration is difficult.

In Figure 6.1, structures of promising light modulator and optical switch candidates are shown, such as the variable well optical IC (VWOIC) [1–4], the waveguide prism deflector (WPD) optical switch [2–10], the total internal reflection (TIR) optical switch, the digital optical switch, the ring resonator optical switch, and the photonic crystal. The VWOIC, WPD optical switch, TIR optical switch, and digital optical switch are phase-insensitive flux-controlled devices that are suitable for large-scale integration.

Examples of EO material candidates for light modulators and optical switches are listed in Table 6.1. Materials having the Pockels effect are $LiNbO_3$ (LN), III–V semiconductor quantum dot (QD) [11,12], styrylpyridinium cyanine dye (SPCD) [13], which is a molecular crystal, and polymer MQD [14–16], which is an artificial material consisting of polymer wires with MQDs. The EO property of polymer MQDs is theoretically predicted in Chapter 5 [16]. EO material having the Kerr effect is lead lanthanum zirconate titanate (PLZT) [17]. The polymer MQD might also be expected to exhibit a large Kerr effect, although precise analysis was not done. Electric-field-induced refractive index change, Δn, is calculated by the following expressions.

$$\text{Pockels Effect: } \Delta n = \frac{1}{2} n^3 rE \tag{6.1}$$

$$\text{Kerr Effect: } \Delta n = \frac{1}{2} n^3 RE^2 \tag{6.2}$$

Here, n and E represent the refractive index of the EO material and electric field in the materials. r and R are, respectively, the EO coefficients for the Pockels effect and the Kerr effect. Figure 6.2 shows the expected Δn that is induced when 1 V is applied between electrodes on the top and bottom surfaces of 1-μm-thick EO materials. It is found that III-V semiconductor QD and SPCD exhibit large Δn as compared with

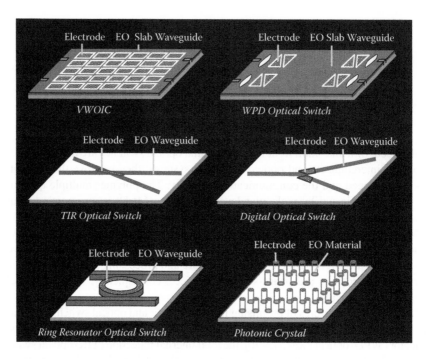

FIGURE 6.1 Structures of light modulators and optical switches.

TABLE 6.1

Typical Properties of EO Material Candidates for Light Modulators and Optical Switches

Pockels Effect

	LN	**III-V QD**	**SPCD**	**Polymer MQD**
n	2.2	3.5	1.55	~2
r (pm/V)	31	26[11]	430[13]	1000 *[16]

*Theoretically Predicted Value

Kerr Effect

	PLZT
n	2.48
R (m²/V²)	$5.7 \times 10^{-16[17]}$

lithium niobate (LN). PLZT and polymer MQD exhibit extremely large Δn. Δn of more than 0.01 is theoretically expected with a few volts applied.

In this chapter, the design and predicted performance of VWOICs and WPD optical switches previously proposed by the author are reviewed. The VWOIC and the WPD optical switch are especially superior as EO switches for integrated polymer optical circuits with a large number of elements. Ring resonator optical switches that

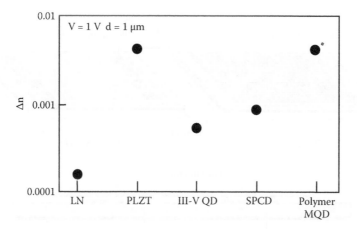

FIGURE 6.2 Expected electric-field-induced refractive index change Δn for typical EO materials. The solid circle with an asterisk (*) is the theoretically predicted value diselectrode seperation.

are expected as ultrasmall optical switches are briefly described. In addition, bandwidth limit in photonic crystal waveguides is discussed.

Since light modulators and optical switches have the same structures in many cases, light modulators are discussed along with optical switches in the following text.

6.1 VARIABLE WELL OPTICAL ICS (VWOICS) AND WAVEGUIDE PRISM DEFLECTORS (WPDS)

Structures of the VWOIC [1–4] and the WPD optical switch [2–10] are schematically shown in Figure 6.3. In the former, rectangular electrodes are formed on an EO slab waveguide. In the latter, prism-shaped cascading electrodes are formed on an EO slab waveguide. The slab waveguides consist of three layers (clad/core/clad) placed between counter electrodes and rectangular or prism-shaped electrodes. The clad is not drawn in the figure. The core is made of EO materials and the refractive index changes by electric fields. Interface films, in which thin-film interface ICs are embedded, are stacked on the surface for driving the individual optical switches to perform selective voltage application to the electrodes.

As Figure 6.4 shows, in VWOIC, by selectively applying voltage to the EO waveguide through the patterned electrodes, refractive index change is induced to form potential wells for light waves (see Section 2.1). Then, electric-field-induced dynamic optical waveguides are constructed all over the slab waveguide to switch guided light beams. In WPD optical switches, as shown in Figure 6.5, electric-field-induced dynamic prisms, which deflect guided light beams introduced from input waveguides, are constructed, enabling us to switch the light beams through two-dimensional (2-D) free space region in the slab waveguide like "search lights." Waveguide lenses for light beam collimation are formed to suppress spreading due to diffraction of the propagating light beams in the free space region.

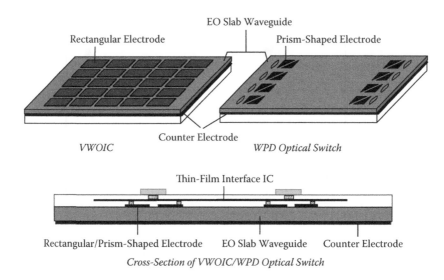

FIGURE 6.3 Structures of the VWOIC and the WPD optical switch.

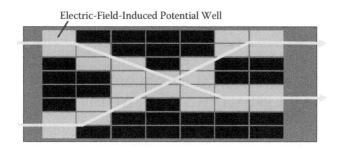

FIGURE 6.4 Operation principle of the VWOIC. From T. Yoshimura, S. Tsukada, S. Kawakami, Y. Arai, H. Kurokawa, and K. Asama, "3D micro optical switching system (3D-MOSS) architecture," *Proc. SPIE* **4653**, 62–70 (2002).

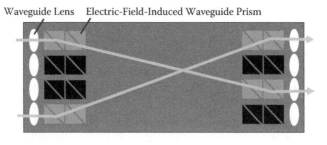

FIGURE 6.5 Operation principle of the WPD optical switch. From T. Yoshimura, S. Tsukada, S. Kawakami, Y. Arai, H. Kurokawa, and K. Asama, "3D micro optical switching system (3D-MOSS) architecture," *Proc. SPIE* **4653**, 62–70 (2002).

Optical switches for optical interconnects within computers and optical switching systems should be high speed, thin, small, and at the same time, should be stable when subjected to the considerable thermal and device shape/configuration disturbance that is present. In ordinary optical switches, such as Mach-Zehnder switches and directional coupler switches, a small phase difference is used for the switching operation. Therefore, the dependence of the device operation on temperature and processing accuracy in device fabrication is large. In fact, in a LiNbO$_3$ matrix optical switch, individual precise adjustment of driving voltage is required for each element in the 8 × 8 scale.

On the contrary, VWOICs and WPD optical switches are phase-insensitive flux-controlled devices. That is, electric-field-induced refractive index differences in the slab waveguide are used to switch the optical path directly. The refractive index difference is relatively insensitive to the temperature changes compared with the refractive index itself. Furthermore, for processing accuracy, as far as the edge angles of electrodes are concerned, the size of each part does not largely affect the switching characteristics. Therefore, VWOICs and WPD optical switches are expected to exhibit stable operation compared with ordinary optical switches, and to be suitable for large-scale optical systems such as integrated optical interconnects within boxes and three-dimensional micro-optical switching systems (3D-MOSSs) described in Chapter 8.

6.1.1 DESIGN OF VWOIC

The switching operation was simulated by the beam propagation method (BPM) for VWOIC with 12 × 7 electrodes of 2 μm × 100 μm. The results are shown in Figure 6.6 for a wavelength of 1.3 μm. The refractive index change induced by voltage application is assumed to be 0.005. Two adjacent electrodes are simultaneously selected for voltage application, which corresponds to 4-μm-wide waveguide construction. Selected regions are shifted by 2 μm along a direction perpendicular to

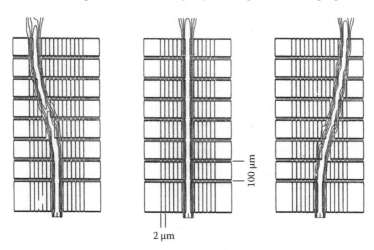

FIGURE 6.6 Switching operation of VWOIC simulated by the BPM. From T. Yoshimura, S. Tsukada, S. Kawakami, Y. Arai, H. Kurokawa, and K. Asama, "3D micro optical switching system (3D-MOSS) architecture," *Proc. SPIE* **4653**, 62–70 (2002).

the light propagation direction in order to achieve quasicontinuous index variation. It can be seen that a light beam is switched in a length of 700 μm.

6.1.2 Design of WPD Optical Switch Utilizing the Pockels Effect

6.1.2.1 Simulation Procedure

When designing the WPD optical switches, the following two are the objectives.

1. Reduction of the prism size and the interchannel pitch
2. Suppression of the beam spreading accompanying the free space propagation in the slab waveguide

These two objectives have a reciprocal relationship. If the prism size is reduced, the prism aperture decreases and the diffraction angle increases. Consequently, the beam spreading increases. So it is important to find a practical solution for the WPD optical switch structure within this trade-off.

As shown in Figure 6.7, the prism electrode is formed by a pair. When a −V voltage is applied to one electrode and a +V voltage to the other, prism-shaped refractive index changes of −Δn and +Δn are induced in the slab waveguide under the electrodes. As a result, a waveguide prism is produced, and the light beams are deflected. The deflection angle θ_d is given by the following equation from the incident angle θ_1 and the transmission angle θ_2.

$$\theta_d = \theta_1 - \theta_2 \tag{6.3}$$

By lining up multiple prism electrode pairs in a cascade, a larger deflection angle can be obtained. The polarity of the voltage pattern of the prism-shaped electrodes at the output is reversed. Thus, the deflected light beams return in the horizontal direction and can smoothly couple to the output waveguide. The light beams introduced from the input waveguide spread as they propagate in the slab waveguide.

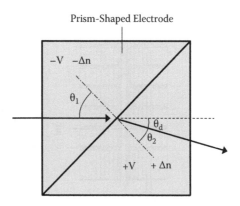

FIGURE 6.7   Detail of the electric-field-induced waveguide prism in a slab waveguide with the Pockels effect. From M. Ninomiya, Y. Arai, T. Yoshimura, H. Kurokawa, and K. Asama, "Characteristics evaluation for waveguide-prism-deflector type micro optical switches (WPD-MOS)," *Electronics and Communications in Japan, Part 2* **86**, 38–48 (2003).

The switching characteristics are simulated by the 2-D analysis of wide-angle approximate BPM using the finite difference method. The propagating light beam has a wavelength λ of 1.3 μm and is in transverse electric (TE) mode. The Δz step in the light beam propagation direction is set to 0.5 μm and the Δx step in the perpendicular direction is set to $\lambda/30$. The core refractive index of the input and output waveguides is equal to the core refractive index of the slab waveguide, that is, the refractive index of the EO material with no applied voltage. The difference in the refractive index between the core and the clad is set to 0.03. The electric field distribution of the input light beam is that of the input waveguide fundamental mode. The integrated intensity of the input light beam is 1.0.

For the present simulation, EO material is assumed to be a 400-nm-thick SPCD film with $r = 430$ pm/V and $n = 1.55$. When 6.6 V is applied to the 400-nm-thick EO waveguide, a refractive index change of about $\Delta n = 0.013$ is theoretically expected.

6.1.2.2 Structural Model

Figure 6.8 is a top view of a WPD optical switch model with a SPCD slab waveguide. The pitch of the input and output waveguides with an aperture of D is set to P μm, and the number of steps of the prism electrode pair to N. The prism electrode has a right-angle triangle shape with a length and an aperture of D_p.

The distance L_d between the prisms required for switching is estimated. For convenience, L_d is defined as the distance to the center of the multistage prisms at the input and the output. The refractive index n_1 in the negative voltage application unit and the refractive index n_2 of the positive voltage application unit are expressed as follows using the electric-field-induced refractive index change Δn:

$$n_1 = n_0 - \Delta n \tag{6.4}$$

$$n_2 = n_0 + \Delta n \tag{6.5}$$

where n_0 is the refractive index when no voltage is applied. From the refraction rule at the prism electrode boundary, the relationship between the incident angle θ_1 and the transmission angle θ_2 is given by

$$\theta_2 = \sin^{-1}\left(\frac{n_1}{n_2}\sin\theta_1\right) \tag{6.6}$$

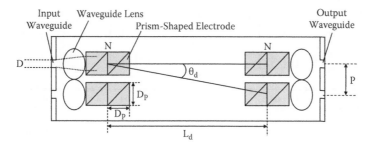

FIGURE 6.8 Top view of WPD optical switch model with SPCD slab waveguide. From M. Ninomiya, Y. Arai, T. Yoshimura, H. Kurokawa, and K. Asama, "Characteristics evaluation for waveguide-prism-deflector type micro optical switches (WPD-MOS)," *Electronics and Communications in Japan, Part 2* **86**, 38–48 (2003).

The deflection angle θ_d is determined from Equations (6.6) and (6.3). For an N-stage cascade, N iterations of the calculation in Equation (6.6) determines the final θ_d. L_d in the zeroth approximation is expressed in terms of θ_d and P by the following:

$$L_d = P/\tan\theta_d \qquad (6.7)$$

The diffraction angle of the light beams increases with decreasing D. The beam diameter must be less than or equal to the waveguide pitch P. Therefore, restrictions develop in L_d depending on D.

6.1.2.3 Simulated Performance

6.1.2.3.1 2 × 2 WPD Optical Switch

From preliminary tests, $N = 4$ and $V = 11$ V was extracted as the appropriate condition for WPD optical switches with SPCD. A 2 × 2 WPD optical switch structure is created with this condition as shown in Figure 6.9. Here, $P = 20$ μm, $D = 4$ μm, and $D_p = 16$ μm. The diameter and the refractive index of the lens are 40 μm and 2.0, respectively. The optical switching length is 390 μm.

Figure 6.10 shows the Input 1 –> Output 1 switching and Input 2 –> Output 1 switching simulated by the BPM. Light beam deflection by the waveguide prism and smooth switching operation are observed. For Input 1 –> Output 1, the crosstalk was −12 dB and the insertion loss, 0.44 dB. For Input 2 –> Output 1, the crosstalk was −14 dB and the insertion loss, 0.96 dB. Thus, the overall WPD optical switch characteristics were the crosstalk $\leq -12dB$, the insertion loss $\leq 0.96dB$.

6.1.2.3.2 3 × 3 WPD Optical Switch

Figure 6.11 shows a 3 × 3 WPD optical switch structure. Figure 6.12 shows the voltage application patterns, and Figure 6.13 shows the switching operation simulated by the BPM for Input 1 –> Output 3 and Input 1 –> Output 2. The crosstalk was ≤ -10 dB, insertion loss was ≤ 1.2 dB, and the length was 400 μm. Thus, the possibility of a 3 × 3 optical switch with a length of less than 1 mm was demonstrated.

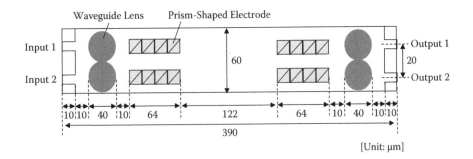

FIGURE 6.9 Typical structure of 2 × 2 WPD optical switch with SPCD slab waveguide. From M. Ninomiya, Y. Arai, T. Yoshimura, H. Kurokawa, and K. Asama, "Characteristics evaluation for waveguide-prism-deflector type micro optical switches (WPD-MOS)," *Electronics and Communications in Japan, Part 2* **86**, 38–48 (2003).

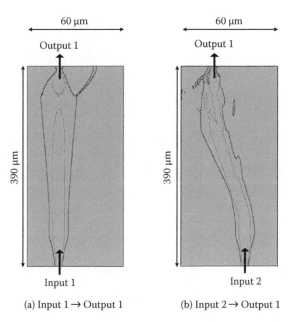

(a) Input 1 → Output 1   (b) Input 2 → Output 1

FIGURE 6.10   Switching operation of 2 × 2 WPD optical switch with SPCD slab waveguide simulated by the BPM. From M. Ninomiya, Y. Arai, T. Yoshimura, H. Kurokawa, and K. Asama, "Characteristics evaluation for waveguide-prism-deflector type micro optical switches (WPD-MOS)," *Electronics and Communications in Japan, Part 2* **86**, 38–48 (2003).

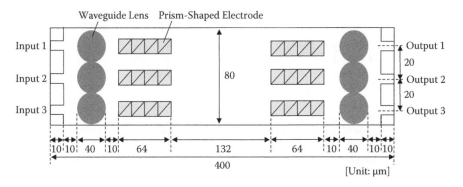

FIGURE 6.11   Structure of 3 × 3 WPD optical switch with SPCD slab waveguide. From M. Ninomiya, Y. Arai, T. Yoshimura, H. Kurokawa, and K. Asama, "Characteristics evaluation for waveguide-prism-deflector type micro optical switches (WPD-MOS)," *Electronics and Communications in Japan, Part 2* **86**, 38–48 (2003).

6.1.2.3.3   *Switching Time*

The switching time τ of a WPD optical switch is given by the following equation,

$$\tau = R_{ON}C \, [s], \tag{6.8}$$

where the ON resistance of the drive circuit is R_{ON} [Ω] and the capacitance between the prism electrode and the counter electrode is C [F].

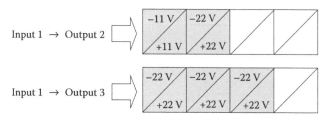

Input 1 → Output 2

Input 1 → Output 3

FIGURE 6.12 Voltage application patterns to prism-shaped electrodes in 3 × 3 WPD optical switch. From M. Ninomiya, Y. Arai, T. Yoshimura, H. Kurokawa, and K. Asama, "Characteristics evaluation for waveguide-prism-deflector type micro optical switches (WPD-MOS)," *Electronics and Communications in Japan, Part 2* **86**, 38–48 (2003).

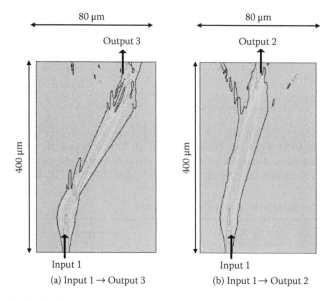

FIGURE 6.13 Switching operation of 3 × 3 WPD optical switch with SPCD slab waveguide simulated by the BPM. From M. Ninomiya, Y. Arai, T. Yoshimura, H. Kurokawa, and K. Asama, "Characteristics evaluation for waveguide-prism-deflector type micro optical switches (WPD-MOS)," *Electronics and Communications in Japan, Part 2* **86**, 38–48 (2003).

When the area of the prism-shaped electrode is S [m^2], the distance between electrodes is d [m], and the dielectric constant of the EO material is ε, C is given by

$$C = \varepsilon S/d \ [F].$$
(6.9)

By substituting $S = (1/2) \times (16 \times 10^{-6}) \times (16 \times 10^{-6}) = 128 \times 10^{-12} \ m^2$, $d = 4 \times 10^{-7}$ m, and the assumed dielectric constant of SPCD, $\varepsilon \approx 4 \ \varepsilon_0$, into Equation (6.9), $C = 11$ fF is obtained. Then, from Equation (6.8), τ is calculated to be 1.1 ps for $R_{ON} = 100 \ \Omega$. We see that the speed of the switch is sufficiently fast.

Power consumption *Power* is given by

$$Power = FCV^2$$
(6.10)

where F is switching rate. When $F = 5 \times 10^9$ 1/s and $V = 11$ V, power consumption is calculated to be 6.9 mW.

6.1.3 Design of WPD Optical Switch Utilizing the Kerr Effect

As Figure 6.2 shows, PLZT is expected to exhibit a large electric-field-induced refractive index change due to its large Kerr coefficient. In the present subsection, design, simulated switching operation, and a preliminary experimental result of WPD optical switches utilizing PLZT thin films are reviewed.

6.1.3.1 Simulation Procedure

A top view of a WPD optical switch model with PLZT slab waveguide is shown in Figure 6.14. The same voltage is applied to the input-side electrodes and output-side electrodes in order to make the deflected light path parallel to the output waveguides.

Light wavelength λ is assumed to be 1.3 μm. Input and output waveguides consist of a channel waveguide with a core width D of 4 μm, a core refractive index n_{Core} of 1.52, and a clad refractive index n_{Clad} of 1.50, which permits the propagation of the fundamental mode with $m = 0$ and the first excited mode with $m = 1$. Channel distance P is 20 μm. Waveguide lens diameter D_{Lens} is 40 μm. The lens expands the light beam size to D_B. Right-angle prism-shaped electrodes with an aperture D_P of 16 μm and length $2D_P$ are formed on the slab waveguide with a separation d of 400 nm from a counter electrode beneath the PLZT film. Two-step prism cascading configuration is considered.

Reflections from the integrated components such as waveguide lenses were not taken into account. Anti-reflective (AR) coating on the component edge surfaces using atomic layer deposition (ALD) or MLD, which can grow ultrathin conformal films on surfaces with three-dimensional structures such as trenches, might make the model viable.

6.1.3.2 Structural Model

In order to determine a rough sketch of WPD optical switch structure, an estimation for EO slab waveguide length L_{EO} is carried out. A detail of the electric-field-induced waveguide prism is shown in Figure 6.15. The refractive index change Δn is induced in the EO slab waveguide by a driving voltage V according to Equation (6.2). Waveguide prism refractive index n_p is given by

$$n_p = n_o + \Delta n \tag{6.11}$$

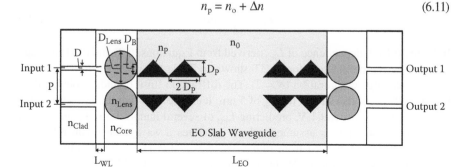

FIGURE 6.14 Top view of WPD optical switch model with PLZT slab waveguide.

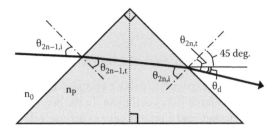

FIGURE 6.15   Detail of the electric-field-induced waveguide prism in a slab waveguide with the Kerr effect. From T. Yoshimura, M. Ojima, Y. Arai, N. Fujimoto, and K. Asama, "Simulation on cross-talk reduction in multi-mode-waveguide-based micro optical switching systems using single mode filters," *Appl. Opt.* **43**, 1390–1395 (2003).

Here, n_0 is the refractive index of the EO slab waveguide with no applied voltage. Deflection angle θ_d can be calculated using the following equations:

$$n_0 \sin\theta_{2n-1,i} = n_P \sin\theta_{2n-1,t}$$

$$n_P \sin\theta_{2n,i} = n_0 \sin\theta_{2n,t}$$

$$\theta_{2n,i} = \pi/2 - \theta_{2n-1,t}$$

$$\theta_{2n+1,i} = \pi/2 - \theta_{2n,t}$$

$$(n = 1, 2, \cdots, N)$$

$$\theta_d = \theta_{2N,t} - \pi/4 \tag{6.12}$$

As the 0th approximation, neglecting the size of prism-shaped electrodes, L_{EO} is given by

$$L_{\text{EO}} = P/\tan\theta_d. \tag{6.13}$$

The Diffraction angle of light beams propagating in the slab waveguide is expressed as $\theta = \lambda/(\pi n_0 D_B)$. The light beam width should be less than P at the end of the EO slab waveguide after propagation. Then, L_{EO} should satisfy

$$L_{\text{EO}} < \frac{P}{2}/\tan\theta. \tag{6.14}$$

In Figure 6.16, V dependence of L_{EO} derived from Equations (6.2) and (6.11) through (6.13) using the parameters for PLZT shown in Table 6.1 is shown for a two-step prism cascading configuration ($N = 2$). The diffraction limit boundary obtained by Equation (6.14) is also shown for D_B of 5 µm. It can be seen that the lowest possible driving voltage is around 1 V, predicting L_{EO} of several hundred µm. In the present simulation, $V = 1.2$ V is assumed, which generates a waveguide prism refractive index $n_{\text{P(On)}}$ of 2.52 derived from Equations (6.2) and (6.11). The waveguide prism refractive index $n_{\text{P(Off)}}$ for $0 - V$ application is equal to n_0 of 2.48.

Based on the parameters mentioned above, the waveguide lens refractive index n_{Lens}, the distance between the waveguide lens and the input/output waveguide L_{WL},

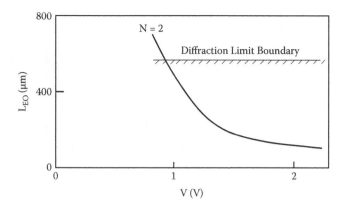

FIGURE 6.16   V dependence of L_{EO} and diffraction limit for L_{EO}. From T. Yoshimura, M. Ojima, Y. Arai, N. Fujimoto, and K. Asama, "Simulation on cross-talk reduction in multi-mode-waveguide-based micro optical switching systems using single mode filters," *Appl. Opt.* **43**, 1390–1395 (2003).

and L_{EO} are optimized. Figure 6.17 shows beam propagation in the slab waveguide simulated by BPM for n_{Lens} of 1.8, 2.0, and 2.5. For $n_{Lens} = 1.8$, collimation of the light beam in the EO slab waveguide is insufficient, while for n_{Lens} of 2.5 the focusing effect is too strong, and large beam size expansion occurs. n_{Lens} of 2.0 is a proper condition to produce collimated light beams, resulting in smooth coupling to the output waveguide at a distance of 310 μm from the input waveguide. Figure 6.18(a) shows L_{WL} dependence of transmission efficiency from the input waveguide into the output waveguide. The optimum L_{WL} is found to be 20 μm. For L_{EO}, as shown in Figure 6.18(b), the optimum is 190–200 μm. From a viewpoint of miniaturization of optical switches, 190 μm should be the optimum L_{EO}. Thus, the total length of the WPD optical switch is calculated to be 310 μm by adding $2(D_{Lens} + L_{WL})$ to L_{EO}.

Design parameters determined for the WPD optical switch are summarized in Table 6.2 and the typical structure of a WPD optical switch is shown in Figure 6.19. The slab waveguide is divided into two regions. One is a region with waveguide lenses. The waveguide lens refractive index of 2.00 is available by using Ti_xSi_yO, for example. The other is a region with prisms where the entire slab waveguide is made of 400-nm-thick PLZT. Right-angle prism-shaped electrodes with an area of 16×16 μm^2 are formed on the slab waveguide. The size of the main switch part is 310 μm $\times$ 20 μm. When channel separation is widened to 60 μm by bending waveguides of 2.5-mm radius at the input and output ports, the total switch length L_{SW} and width W_{SW} become 1190 μm and 100 μm, respectively.

6.1.3.3   Simulated Performance

Figure 6.20 shows the switching operation of the WPD optical switch simulated by the BPM. When no voltages are applied to the prism-shaped electrodes, a light beam propagates from Input 1 to Output 1, achieving a *bar* condition. By 1.2-V application to the electrodes, a light beam propagates from Input 1 to Output 2, achieving a *cross* condition. For both the bar and cross conditions, cross-talk is less than −20

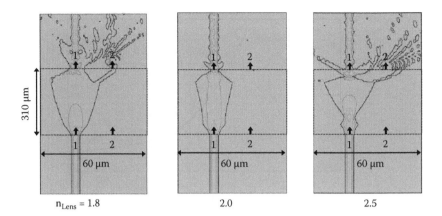

FIGURE 6.17 Beam propagation in the slab waveguide with waveguide lenses simulated by the BPM. From T. Yoshimura, M. Ojima, Y. Arai, N. Fujimoto, and K. Asama, "Simulation on cross-talk reduction in multi-mode-waveguide-based micro optical switching systems using single mode filters," *Appl. Opt.* **43**, 1390–1395 (2003).

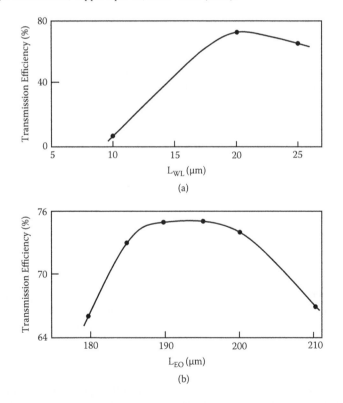

FIGURE 6.18 Dependence of transmission efficiency into the output waveguide on (a) L_{WL} and (b) L_{EO}. From T. Yoshimura, M. Ojima, Y. Arai, N. Fujimoto, and K. Asama, "Simulation on cross-talk reduction in multi-mode-waveguide-based micro optical switching systems using single mode filters," *Appl. Opt.* **43**, 1390–1395 (2003).

TABLE 6.2
Design Parameters for the WPD Optical Switch

Optical signal

Wavelength: λ	1.3 µm

Channel Waveguide

Width: D	4 µm
Channel separation: P	20 µm
Core refractive index: n_{Core}	1.52
Clad refractive index: n_{Clad}	1.50

Waveguide Lens

Lens-Input/Output waveguide distance: L_{WL}	20 µm
Diameter: D_{Lens}	40 µm
Refractive index: n_{Lens}	2.0

EO Slab Waveguide

Length: L_{EO}	190 µm
Electrode separation: d	400 nm
Driving voltage: V	1.2 V
Cascaded prism count: N	2
Kerr coefficient: R_{Kerr}	5.7×10^{-16} m²/V²
Refractive index: n_0	2.48
Prism refractive index (Off): $n_{P(Off)}$	2.48
Prism refractive index (On): $n_{P(On)}$	2.52

Structure

Total Length:	310 µm

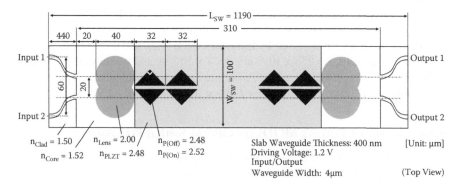

FIGURE 6.19 Typical structure of WPD optical switch with PLZT slab waveguide. From T. Yoshimura, M. Ojima, Y. Arai, and K. Asama, "Three-dimensional self-organized micro optoelectronic systems for board-level reconfigurable optical interconnects: Performance modeling and simulation," *IEEE J. Select. Topics in Quantum Electron.* **9**, 492–511 (2003).

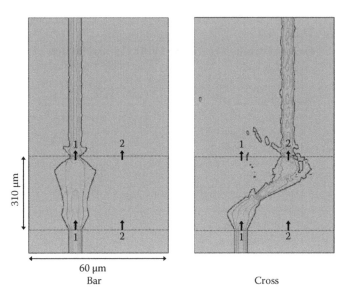

60 μm

310 μm

Bar   Cross

FIGURE 6.20 Switching operation of WPD optical switch with PLZT slab waveguide simulated by the BPM. From T. Yoshimura, M. Ojima, Y. Arai, N. Fujimoto, and K. Asama, "Simulation on cross-talk reduction in multi-mode-waveguide-based micro optical switching systems using single mode filters," *Appl. Opt.* **43**, 1390–1395 (2003).

dB. Transmission efficiency is 80% and 69% for bar and cross, respectively. The maximum insertion loss within the WPD optical switch is 1.7 dB including losses in bending waveguides.

Switching speed and power consumption can be estimated based on a charge–discharge model for a prism-shaped capacitor as the 0th approximation. A waveguide prism deflector is regarded as a capacitor with an area of S, a thickness of d, and a dielectric constant of ε. The capacitance C, switching speed τ, and power consumption *Power* are respectively given by Equations (6.9), (6.8), and (6.10). In the model shown in Figure 6.19, $S = 256 \times 10^{-12}\ m^2$, $d = 4 \times 10^{-7}\ m$, and the dielectric constant of PLZT $\varepsilon = 1000\varepsilon_0$. Then, C, τ, and *Power* are respectively calculated to be 5.7 pF, 570 ps for $R_{ON} = 100\ \Omega$, and 41 mW for $F = 5 \times 10^9\ 1/s$ and $V = 1.2$ V.

6.1.4 IMPACT OF POLYMER MQDS ON OPTICAL SWITCH PERFORMANCE

As Table 6.1 and Figure 6.2 show, the polymer MQD [14–16] is theoretically expected to exhibit large EO effects and a low dielectric constant, say, around $4\varepsilon_0$, which will improve switching performance of optical switches. Figure 6.21 shows EO material dependence of switching operation of WPD optical switches simulated by the BPM. When the DAAD-type polymer MQD described in Chapter 5 is used, large deflection angles induced by small driving voltages are expected.

The dielectric constant for electrical signals is expected to be much smaller in the polymer MQD than in PLZT because the EO effect in the polymer wires mainly arises from electronic polarization while that in PLZT arises from atomic

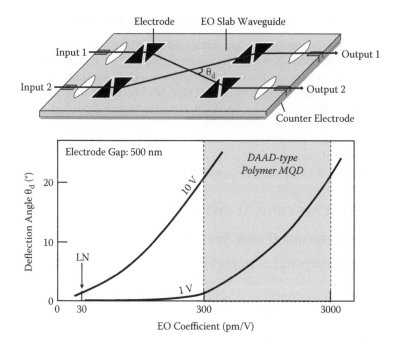

FIGURE 6.21 Improvement of WPD optical switches by utilizing the polymer MQD simulated by the BPM.

polarization. This feature of the polymer MQD is desirable to achieve high-speed and low power operations. When ε for the polymer MQD is $4\varepsilon_0$ and PLZT $1000\varepsilon_0$, τ and *Power* for the polymer MQD can be 1/250 of those for PLZT. Thus, the polymer MQD is expected to remove the drawbacks of slow switching time and large power consumption in PLZT. The polymer MQD is an ideal EO material for large-scale integrated optical interconnects and optical switching systems.

6.1.5 FUTURE INTEGRATION ISSUES

The present simulation indicates the possibility of 2×2 and 3×3 WPD optical switches having lengths of 100-µm scales. However, the method for connecting and integrating the driver circuits, optical waveguide systems, and optical switches is a serious problem when building actual systems. The connecting and the integrating will be achieved by using PL-Pack with SORT (described in Section 7.5), which transplants different thin-film devices and materials together into a substrate. EO thin films of the optical switches may be easily integrated with optical waveguides, waveguide lenses, waveguide filters, and waveguide mirrors by PL-PACK with SORT. For integration of EO thin films made of the polymer MQDs, molecular nano duplication (MND) (described in Section 3.7) may be applicable. SOLNET, which is described in Section 7.4, is useful to connect the EO thin films to optical waveguides and other optical devices even when misalignment exists between them.

6.1.6 Experimental Demonstration of WPD Utilizing PLZT

As a preliminary demonstration, we fabricated a WPD using a PLZT slab waveguide using the process shown at the top in Figure 6.22. A patterned indium tin oxide (ITO) transparent electrode was formed on a glass substrate. Then, a thin film of PLZT was applied on it by sol-gel method. The PLZT film thickness was 250 nm. On the film surface an Al counter electrode was deposited by vacuum evaporation. Figure 6.22 shows guided beam deflection by applying 35 V between the ITO and Al electrodes. A deflection angle of 1.1° was observed.

The future challenges are optimization of the PLZT WPD optical switches and replacement of PLZT with the polymer MQD.

6.2 NANO-SCALE OPTICAL SWITCHES

6.2.1 Ring Resonator Optical Switches

For ultrasmall optical switches, although the thermal stability is usually insufficient, ring resonators consisting of high-index contrast (HIC) waveguides might be a possible candidate. In the present section, light beam propagation and switching operation in the ring resonator optical switch are simulated by the finite difference time domain (FDTD) method.

Ring resonator optical switches have been developed by many researchers. A typical model is shown in Figure 6.23. Between two straight waveguides, a ring EO

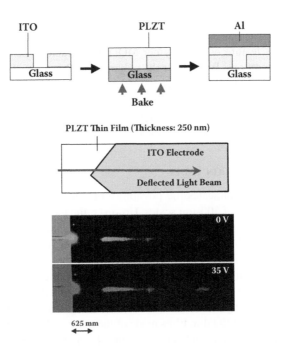

FIGURE 6.22 Light beam deflection in a WPD light modulator using a PLZT slab waveguide.

waveguide is inserted. The core width of these optical waveguides is usually sub-microns. Figure 6.24 shows propagation of a light pulse in a ring resonator optical switch consisting of PLZT. The light polarization is parallel to the z-axis. A light pulse introduced from Input 1 is partially transferred between the straight wave-guides and the ring EO waveguide.

When the device configuration and the refractive index of the optical waveguides are appropriately tuned in a resonance condition, continuous wave (CW) light beam gradually becomes localized into the ring EO waveguide, as can be seen in

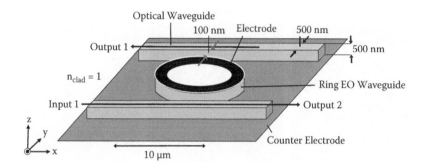

FIGURE 6.23 Model of a ring resonator optical switch.

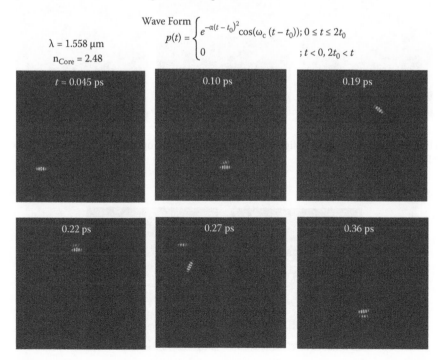

FIGURE 6.24 Propagation of a light pulse in a ring resonator optical switch simulated by the FDTD method. The brightness is automatically adjusted.

Figure 6.25. At the same time, the light beam that was guided to Output 2 is switched to Output 1. When a voltage is applied to the ring EO waveguide, the resonance condition is broken to switch the light beam from Output 2 to Output 1, as shown in Figure 6.26.

The resonance condition varies with the wavelength of the guided beam. Therefore, the ring resonator optical switch acts as a wavelength filter, too. By applying a voltage

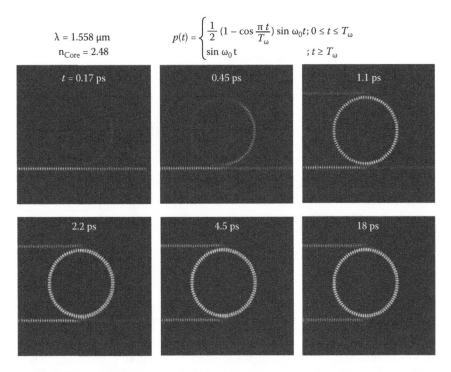

$$
\lambda = 1.558 \ \mu m \qquad\qquad p(t) = \begin{cases} \dfrac{1}{2}(1 - \cos \dfrac{\pi t}{T_\omega}) \sin \omega_0 t; \ 0 \le t \le T_\omega \\ \sin \omega_0 t \qquad\qquad ; t \ge T_\omega \end{cases}
$$

$n_{Core} = 2.48$

FIGURE 6.25 Resonance of light waves in a ring resonator optical switch simulated by the FDTD method.

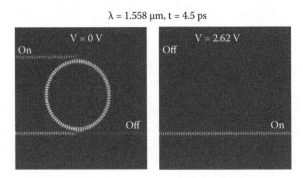

$\lambda = 1.558 \ \mu m, \ t = 4.5 \ ps$

FIGURE 6.26 Switching operation of a ring resonator optical switch simulated by the FDTD method.

to the ring EO waveguide, the resonance wavelength is controlled, enabling tunable wavelength filter operation.

6.2.2 Bandwidth Limit in Photonic Crystal Waveguides

The photonic crystal, which was proposed and developed by Yablonovitch [18], has periodic variations of the refractive index like dielectric multilayer wavelength filters. As the energy bands and energy gap for electrons in crystals that have periodic potential variations, the energy bands and photonic band gap (PBG) for photons appear in photonic crystals as shown in Figure 6.27. When the energy of photons is in the PBG region, light beams cannot exist in the photonic crystals in steady states. When the energy of photons is out of the PBG region, light beams can exist in the photonic crystals.

If some imperfections are created in a photonic crystal, light beams with photon energy within the PBG can exist in the imperfect parts. By creating line-shaped imperfections as shown in Figure 6.27, light beams are confined in the imperfect parts because they cannot exist in the surrounding perfect photonic crystal parts. The line width of the imperfection can be submicrons. Thus, nano-scale waveguides can be constructed in the photonic crystal. By implementation of EO materials into the photonic crystal, ultrasmall optical switches might be available.

Simulation for light beam propagation in a 2-D photonic crystal waveguide was carried out using the FDTD method. The results are shown in Figures 6.28 and 6.29. Holes with a 104-nm radius are arranged with a 521-nm pitch in a medium with a refractive index of 3.38. Light polarization is parallel to the z-axis. For wavelength $\lambda = 1.19$ μm, which corresponds to photon energy out of the PBG, light beams are leaked from the optical waveguide. For wavelength $\lambda = 1.30$ μm, which corresponds to photon energy near the center of the PBG, light beams are strongly confined in the optical waveguide.

Here, however, it should be noted that at the initial stage of light incidence, that is, at time duration (t) of 0.11 ps, confined light intensity is weak, and with time it

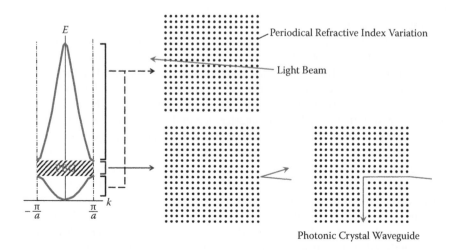

FIGURE 6.27 Band diagrams for a photonic crystal and light beam propagation in it.

$\lambda = 1.19\ \mu m$

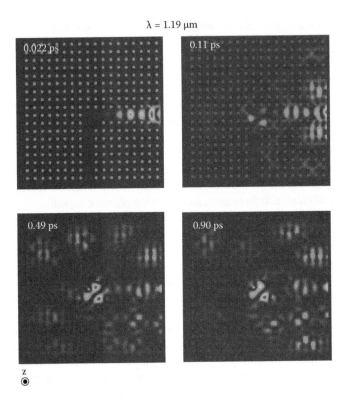

FIGURE 6.28 Light beam propagation in a photonic crystal waveguide for a wavelength out of the PBG simulated by the FDTD method.

gradually increases to reach a saturated state. Figure 6.30 shows a result for transient light beam propagation in the photonic crystal waveguide [19]. The drawing contrast is adjusted to enhance leaked light beams in the surrounding perfect crystal region. It is found that at the initial stage, considerable leakage of light beams into the surrounding perfect crystal region is observed. After 0.6 ps, the leakage is suppressed and strong confinement of light beams in the optical waveguide is accomplished.

A delay before settling waveguide functions arises from the fact that the photonic crystal device uses interference of light beams at a steady state. In the model shown in Figure 6.30, at $t = 0.05$ and 0.1 ps, only a part of reflected light beams are in the photonic crystal, that is, light beam interference in the photonic crystal is not completed yet. At $t = 0.6$ ps, since 0.6 ps corresponds to a light beam propagation distance of ~150 μm that is 17 times the photonic crystal size in the present model, the interference is built up completely, that is, superposition of all the reflected light beams is completed in the entire photonic crystal, to reach the steady state. Thus, the leakage is suppressed, resulting in perfect light beam confinement in the optical waveguide.

Transient responses of light beam confinement in a photonic crystal waveguide are shown in Figure 6.31. For wavelength $\lambda = 1.20$ μm, which corresponds to photon energy at the edge of the PBG, a long delay of ~0.5 ps is observed before the confined light intensity reaches its steady state. For wavelength $\lambda = 1.30$ μm located near the

$\lambda = 1.30$ μm

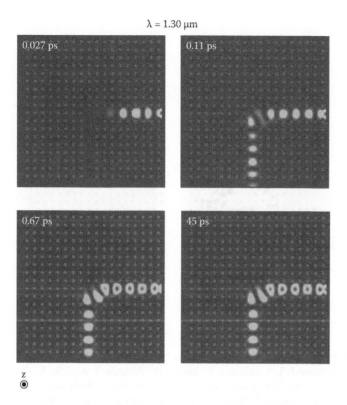

FIGURE 6.29 Light beam propagation in a photonic crystal waveguide for a wavelength within the PBG simulated by the FDTD method.

center of the PBG, on the other hand, the confined light intensity rapidly increases to its saturated value with a short delay of less than 0.2 ps. As can be seen in Figure 6.32, light beam distribution after reaching steady state is wider for $\lambda = 1.20$ μm than for $\lambda = 1.30$ μm, indicating that in the case of $\lambda = 1.20$ μm, the light beams penetrate into the deep part of the surrounding perfect crystal region, and consequently, the interference occurs over a large region as compared to the case of $\lambda = 1.30$ μm. From these results, it is suggested that the delay before the waveguide function appears will cause problems when photonic crystals are used in high-speed systems.

6.2.3 Polymer MQDs in Nano-Scale Optical Switches

As mentioned in Section 6.1.4, the polymer MQD fabricated by MLD is expected to be a promising EO material. It will be a great challenge to realize high-performance optical switches with ring resonator and photonic crystal structures consisting of the polymer MQD.

Figure 6.33 shows a concept of a photonic crystal optical switch with the polymer MQD fabricated by MLD using seed cores, as described in Section 3.4. First, seed cores with a pattern of a photonic crystal waveguide are formed on a substrate surface. Then, by MLD, polymer wires are grown from the seed cores to form EO

λ = 1.30 μm

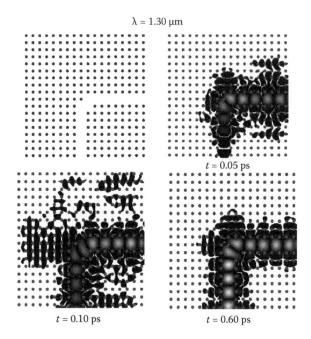

$t = 0.05$ ps

$t = 0.10$ ps $t = 0.60$ ps

FIGURE 6.30 Transient light beam propagation in a photonic crystal waveguide simulated by the FDTD method. T. Yoshimura, Y. Suzuki, N. Shimoda, T. Kofudo, K. Okada, Y. Arai, and K. Asama, "Three-dimensional chip-scale optical interconnects and switches with self-organized wiring based on device-embedded waveguide films and molecular nanotechnologies," *Proc. SPIE* **6126**, 612609-1-15 (2006).

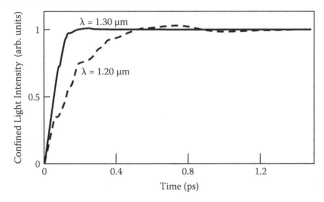

FIGURE 6.31 Transient responses of light beam confinement in a photonic crystal waveguide simulated by the FDTD method.

rods of the polymer MQD. By applying voltages to the rods, some optical switching operation might occur. For the ring resonator optical switches, the same process is applicable by forming seed cores with ring waveguide patterns.

Connection and integration of the nano-scale optical switches can be done by PL-Pack with SORT, MND, and SOLNET.

λ = 1.30 μm

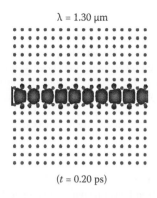

(t = 0.20 ps)

λ = 1.20 μm

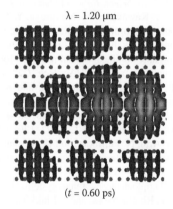

(t = 0.60 ps)

FIGURE 6.32 Light beam distribution in photonic crystal waveguides.

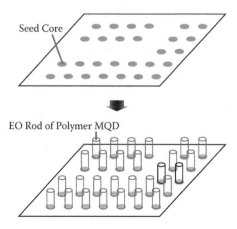

FIGURE 6.33 Concept of a photonic crystal optical switch with the polymer MQD fabricated by MLD.

REFERENCES

1. T. Yoshimura, S. Tsukada, S. Kawakami, Y. Arai, H. Kurokawa, and K. Asama, "3D micro optical switching system (3D-MOSS) packaging design," 2001 IEEE LEOS Annual Meeting Conference Proceedings, San Diego, California, 499–500 (2001).

2. T. Yoshimura, S. Tsukada, S. Kawakami, Y. Arai, H. Kurokawa, and K. Asama, "3D micro optical switching system (3D-MOSS) architecture," *Proc. SPIE* **4653**, from Photonics West 2002, San Jose, California, 62–70 (2002).

3. T. Yoshimura, Waveguide-type electro-optic devices, Japanese Patent Tokukai Hei 4-181231 (1992) [in Japanese].

4. T. Yoshimura, Optical integrated circuit devices, Japanese Patent Tokukai Hei-204633 (1992) [in Japanese].

5. T. Yoshimura, M. Ojima, Y. Arai, and K. Asama, "Three-dimensional self-organized micro optoelectronic systems for board-level reconfigurable optical interconnects: performance modeling and simulation," *IEEE J. Select. Topics in Quantum Electron.* **9**, 492–511 (2003).

6. T. Yoshimura, S. Tsukada, S. Kawakami, M. Ninomiya, Y. Arai, H. Kurokawa, and K. Asama, "Three-dimensional micro-optical switching system architecture using slab-waveguide-based micro-optical switches," *Opt. Eng.* **42**, 439–446 (2003).

7. T. Yoshimura, "Design of organic nonlinear optical materials for electro-optic and all-optical devices by computer simulation," *FUJITSU Sc. Tech. J.* **27**, 115–131 (1991).

8. M. Ninomiya, Y. Arai, T. Yoshimura, H. Kurokawa, and K. Asama, "Characteristics evaluation for waveguide-prism-deflector type micro optical switches (WPD-MOS)," *Electronics and Communications in Japan, Part 2* **86**, 38–48 (2003) [Translated from *Denshi Joho Tsushin Gakkai Ronbunshi* **J85-C**, 1192–1201 (2002)].

9. T. Yoshimura, M. Ojima, Y. Arai, N. Fujimoto, and K. Asama, "Simulation on cross-talk reduction in multi-mode-waveguide-based micro optical switching systems using single mode filters," *Appl. Opt.* **43**, 1390–1395 (2003).

10. T. Yoshimura, and Y. Arai, 3D optoelectronic micro system, U.S. Patent 7,387,913 (2008).

11. O. Qasaimeh, K. Kamath, P. Bhattacharya, and J. Phikkips, "Linear and quadratic electro-optic coefficients of self-organized In$_{0.4}$Ga$_{0.6}$As/GaAs quantum dots," *Appl. Phys. Lett.* **72**, 1275–1277 (1998).

12. T. Yoshimura and T. Futatsugi, Non-linear optical device using quantum dots, U.S. Patent 6,294,794 (2001).

13. T. Yoshimura, "Characterization of the electro-optic effect in styrylpyridinium cyanine dye thin-film crystals by an AC modulation method," *J. Appl. Phys.* **62**, 2028–2032 (1987).

14. T. Yoshimura, S. Tatsuura, W. Sotoyama, A. Matsuura, and T. Hayano, "Quantum wire and dot formation by chemical vapor deposition and molecular layer deposition of one-dimensional conjugated polymer," *Appl. Phys. Lett.* **60**, 268–270 (1992).

15. T. Yoshimura, A. Oshima, D. Kim, and Y. Morita, "Quantum dot formation in polymer wires by three-molecule molecular layer deposition (MLD) and applications to electro-optic/photovoltaic devices," *ECS Transactions* **25**, No. 4 "Atomic layer deposition applications 5," from the Vienna, Austria Meeting, 15–25 (2009).

16. T. Yoshimura, "Enhancing second-order nonlinear optical properties by controlling the wave function in one-dimensional conjugated molecules," *Phys. Rev.* **B 40**, 6292–6298 (1989).

17. T. Aizawa, Furuuchi Chemical Corporation Technical Articles for PLZT Shutter Arrays.

18. E. Yablonovitch, "Inhibited spontaneous emission in solid state physics and electronics," *Phys. Rev. Lett.* **58**, 2059–2062 (1987).

19. T. Yoshimura, Y. Suzuki, N. Shimoda, T. Kofudo, K. Okada, Y. Arai, and K. Asama, "Three-dimensional chip-scale optical interconnects and switches with self-organized wiring based on device-embedded waveguide films and molecular nanotechnologies," *Proc. SPIE* **6126**, from Photonics West 2006, San Jose, California, 612609-1-15 (2006).

7 Organic Photonic Materials, Devices, and Integration Processes

Although artificial thin-film organic materials fabricated by molecular layer deposition (MLD) and devices made of artificial materials are the final goal, sometimes, other kinds of thin-film organic materials and devices are still useful. Furthermore, in order to construct real optoelectronic (OE) systems, integration processes of the thin-film materials and devices are important.

In the present chapter, these materials and devices, and the integration processes are presented.

7.1 ELECTRO-OPTIC (EO) MATERIALS

Nonlinear optical devices such as EO devices, all-optical devices, and optical integrated circuits (ICs) consisting of these devices are key in optical communication, optical interconnects, and other OE systems. However, application areas of nonlinear optical devices are limited due to a lack of high-performance nonlinear optical materials.

Organic materials with π-conjugated systems have attracted interest as one of the candidates for nonlinear optical materials. The nonlinearity, as mentioned in Chapter 5, is predicted to be superior to that of conventional inorganic materials, such as $LiNbO_3$ (LN) and the multiple quantum wells (MQWs) of compound semiconductors. Low dielectric constant characteristics are also an advantage of organic materials, especially for EO devices, enabling high speed and low power operation. Despite much work on organic materials, however, few notable improvements in optical nonlinearities have been made yet. It is important for the photonics world to make a breakthrough in the nonlinear optical properties of organic materials and to clarify the potential of the organic materials for EO devices.

The EO coefficients r of organic EO materials are compared in Figure 7.1. The materials can be classified into poled polymers [1], molecular crystals [2,3], and conjugated polymers [4,5]. The most promising EO material is the conjugated polymer [4,5]. As described in Chapter 5, polydiacetylene (PDA) wires with long π-conjugated systems are regarded as natural one-dimensional systems (quantum wires), and are more favorable to induce a larger oscillator strength than two- or three-dimensional systems. When multiple quantum dot (MQD) structures are introduced into the wires, the wires have long variable wavefunctions offering a greatly enhanced Pockels effect that is about 100 times r (LN). However, the conjugated EO polymers are a kind of far-reaching material. The second-best solutions are the molecular crystals and the poled polymers.

EO Materials	LiNbO$_3$	Poled Polymer	Molecular Crystal	Conjugated Polymer
r(pm/V)	~30	~50	~400	>1000 (Simulation)
Processability	Good	Good	Poor	Good
Development Phase	Commercially Available	Test Vehicle	Material Preparation	Material Design

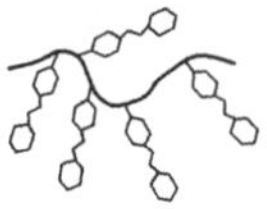

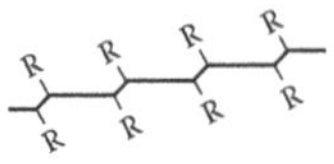

FIGURE 7.1 Organic nonlinear optical materials for EO devices.

For the molecular crystals, in 1987, using the AC modulation method, we found that the styrylpyridinium cyanine dye (SPCD) thin-film crystal exhibits an *r* of 430 pm/V, about 14 times that of LN [6]. This was the largest value ever observed in organic materials. The result demonstrated the prospect for applications of organic materials to EO devices. For the poled polymers, although *r* is usually smaller than that of the molecular crystals, they can be easily formed by spin-coating into thin films with high optical quality.

In the present section, various types of molecular crystals and poled polymers for EO materials are reviewed, and some examples of applications to polymer optical waveguides and optical switches are shown.

7.1.1 CHARACTERIZATION PROCEDURE FOR THE POCKELS EFFECT IN ORGANIC THIN FILMS

We developed the AC modulation method for characterization of the Pockels effect of organic thin films [6]. The schematic diagram of the experimental setup is shown in Figure 7.2(a). In order to apply an electric field to the thin films, a substrate, on which slit-type electrodes are formed, is pressed against the thin film. This method makes it possible to form electrodes on a small flake of organic thin film, say, with μm-order sizes. AC voltage from a function generator is applied between the electrodes. Linearly polarized light from a laser passes through the thin film and the electrode gap, and the light transmitted through the analyzer is detected by a photodetector.

In Figure 7.2(b) polarization states corresponding to positions I, II, and III, described in Figure 7.2(a), are schematically illustrated. The z and y components of the electric field of the laser (E_z and E_y) at positions I and II and the component along the analyzed polarization direction E_θ at position III can be written as follows:

Position I:

$$E_z = E_0 \cos\varphi \cos\omega t$$

$$E_y = E_0 \sin\varphi \cos\omega t$$

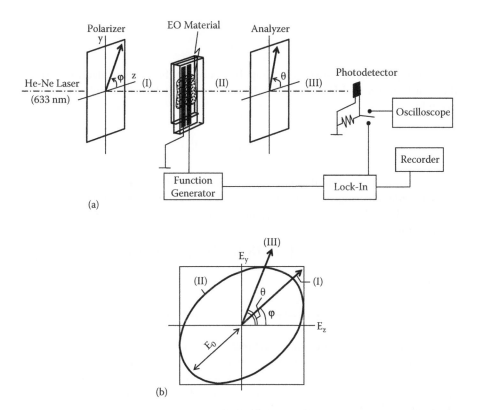

(a)

(b)

FIGURE 7.2   (a) Schematic diagram of experimental setup for the AC modulation method. (b) Polarization states corresponding to positions I, II, and III in (a). From T. Yoshimura, "Characterization of the EO effect in styrylpyridinium cyanine dye thin-film crystals by an ac modulation method," *J. Appl. Phys.* **62**, 2028–2032 (1987).

Position II:

$$E_z = E_0 \cos\varphi \cos\omega t$$

$$E_y = E_0 \sin\varphi \cos(\omega t - \delta)$$

Position III:

$$E_\theta = E_{\theta 0}(\delta)\cos(\omega t + \alpha)$$

$$E_{\theta 0} = E_0 \left[(\cos\varphi\cos\theta + \sin\varphi\sin\theta\cos\delta)^2 + (\sin\varphi\sin\theta\sin\delta)^2 \right]^{1/2}, \qquad (7.1)$$

where φ is the angle between the polarization direction of incident light and the z-axis, θ is the angle between the analyzed polarization direction and the z-axis, and δ is the phase retardation arising from birefringence of the thin film. Then, the detector output V_{out} can be written as

$$V_{out} = const[E_{\theta 0}]^2.$$

δ varies depending on the voltage V_A applied between the slit-type electrodes since the thin film has an EO effect. The detector output at $V_A = 0$, denoted by V_{dc}, is given by

$$V_{dc} = const[E_{\theta 0}(\delta_0)]^2,\tag{7.2}$$

where δ_0 is the phase retardation at $V_A = 0$. When V_A is increased from 0 to V, the phase retardation changes from δ_0 to $\delta_0 + \Delta\delta$. $\Delta\delta$ is the EO phase retardation induced by $V_A = V$. Then the detector output change V_{ac} is induced by applying $V_A = V$ and is given as follows.

$$V_{ac} = const\left\{ [E_{\theta 0}(\delta_0)]^2 - [E_{\theta 0}(\delta_0 + \Delta\delta)]^2 \right\}\tag{7.3}$$

From Equations (7.2) and (7.3),

$$V_{ac}/V_{dc} = \left\{ [E_{\theta 0}(\delta_0)]^2 - [E_{\theta 0}(\delta_0 + \Delta\delta)]^2 \right\} / [E_{\theta 0}(\delta_0)]^2\tag{7.4}$$

is obtained. Substituting Equation (7.1) into Equation (7.4), an expression for V_{ac}/V_{dc} related to the four parameters, that is, φ, θ, δ_0, and $\Delta\delta$ is obtained. Here φ and θ are known, and δ_0 is obtained from the angular dependence of V_{dc}. Therefore, by measuring V_{ac} experimentally with the use of the setup illustrated in Figure 7.2(a), the EO phase retardation $\Delta\delta$ can be determined from Equation (7.4).

Once $\Delta\delta$ is obtained, the figure of merit of the EO phase retardation F is estimated according to the formula

$$F = (\lambda_0 \Delta\delta\, D)/(2\pi V_A L)\tag{7.5}$$

where λ_0 is the wavelength of the light in vacuum, D is the distance between the slit-type electrodes, and L is the film thickness.

The AC modulation method exhibits a high sensitivity for estimating EO effects since the lock-in technique is used. Therefore, the method is effective for characterizing thin-film EO materials.

7.1.2 MOLECULAR CRYSTALS

7.1.2.1 SPCD

A lot of studies have been done on second harmonic generation (SHG), which is one of the second-order nonlinear optical phenomena for molecular crystals. Some materials such as 2-methyl-4-nitroaniline (MNA) have been found to show extremely large SHG activity [2].

Molecular crystals have also been known to show the Pockels effect, which, as well as SHG, is involved in second-order nonlinearity. Lipscomb et al. studied the

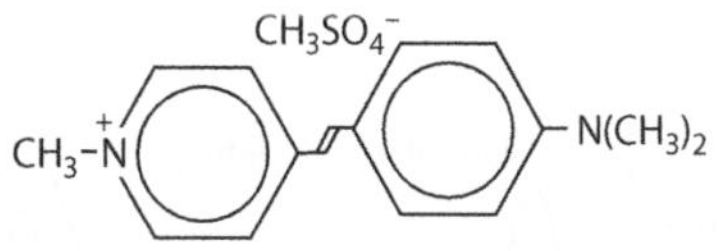

FIGURE 7.3 Styrylpyridinium cyanine dye (SPCD). From T. Yoshimura, "Characterization of the EO effect in styrylpyridinium cyanine dye thin-film crystals by an ac modulation method," *J. Appl. Phys.* **62**, 2028–2032 (1987).

Pockels effect in an MNA single crystal and estimated the EO coefficient r_{11} to be 67 pm/V [3], which is about 2 times as large as r_{33} in LiNbO$_3$.

In the present subsection, first, the Pockels effect of SPCD molecular crystals is reported [6]. SPCD has a molecular structure schematically shown in Figure 7.3. Referring to an SHG measurement, SPCD shows extremely large SHG activity, about one order of magnitude larger than that in MNA [7]. Consequently, it is expected that in SPCD the Pockels effect would be several times larger than in MNA.

According to this guideline, we prepared SPCD thin film crystals and characterized the EO property of them. The characterization was performed by the AC modulation method described in Section 7.1.1. The accuracy of the method was confirmed by comparing the measured value with the value reported in other articles using LiNbO$_3$ crystal as a standard sample.

7.1.2.1.1 Crystal Growth

An SPCD thin-film crystal was grown as follows. Two quartz substrates (55 × 55 × 1 mm^3) were paired in a methanol solution of SPCD (Japanese Research Institute for Photosensitizing Dyes Co. Ltd.). The solution was inserted between the two substrates by capillary effect. After several days the methanol was evaporated naturally, and an SPCD thin film was grown between the substrates. The two substrates were separated carefully, and the thin film was left on one of the substrates. The film size typically ranges from 1–10 mm long, < 1 mm wide, and 3–10 μm thick. Observation of the SPCD thin film between crossed polarizers revealed that the SPCD thin film exhibits birefringence. When the thin film was rotated on a stage, some of the thin films showed complete and uniform extinction. These results indicate that the SPCD thin film is a single crystal. Since the SPCD crystal, as in most organic crystals, is not durable against mechanical, thermal, and chemical attack, we treated the crystal carefully during measurements.

7.1.2.1.2 Optical Properties

In Figure 7.4 a schematic shape of SPCD thin-film crystal is illustrated with a roughly estimated angle of each corner, as well as the absorption spectra of the crystal. Two optical axes are denoted by z and y. As can be seen in the figure, the light absorption of the SPCD thin-film crystal increases rapidly in the range above 2 eV, and the crystal shows strong dichroism. The absorption is greater for polarization along the z-axis than along the y-axis. This suggests that the molecular dipole moments of SPCD tend to align along the z-axis in the thin-film crystalline state.

Photoluminescence spectra of an SPCD thin-film crystal, SPCD powder used as the starting material for the crystal growth, and a water solution of SPCD were measured

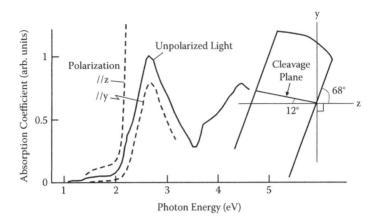

FIGURE 7.4 Schematic shape of SPCD thin-film crystal and its absorption spectra for polarized light. From T. Yoshimura, "Characterization of the EO effect in styrylpyridinium cyanine dye thin-film crystals by an ac modulation method," *J. Appl. Phys.* **62**, 2028–2032 (1987).

with a He-Cd laser (325 nm) excitation light source at room temperature. The luminescence spectrum of the SPCD powder appears in the low-energy side (peak position of 1.5 eV) compared with that of the SPCD thin-film crystal (peak position of 2.05 eV), and the luminescence intensity for the powder was about 1/1000 of that for the crystal. This suggests that the electronic state in the crystal is largely different from the state in the powder. The spectrum of the SPCD crystal is relatively close to that of the water solution. This implies that in the crystal the electronic state of the SPCD molecule is like the state in water, that is, the interaction between SPCD molecules is weak.

The refractive index of the SPCD thin-film crystal was estimated to be $n_z = 1.55$ and $n_y = 1.31$ by using the prism coupler method proposed by Wei and Westwood [8] and a differential interferometer. The result of $n_z > n_y$ is consistent with the strong dichroism of the crystal and indicates again that the molecular dipole moments align along the z-axis.

7.1.2.1.3 Measurement

In order to apply an electric field to SPCD thin-film crystals, a substrate, on which slit-type NiCr electrodes (distance between the electrodes, 5 μm; electrode width, 100 μm; thickness 80 nm) are formed, was pressed against the thin-film crystal. Under application of AC voltage from a function generator between the electrodes, linearly polarized light at 633 nm from a He-Ne laser passed through the SPCD thin-film crystal and the electrode gap as shown in Figure 7.2. The light transmitted through the analyzer was detected by a solar cell.

7.1.2.1.4 Pockels Effect

As mentioned in Section 7.1.1, in order to estimate $\Delta\delta$ and F, V_{dc} and V_{ac} must be measured. V_{dc} can be obtained straightforwardly by an oscilloscope. From the angular dependence of V_{dc}, δ_0 is obtained. The result is shown in Figure 7.5 for a 5-μm-thick

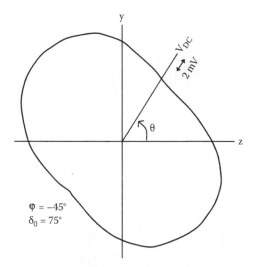

FIGURE 7.5 V_{dc} as a function of θ for a 5-μm-thick SPCD thin-film crystal. From T. Yoshimura, "Characterization of the EO effect in styrylpyridinium cyanine dye thin-film crystals by an ac modulation method," *J. Appl. Phys.* **62**, 2028–2032 (1987).

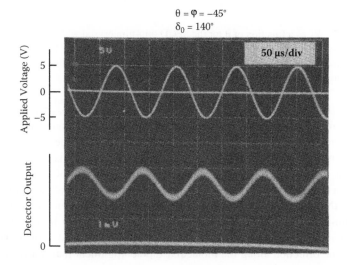

FIGURE 7.6 Light modulation introduced in the SPCD thin-film crystal by an 8-kHz sine wave. From T. Yoshimura, "Characterization of the EO effect in styrylpyridinium cyanine dye thin-film crystals by an ac modulation method," *J. Appl. Phys.* **62**, 2028–2032 (1987).

SPCD thin-film crystal. In Figure 7.6, a typical example of the light modulation introduced in the SPCD thin-film crystal by an 8-kHz sine wave is shown. V_{ac} is superposed on V_{dc}. In this work for detecting V_{ac}, the lock-in technique was used. Figure 7.7 shows V_{ac}-versus-θ characteristics of the SPCD thin-film crystal for an electric field applied along the z-axis. V_{ac} becomes zero at $\theta = 0°$, $90°$, and $180°$,

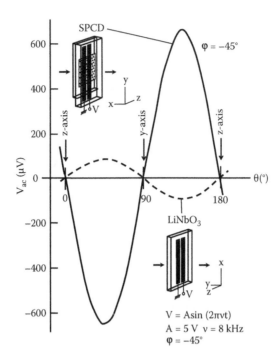

FIGURE 7.7 V_{ac} as a function of θ for the SPCD thin-film crystal with electric field applied along the z-axis (solid line) and for a LiNbO$_3$ crystal (dotted line). From T. Yoshimura, "Characterization of the EO effect in styrylpyridinium cyanine dye thin-film crystals by an ac modulation method," *J. Appl. Phys.* **62**, 2028–2032 (1987).

that is, in the direction of the optical axes (z-axis and y-axis), and has its maximum or minimum at $\theta = 45°$ and $135°$. This behavior can be completely explained by Equations (7.3) and (7.1) and indicates that the rotation of the optical axes does not occur in the SPCD thin-film crystal when the electric field is applied along the z-axis. For comparison, V_{ac}-versus-θ characteristics for a LiNbO$_3$ crystal are also shown by a dotted line. The SPCD thin-film crystal exhibits V_{ac} about 5 times larger than the LiNbO$_3$ crystal. When the electric field was applied along the y-axis of the SPCD thin-film crystal, a maximum value of V_{ac} was reduced to 1/3 of that with the electric field along the z-axis. This means that in the SPCD crystal, the EO effect is larger for the direction along the molecular dipole alignment than for the direction perpendicular to the dipole alignment.

The EO phase retardation $\Delta\delta$ is obtained by substituting the experimental values, that is, $\varphi = \theta = -45°, \delta_0 = 75°, V_{dc} = 14\ mV$, and $V_{ac} = 0.65\ mV$ into Equations (7.1) and (7.4). The result is shown in Figure 7.8. The SPCD thin-film crystal shows $\Delta\delta = 7°$ at the applied voltage of 20 V, which corresponds to the half-wave voltage V_π of 500 V, while the LiNbO$_3$ crystal shows $\Delta\delta = 1.5°$ at 20 V, corresponding to $V_\pi = 2400$ V.

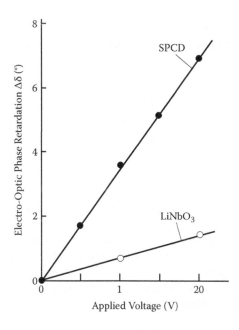

FIGURE 7.8 Dependence of EO phase retardation on applied voltage for the SPCD thin-film crystal with electric field applied along the z-axis (filled circles) and for the LiNbO₃ crystal (open circles). From T. Yoshimura, "Characterization of the EO effect in styrylpyridinium cyanine dye thin-film crystals by an ac modulation method," *J. Appl. Phys.* **62**, 2028–2032 (1987).

The figure of merit of EO phase retardation of SPCD and the standard sample of LiNbO₃ can be estimated by using Equation (7.5) and the results in Figure 7.8. Here in the estimation of the figure of merit for LiNbO₃ crystal, an approximation of $L = D$ was made, considering that the electric field between the electrodes penetrates from the surface into the crystal to the same extent as the electrode distance D. The estimated value of the figure of merit in the LiNbO₃ crystal is 1.2×10^{-10} m/V and nearly equals the value of 0.95×10^{-10} m/V reported by Lipscomb et al. [3]. This confirms the accuracy of the AC modulation method and the approximation done previously. For the SPCD thin-film crystal a large figure of merit of EO phase retardation, $F = 6.5 \times 10^{-10}$ m/V, is obtained. The value is about 5 times as large as that of the LiNbO₃ crystal and 2 times as large as 2.7×10^{-10} m/V of the MNA crystal.

In addition it should be noted that the EO effect can be detected only by applying AC voltage. The effect cannot be observed by applying DC voltage. The reason is that in SPCD, the EO effect decays rapidly under a constant voltage application. The origin of the decay could be that some mobile ions cancel the applied voltage between the electrodes.

7.1.2.1.5   EO Coefficient

Since the SPCD crystal has an orthorhombic structure, the EO coefficient matrix is described as follows.

$$
\begin{array}{ccc}
0 & 0 & r_{13} \\
0 & 0 & r_{23} \\
0 & 0 & r_{33} \\
0 & r_{42} & 0 \\
r_{51} & 0 & 0 \\
0 & 0 & 0
\end{array}
$$

According to this matrix, rotation of optical axes does not occur only when the electric field is applied along the principal z-axis of the index ellipsoid. As described previously, no rotation of optical axes is observed when an electric field is applied along the optical z-axis, indicating that the optical and principal z-axes coincide in the SPCD thin-film crystal.

EO coefficient r_{33} in the SPCD crystal and the figure of merit of EO phase retardation can be related as follows.

$$F = |(n_z^3 r_c)/2|, \tag{7.6}$$

$$r_c = r_{33} - (n_y/n_z)^3 r_{23} \tag{7.7}$$

By substituting $F = 6.5 \times 10^{-10}$ m/V and $n_z = 1.55$ into Equation (7.6), r_c is found to be 4.3×10^{-10} m/V. Usually, the coefficient corresponding to the direction along the molecular dipole alignment tends to be the largest [3]. Then, by neglecting r_{23} in Equation (7.7), $r_{33} \approx r_c = 430$ pm/V is obtained. This value is extremely large compared with $r_{33} = 32$ pm/V of the LiNbO$_3$ and $r_{11} = 67$ pm/V of the MNA crystal [3]. The results led us to believe that organic solids would be promising as EO materials.

Finally, the relationship between the EO coefficient r_{33} and the SHG efficiency of the SPCD crystal is discussed. In organic nonlinear materials it is known that the EO coefficient r and SHG efficiency η are related to the second-harmonic coefficient d and the refractive index n as follows [2,3].

$$r \propto d/n^4$$

$$\eta \propto d^2/n^3$$

Here, dispersion is ignored for simplicity. From these relationships,

$$r \propto \eta^{0.5}/n^{2.5}$$

is obtained, and then we have

$$r_{33}(SPCD)/r_{11}(MNA) = [\eta(SPCD)/\eta(MNA)]^{0.5} \times [n(MNA)/n(SPCD)]^{2.5}. \tag{7.8}$$

Using this expression, we can calculate the relative ratio of r from the SHG efficiencies. By substituting $n(\text{SPCD}) = 1.55$ and $n(\text{MNA}) = 2$ at 633 nm [2], and $\eta(SPCD)/\eta(MNA) = 10[7]$ into Equation (7.8), we have $r_{33}(SPCD)/r_{11}(MNA) = 6$ as the expected value based on the SHG measurement. The ratio of 6 agrees well with the measured value of $r_{33}(SPCD)/r_{11}(MNA) = 4.3 \times 10^{-10}/0.67 \times 10^{-10} = 6.4$. This coincidence again supports the accuracy of the AC modulation method.

7.1.2.2 MNA

Next, relationships between the Pockels effect and the growth conditions of an organic crystal are briefly noted using 2-methyl-4-nitroaniline (MNA) as an example [9].

Figure 7.9 shows the crystal growth apparatus. It has temperature control accuracy of 0.1°C. In Figure 7.10 a temperature-time profile for MNA thin-film crystal growth

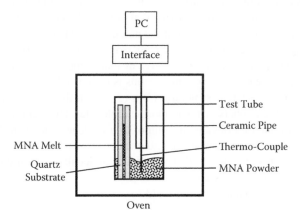

FIGURE 7.9   Crystal growth apparatus with automatic temperature control.

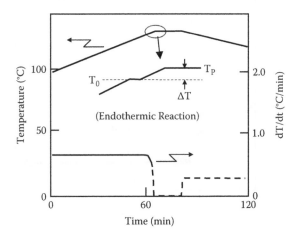

FIGURE 7.10   Temperature-time profile for MNA thin-film crystal growth. From Y. Kubota and T. Yostrimura, "Endothermic reaction aided crystal growth of 2-methyl-4-nitroaniline and their electro-optic properties," *Appl. Phys. Lett.* **53**, 2579–2580 (1988).

is plotted. Two paired quartz substrates ($15 \times 10 \times 0.5$ mm^3) and MNA powder (3 mg) were set in a test tube and heated in an oven. MNA temperature changes were monitored by a computer. Near the melting point (T_M) of MNA, the temperature change rate dT/dt decreases rapidly due to an endothermic reaction, and the MNA powder begins to melt. The temperature, at which dT/dt started to decrease (T_0), was detected, and feedback control was used to maintain the melt temperature at $T_p = T_0 + \Delta T$. The melt was introduced into the gap space between the substrates by capillary action. MNA thin-film crystals were grown between the substrates by slow cooling. The two substrates were carefully separated, leaving the thin film on one of the substrates. The film is typically 2–3 mm in length and width, and 5–10 µm in thickness.

The figure of merit for EO phase retardation (F) of the MNA thin-film crystals was measured by the AC modulation method. In Figure 7.11 the T_p dependence of F is shown. As T_p decreases toward T_M, F increases. When the crystal is prepared so that $T_M \approx T_p$, the maximum F is twice that of the LiNbO$_3$ crystal. The reproducibility of this value is good as a result of the automatic control of the crystal growth. With annealing (at 118°C for 120 min), F increases from 1.8×10^{-10} to 2.4×10^{-10} m/V. This value is about three times as large as that of the LiNbO$_3$ crystal.

We expect that one of the following reasons is responsible for the variation in F: (1) a change in the molecular structure, (2) a change in the crystal structure, (3) a change in the interaction between neighboring molecules. Differential thermal analysis on MNA revealed no obvious decomposition or degeneration in the molecule. X-ray diffraction analysis showed no T_p dependence of the peak position and the half-width. These results suggest that significant molecular structure degeneration and crystal structure changes did not occur for various values of T_p, and that the variation in F might be caused by reason (3).

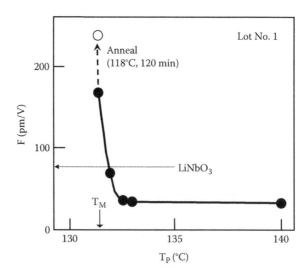

FIGURE 7.11 T_p dependence of the figure of merit for electro-optic phase retardation (F). From Y. Kubota and T. Yostrimura, "Endothermic reaction aided crystal growth of 2-methyl-4-nitroaniline and their electro-optic properties," *Appl. Phys. Lett.* **53**, 2579–2580 (1988).

In Figure 7.12 a relationship between F and a peak position of the photoluminescence spectra (E_{peak}) is shown. As the peak position shifts to the lower-energy side, F increases. This suggests that changes in F might be attributed to changes in the electronic state of the molecule. The molecular second-order nonlinear susceptibility expressed by Equation (5.23) can be written as follows for the two-level model.

$$\beta = -\frac{3e^3}{2\hbar^2}\frac{\omega_{eg}^2}{\left(\omega_{eg}^2 - \omega^2\right)\left(\omega_{eg}^2 - 4\omega^2\right)}r_{ge}^2\Delta r_e \tag{7.9}$$

Here, r_{ge} is the transition dipole moment from the ground state to the excited state, Δr_e is the difference in the dipole moment between the excited state and the ground state, and $\hbar\omega_{ge}$ is the energy gap between the two states. Since the absorption spectra measurements showed little dependence of the energy gap on T_p, the change of dipole moments would cause T_p dependence on F. When crystals are prepared so that some amount of impurity is produced by decomposition, molecular rotations might occur. As a result, the interaction between donors and acceptors of neighboring molecules changes, decreasing the total dipole moment. Consequently, F decreases [10].

Figure 7.13 shows electron distributions in MNA molecules for three arrangements: isolated molecule, three molecules with rotating by $\theta = 0°$, and three molecules

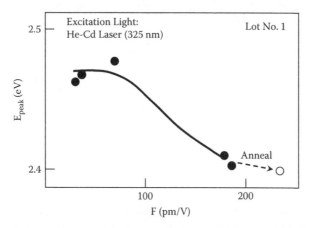

FIGURE 7.12 Relationship between F and a peak position of the photoluminescence spectra. From Y. Kubota and T. Yostrimura, "Endothermic reaction aided crystal growth of 2-methyl-4-nitroaniline and their electro-optic properties," *Appl. Phys. Lett.* **53**, 2579–2580 (1988).

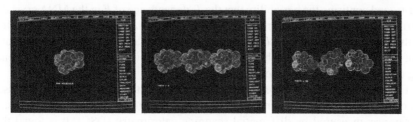

FIGURE 7.13 Influence of arrangement of MNA molecules on electron distributions.

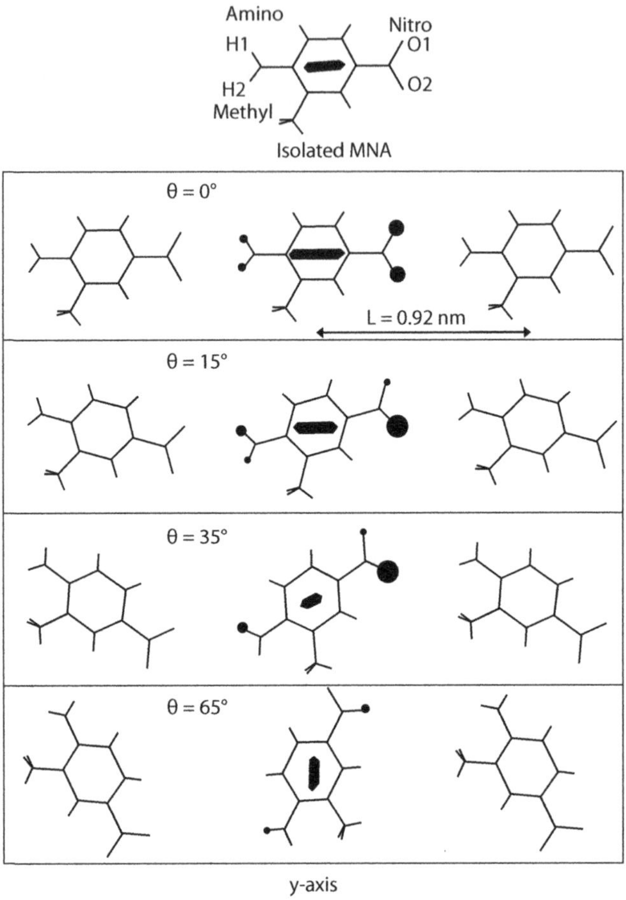

FIGURE 7.14 Influence of arrangement of MNA molecules on β.

with rotating by θ = 35°. The molecular orbitals for each arrangement were calculated by Austin Model 1 (AM1). From Equation (7.9), β of the central molecule was calculated and is shown in Figure 7.14 with bars. When θ increases from 0 to 35°, β decreases. With further increases in θ to 65°, β increases. This result suggests that the interaction between molecules in solids greatly affects the second-order nonlinearity, and confirms the possibility that the T_P dependence of F in the MNA crystal is attributed to the change in the interaction between MNA molecules. Also note that, in the configuration of θ = 0°, F is larger than in the isolated molecule. This indicates that enhancement of β occurs by interaction between neighboring molecules depending on the arrangements of those molecules.

Figure 7.15 shows the dependence of F on the purity of the MNA powder used for crystal growth. Here, F is normalized with respect to that of the $LiNbO_3$ crystal. Purity was measured by differential scanning calorimetry (DSC). It is found that F drastically decreases with decreases in the purity of MNA powder to less than

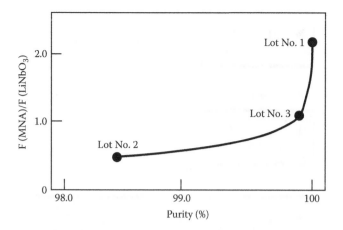

FIGURE 7.15   Dependence of F on the purity of the MNA powder. From Y. Kubota and T. Yostrimura, "Endothermic reaction aided crystal growth of 2-methyl-4-nitroaniline and their electro-optic properties," *Appl. Phys. Lett.* **53**, 2579–2580 (1988).

99.9%. The origin of the purity dependence of F might be similar to that of the T_p dependence mentioned previously. That is, a small number of impurity molecules will cause the rotation of a large number of molecules.

7.1.3   POLED POLYMERS AND OPTICAL SWITCHES

7.1.3.1   EO Effects in Poled Polymers

Poled polymers are obtained by applying a poling process to spin-coated polymer films containing nonlinear optical molecules. The typical fabrication process of poled polymers is shown in Figure 7.16. At temperature T_p above glass transition temperature T_g, an electric field is applied to the polymer film. Then, the nonlinear optical molecules in the film rotate to align in the electric field direction. By cooling to room temperature, the aligned orientations of molecules are fixed in the film to generate second-order optical nonlinearity.

EO coefficient r of poled polymers is expressed by the following formula.

$$r = -2Nf^2 f_0^2 \beta\mu E_p / 5n^4 k_B T_p \qquad (7.10)$$

Here, β, μ and N are molecular second-order nonlinear susceptibility, electric dipole moment, and concentration of nonlinear optical molecules, respectively. E_p is the poling electric field, T_p is the poling temperature, and k_B is the Boltzman constant. f and f_0 are given by the following expressions using dielectric constant ε and refractive index n and n_∞ .

$$f = (n^2 + 2) / 3, \quad f_0 = \varepsilon(n_\infty^2 + 2) / (n_\infty^2 + 2\varepsilon) \qquad (7.11)$$

From Equation (7.10), it is found that r increases by increasing β, μ, N, and E_p, and by decreasing T_p. Figure 7.17 summarizes the influence of β and T_p on r [11]. With decreasing T_p, thermal disturbance during poling is suppressed to improve molecular

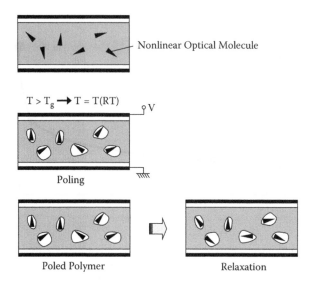

FIGURE 7.16 Fabrication process of poled polymers.

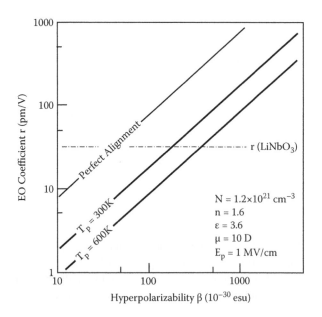

FIGURE 7.17 Influence of β and T_p on r in poled polymers.

alignment along the electric fields, resulting in an increase in r. However, low T_p implies low T_g of the polymer, which causes decay of the Pockels effect due to relaxation of aligned orientations of molecules to random orientations (Figure 7.16). This results in a decrease in thermal stability. Thus, in poled polymers, resolving the trade-off relationship between r and thermal stability is the most important issue.

7.1.3.2 EO Polyimide

Poled polymers with second-order nonlinear optical properties are suitable materials for EO devices because of their low dielectric constants and waveguide processability. In order to improve the thermal stability after poling, polyimide was developed with nonlinear optical molecules of azobenzene dyes attached as side chains [12]. The polymer system with side chains is expected to offer high EO coefficients because of the high-density nonlinear optical molecules and high thermal stability because of the polyimide backbones.

The chemical reaction for side-chain polyimide synthesis is shown in Figure 7.18. As shown in Figure 7.19, obvious thermal decomposition does not occur below 250°C since the absorption peak at 480 nm, which is attributed to the azobenzene dye, is preserved.

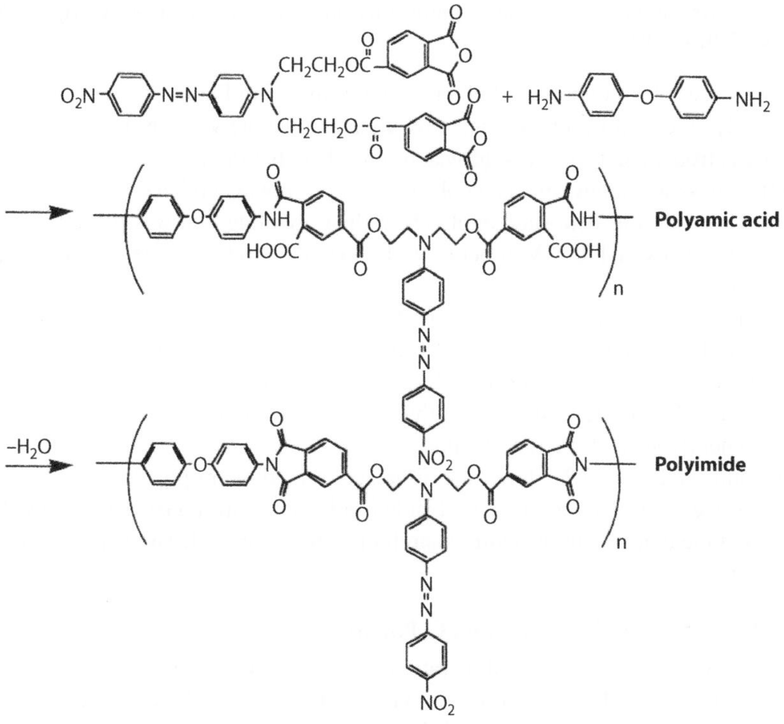

FIGURE 7.18 Chemical reaction for the side-chain polyimide synthesis. From W. Sotoyama, S. Tatsuura, and T. Yoshimura, "Electro-optic side-chain polyimide system with large optical nonlinearity and high thermal stability," *Appl. Phys. Lett.* **64**, 2197–2199 (1994).

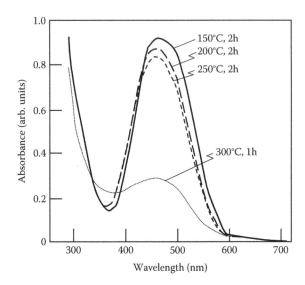

FIGURE 7.19 Absorption spectra of the side-chain polyimide after baking at 150, 200, 250, and 300°C. From W. Sotoyama, S. Tatsuura, and T. Yoshimura, "Electro-optic side-chain polyimide system with large optical nonlinearity and high thermal stability," *Appl. Phys. Lett.* **64**, 2197–2199 (1994).

The process for EO polyimide fabrication is shown in Figure 7.20. By spin coating, a polyamic acid film was formed on a glass substrate with an indium tin oxide (ITO) electrode. Next, the solvent was removed by baking at 100°C. The polymer film thickness was a few μm. The polyamic acid film was imidized at 250°C for 2h. An Au electrode was deposited on it. Then, the polyimide film was poled by applying electric fields of 170 MV/m between the ITO electrode and the Au electrode at 180°C for 1 h.

The EO coefficient r_{33}, measured by the reflection technique developed by Teng and Man [13], was 10.8 pm/V at 1.3 μm in wavelength. The thermal stability of the EO efficiency of the EO polyimide was tested by keeping the sample at 120°C. As shown in Figure 7.21, after more than 200 h, the EO coefficient of the sample retained nearly 90% of the initial value.

In addition, Figure 7.22 shows the influence of nonlinear optical molecule structures on decay of EO coefficients of poled polymers with epoxy backbones. It is found that the diacetylene structure is preferable to suppress decay than the azobenzene structure.

7.1.3.3 Optical Switches Using EO Polyimide

Figure 7.23 shows a schematic illustration and switching operation of a directional coupler optical switch using the EO polyimide. Figure 7.24 shows a photograph of the optical switch. The devices are fabricated on a conductive Si wafer, which is used as the lower electrode. In this device, the clad part consists of the EO polyimide and the core consists of a passive fluorinated polyimide. When applied voltage V is 0 V, a light beam is guided to one branch of the optical switch. When 500 V is applied,

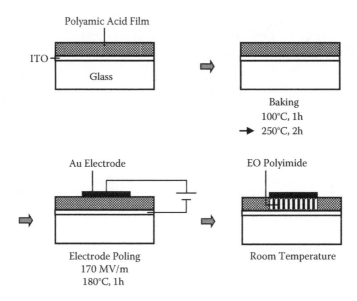

FIGURE 7.20   Process for EO polyimide fabrication.

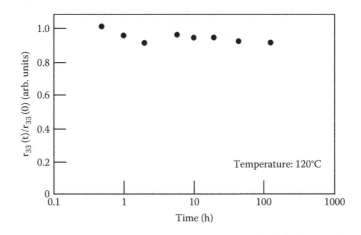

FIGURE 7.21   Relationship between heating time (at 120°C) and EO coefficient of the EO polyimide. From W. Sotoyama, S. Tatsuura, and T. Yoshimura, "Electro-optic side-chain polyimide system with large optical nonlinearity and high thermal stability," *Appl. Phys. Lett.* **64**, 2197–2199 (1994).

the light beam is picked up to another branch, showing the switching operation. The drive voltage is extremely high. This is attributed to the fact that only a small fraction of the light beam feels the refractive index change that is induced by the EO effect because the EO materials do not exist in the core region, where most of the light beam's electric fields are confined, but exist in the surrounding clad region.

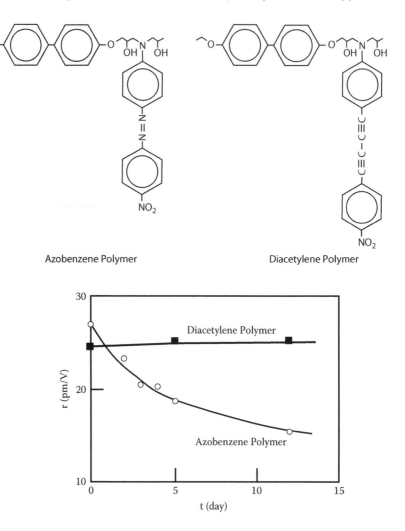

FIGURE 7.22 Influence of nonlinear optical molecule structures on decay of EO coefficients of poled polymers.

7.1.3.4 3-D Optical Switches

To fabricate optical integrated circuits for high-density optical wiring, small light modulators and optical switches in three-dimensional (3-D) optical circuits are required. The vertical directional coupler consisting of two stacked EO polymer waveguides [14] is a candidate for the 3-D optical switch.

The 3-D optical switch with a directional coupler structure is constructed by stacking two polymer waveguides and poling them to create the EO effect in them. The directional coupler has a certain coupling length, which varies according to the electric-field-induced refractive index change of the EO polymer waveguides. Consequently, a laser beam introduced into one of the waveguides is switched between the two waveguides by voltage application.

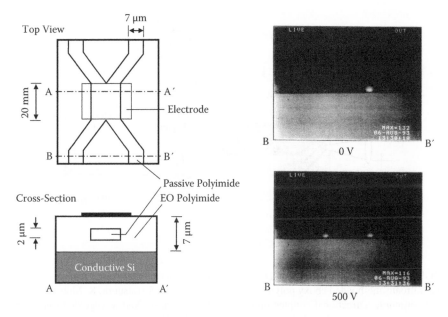

FIGURE 7.23 Schematic illustration and switching operation of an optical switch using the EO polyimide.

FIGURE 7.24 Photograph of the optical switch using the EO polyimide.

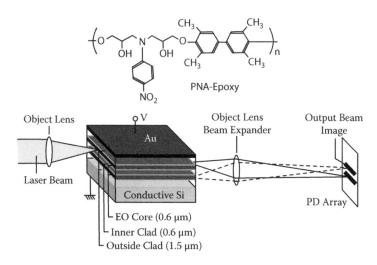

PNA-Epoxy

FIGURE 7.25   Structure of the 3-D optical switch, experimental setup, and structure of paranitroaniline-bonded epoxy polymer. From W. Sotoyama, S. Tatsuura, K. Motoyoshi, and T. Yoshimura, "Directional-coupled optical switch between stacked waveguide layers using electro-optic polymer," *Jpn. J. Appl. Phys.* **31**, L1180–L1181 (1992).

In Figure 7.25, the structure of the 3-D optical switch and experimental setup are shown, as well as the structure of paranitroaniline-bonded epoxy polymer (PNA-epoxy) used for the EO core layer. The refractive index of the PNA-epoxy layer is 1.64 before poling. Polyvinylalcohol (PVA) with a refractive index of 1.53 is used for the clad layers. These layers are stacked on a conductive silicon wafer by spin coating. Each waveguide layer is considered a single-mode slab waveguide having the same effective index.

A He-Ne laser beam (633 nm) was coupled into the stacked waveguide through the cleaved edge. The scattered light from the top of the waveguide was observed as bright and dark stripes perpendicular to the direction of the propagating light beam, which indicates that the light beam is transfered between the upper and the lower waveguides alternately according to the stripe interval. The stripe interval was about 0.3 mm for the transverse electric (TE)-mode wave, being consistent with the calculated coupling length of 0.34 mm.

A 100-nm-thick planar Au electrode was deposited on top of the stacked layers. The PNA-epoxy core layers were poled by applying 200 V between the substrate and the top Au electrode at 60°C for 5 minutes. After poling, the sample was cleaved for a 5-mm waveguide length, and optical switching measurement was performed in the experimental setup shown in Figure 7.25. A He-Ne laser beam focused by a 100× object lens was coupled to the waveguide in the transverse magnetic (TM)-mode. The exiting beam was collected by a 50× object lens and a 10× beam expander and detected by a photodiode array.

Figure 7.26 shows examples of the output beam line images from the 3-D optical switch detected by the photodiode (PD) array. Two parallel lines spaced about 1 mm

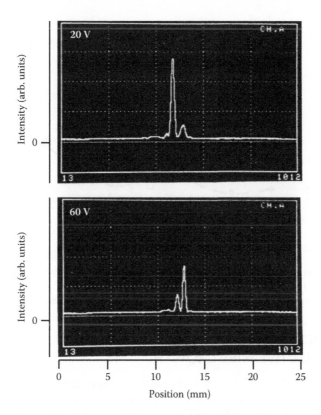

FIGURE 7.26 Output beam line images from the 3-D optical switch detected by the PD array at applied voltage of 20 V and 60 V. Horizontal axis represents position on the PD array. From W. Sotoyama, S. Tatsuura, K. Motoyoshi, and T. Yoshimura, "Directional-coupled optical switch between stacked waveguide layers using electro-optic polymer," *Jpn. J. Appl. Phys.* **31**, L1180–L1181 (1992).

apart appear, corresponding to beams from the upper waveguide and the lower waveguide. As Figure 7.27 shows, the intensity pattern of these lines varies when voltage is applied between the substrate and the top electrode, indicating that the light beam switching between layered waveguides is voltage dependent. The switching voltage obtained from the intensity versus voltage characteristics is about 30 V.

It should be noted that an optical switch consisting of channel waveguides can be easily constructed in principle using a stripe electrode of a few μm wide. Figure 7.28 shows an example of stacked channel waveguides [15].

For the fabricated device, the switching voltage was 30 V, which is too high for optical interconnect applications, because of the small EO coefficient of the PNA-epoxy polymer, $r_{33} = 3$ pm/V. The switching voltage can be reduced by improving the polymer, for example, by implementation of the polymer multiple quantum dots (MQDs) fabricated by MLD.

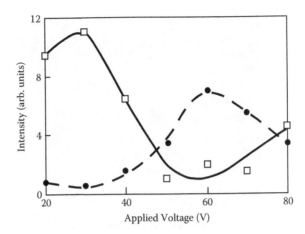

FIGURE 7.27 Output beam line intensity as a function of applied voltage. Solid circle represents intensity of one line and rectangle the other. From W. Sotoyama, S. Tatsuura, K. Motoyoshi, and T. Yoshimura, "Directional-coupled optical switch between stacked waveguide layers using electro-optic polymer," *Jpn. J. Appl. Phys.* **31**, L1180–L1181 (1992).

7.2 OPTICAL WAVEGUIDES FABRICATED BY SELECTIVE WIRE GROWTH

As mentioned in Section 7.1.3, the trade-off relationship between r and thermal stability is the most serious problem in poled polymers. One of the ways to resolve it is to apply selectively aligned polymer wire growth to EO polymer fabrication. This section covers several examples of EO waveguide formation by selectively aligned polymer wire growth utilizing electric fields and anisotropic surface structures.

7.2.1 EO WAVEGUIDES FABRICATED BY ELECTRIC-FIELD-ASSISTED GROWTH

When molecules with sufficiently high reactivities are used in electric-field-assisted growth (described in Section 3.6.3), polymer films could be fabricated at around room temperature. If electric fields are selectively applied to the polymer wire growth regions, nonlinear optical molecules are aligned by the applied electric field and simultaneously polymerized, enabling poled polymer fabrication below the glass transition temperature. In this method, molecules could move rather freely and disordering by thermal disturbance could be reduced to achieve highly oriented poling. This resolves the problem of the trade-off relationship between r and thermal stability.

Electric-field-assisted growth has additional advantages. In conventional poling, during rotation of the molecules, stress and free volume are generated in the surrounding regions. These cause a force to relax the oriented molecules back to the initial random state as shown in Figure 7.16, which decreases the thermal stability of the poled polymers. On the contrary, in the polymer fabricated by electric-field-assisted growth, such stress is not induced because rotation of molecules occurs before polymerization.

Top View

Cross-Section

FIGURE 7.28   An example of stacked channel waveguides.

Furthermore, since electric-field-assisted growth is carried out in vacuum, lower impurity concentration is expected. By combining electric-field-assisted growth with MLD, precise molecular rotation controllability as well as thickness controllability and conformability of the film are expected.

7.2.1.1   Epoxy-Amine Polymer

Optical waveguides of EO epoxy-amine polymer [BE/AEANP] were fabricated using a reaction between tetramethyl-biphenylepoxy (BE) and 2(2-aminoethylamino)-5-nitropyridine (AEANP), as shown in Figure 7.29 [16,17]. AEANP, which has a pyridine ring connected to a donor (-NH(CH$_2$)-) and an acceptor (-NO$_2$), is a nonlinear optical molecule. AEANP has a dipole moment and is estimated from the MO calculation to have β about a half of β in *p*-nitroaniline. Therefore, AEANP could be aligned

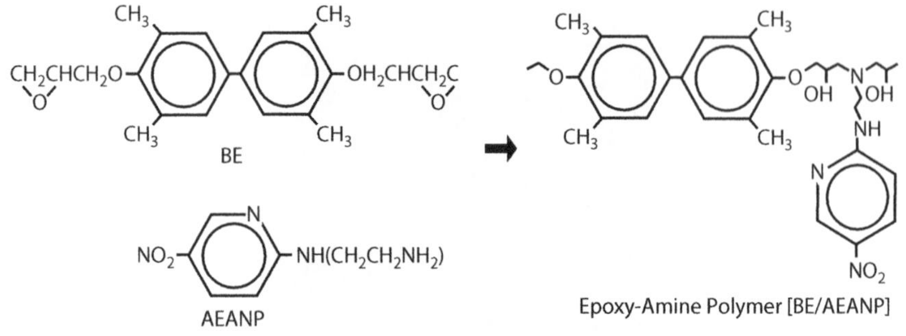

FIGURE 7.29 Reaction for epoxy-amine polymer [BE/AEANP] wire growth.

by electric field application during polymer wire growth in vacuum to produce EO polymer wires.

Epoxy-amine polymer growth was performed with a gas pressure of 5×10^{-3} Pa (background pressure of $< 10^{-4}$ Pa) at a deposition rate of 0.2–0.5 nm/s. The film fabricated at $T_s = 30°C$ was yellow and transparent when the AEANP/BE ratio was 1/1, which is consistent with the polymer structure shown in Figure 7.29.

Figure 7.30 shows infrared spectra of the film and the two molecules. In the film spectrum, the band of the epoxy ring (911 cm^{-1}) becomes very small, that of -NH$_2$ (3367 cm^{-1}) disappears, and that of –OH grows compared to the spectra of AEANP and BE molecules. These changes indicate that polymerization progressed and a polymer film was grown.

To fabricate channel optical waveguides, a thermally oxidized Si wafer with a 1-μm-thick SiO$_2$ clad layer was prepared as the substrate. Al slit-type electrodes with a 10- and 20-μm gap were formed on the substrate. Epoxy-amine polymer [BE/AEANP], 1.3-μm thick, was grown under electric field applied in the gap region of slit-type electrodes. The applied electric field was 0.2 MV/cm and 0.1 MV/cm for the gap of 20 μm and 10 μm, respectively. Figure 7.31 shows a photograph of the sample for epoxy-amine polymer [BE/AEANP] waveguides.

A He-Ne laser beam (633 nm) was introduced into the epoxy-amine polymer [BE/AEANP] film on the gap region to guide the beam. As shown in Figure 7.32, a streak of the guided beam is observed in the gap region, demonstrating that a low-loss channel waveguide is formed in the gap region. Near-field pattern observation revealed that a guided beam with polarization parallel to the y-axis is bright while a guided beam with polarization parallel to the z-axis is very weak, indicating that the guided beam mainly contains polarization parallel to the y-axis. This result is attributed to AEANP alignment induced by the electric field along the y-axis. The AEANP alignment along the y-axis increases the refractive index for polarization parallel to the y-axis and decreases that for polarization perpendicular to the y-axis (i.e., parallel to the z-axis), resulting in selective confinement of the He-Ne laser beam with y-polarization into the channel optical waveguide. The EO coefficient of the channel waveguide evaluated by a Mach-Zehnder interferometer is approximately 0.1 pm/V at 633 nm.

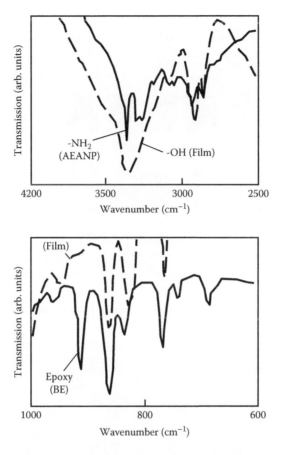

FIGURE 7.30 Infrared spectra of a film and the two molecules. From S. Tatsuura, W. Sotoyama, and T. Yoshimura, "Epoxy-amine polymer waveguide containing nonlinear optical molecules fabricated by chemical vapor deposition," *Appl. Phys. Lett.* **60**, 1158–1160 (1992).

To confirm the effect of an applied electric field during deposition, channel wave-guides were fabricated from the same polymer using electrode poling. After the polymer film deposition, the film was poled with an electric field of 0.2 MV/cm at room temperature and at 66°C. It was found that the EO coefficient of the waveguide formed by electric-field-assisted growth is over 100 times larger than that of the waveguide formed by electrode poling at room temperature and at 66°C.

These results show that molecular alignment and polymerization were accomplished simultaneously during the film deposition and demonstrate the advantage of electric-field-assisted growth.

7.2.1.2 Poly-Azomethine

As mentioned in Section 3.8, conjugated polymers are expected to be high-performance materials for various thin-film photonic/electronic devices. As an

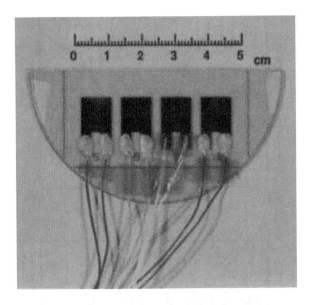

FIGURE 7.31 Photograph of a sample for epoxy-amine polymer [BE/AEANP] waveguides.

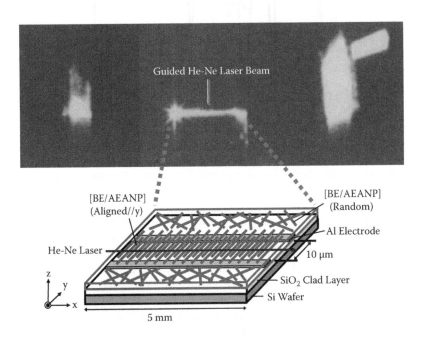

FIGURE 7.32 Guided He-Ne laser beam in a channel waveguide of epoxy-amine polymer [BE/AEANP] grown by electric-field-assisted growth.

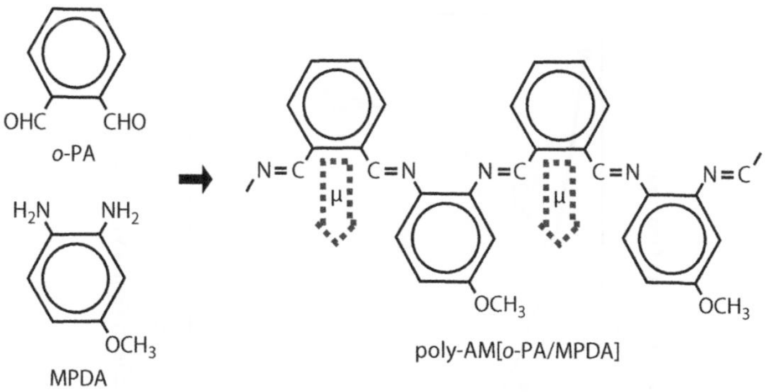

FIGURE 7.33 Reaction for poly-AM [*o*-PA/MPDA] wire growth. From S. Tatsuura, W. Sotoyama, K. Motoyoshi, A. Matsuura, T. Hayano, and T. Yoshimura, "Polyazomethine conjugated polymer film with second-order nonlinear optical properties fabricated by electric-field-assisted chemical vapor deposition," *Appl. Phys. Lett.* **62**, 2182–2184 (1993).

example of a functional conjugated polymer device, EO waveguides of poly-AM [*o*-PA/MPDA] were fabricated by electric-field-assisted growth [18].

Poly-AM [*o*-PA/MPDA] is formed with the reaction shown in Figure 7.33 using *o*-phthalaldehyde (*o*-PA) and 4-methoxy-*o*-phenylenediamine (MPDA). The methoxy group in MPDA acts as a donor. From the MO calculation it is found that in poly-AM [*o*-PA/MPDA] wire dipole moments are generated along a direction perpendicular to the wire, and a large β of 3–5 times that of *p*-nitroaniline is expected.

A thermally oxidized Si wafer with Al slit-type electrodes was prepared as a substrate. The electrodes were about 100-nm thick and the gaps were 5–20 μm wide. Yellow-orange poly-AM [*o*-PA/MPDA] film with a thickness of 0.8–3.3 μm was grown at a molecular gas pressure of $1-10^{-1}$ Pa under an electric field of 0–0.6 MV/cm in the gap region of the slit-type Al electrodes.

Deposition rate dependence of the peak optical density and the energy gap obtained from the absorption spectra of poly-AM [*o*-PA/MPDA] is shown in Figure 7.34. As the deposition rate decreases, the peak optical density increases and the energy gap decreases. The results are attributed to an increase in conjugated lengths of poly-AM [*o*-PA/MPDA] with decreases in the deposition rate. This tendency can be explained because the deposition rate decrease increases the time for *o*-PA and MPDA to migrate on the substrate surface, meet each other, and then combine with strong bonds.

The polymer film's refractive index was 1.68 at 633 nm. A He-Ne laser beam was directly coupled to the poly-AM [*o*-PA/MPDA] film on a thermally oxidized Si wafer with Al slit electrodes. The SiO$_2$ layer was about 1 μm thick. A channel-like beam propagation over 15 mm long was observed in the electrode gap region. The exiting beam's near-field pattern was bright in TE mode and weak in TM mode. This is attributed to the fact that the benzene ring planes in the polymer wires were aligned parallel to the substrate by the applied electric field.

Figure 7.35 shows EO responses in poly-AM [*o*-PA/MPDA] obtained with applied electric fields of 0.5 MV/cm and 0 MV/cm during growth. For this experiment,

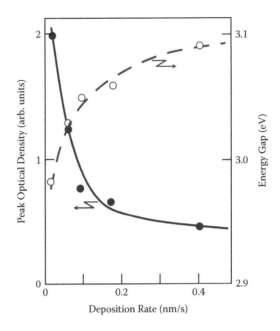

FIGURE 7.34 Deposition rate dependence of the peak optical density and the energy gaps in poly-AM [o-PA/MPDA] grown by electric-field-assisted growth. From S. Tatsuura, W. Sotoyama, K. Motoyoshi, A. Matsuura, T. Hayano, and T. Yoshimura, "Polyazomethine conjugated polymer film with second-order nonlinear optical properties fabricated by electric-field-assisted chemical vapor deposition," *Appl. Phys. Lett.* **62**, 2182–2184 (1993).

quartz substrates with Ni-Cr slit electrodes were used. The EO effect was measured using a Mach-Zehnder interferometer. The poly-AM [o-PA/MPDA] film was placed in one arm of the Mach-Zehnder interferometer with the film plane perpendicular to a He-Ne laser beam. The polarized direction of the laser beam was in the applied electric field direction in the electrode gap region. For 0.5 MV/cm, clear EO responses are observed in a Mach-Zehnder interferometer while no EO responses were observed for 0 MV/cm. The EO response was observed after heating at 80°C for 5 h; this confirms that the conjugated polymer is thermally stable.

These results demonstrate the possibility that conjugated polymer wires can be used for EO waveguides. By combining the process with MLD, the EO effect of the conjugated polymer wires will be optimized.

7.2.2 Conjugated Polymer Waveguides Fabricated on Anisotropic Surface Structures

EO waveguides of poly-AM [TPA/PPDA] were fabricated by selectively-aligned growth on surfaces with anisotropic structures [19]. A schematic illustration of the waveguide structure is shown in Figure 7.36(a). The SiO underlayer was deposited in waveguide core patterns by vacuum evaporation on a glass substrate that was tilted by 45° along the y-axis as mentioned in Section 3.6.2. The obliquely evaporated

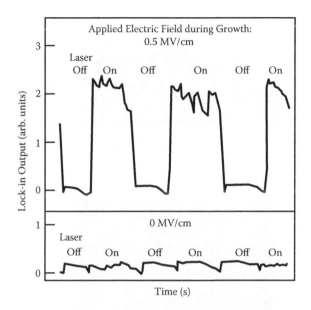

FIGURE 7.35 EO responses in poly-AM [*o*-PA/MPDA] grown by electric-field-assisted growth. From S. Tatsuura, W. Sotoyama, K. Motoyoshi, A. Matsuura, T. Hayano, and T. Yoshimura, "Polyazomethine conjugated polymer film with second-order nonlinear optical properties fabricated by electric-field-assisted chemical vapor deposition," *Appl. Phys. Lett.* **62**, 2182–2184 (1993).

SiO film thickness was 200 nm. A poly-AM [TPA/PPDA] film was grown on the substrate by the carrier gas–type organic chemical vapor deposition (CVD) using terephthalaldehyde (TPA) and *p*-phenylenediamine (PPDA). Growth time was 1 h. The carrier gas flow rate was 4 NL/min. Gas pressure was 100 Pa or more.

Poly-AM deposited all over the substrate surface was selectively aligned along the y-axis on the obliquely evaporated SiO region. In the other region, poly-AM was randomly aligned. Figure 7.36(b) shows a microscope image of the fabricated waveguide. A laser beam of 650 nm in wavelength was introduced into the waveguide of 4-μm core width. It was found that the beam propagates through the waveguide. Thus, waveguide formation using poly-AM [TPA/PPDA] by selectively aligned growth is successfully demonstrated. Again, by combining the process with MLD, the EO effect of the conjugated polymer wires is obtained.

7.2.3 ACCEPTOR SUBSTITUTION INTO CONJUGATED POLYMER WIRES

As shown in Chapter 5, for drastic enhancement of optical nonlinearity of conjugated polymer wires, donor–acceptor substitution into polymer backbones is necessary. We tried a preliminary experiment for the substitution by using TPA for molecule A and 2-nitro-1,4-phenylenediamine (NPDA) for molecule B [5]. As shown in Figure 7.37, NPDA has two $-NH_2$ used for connections with TPA, and $-NO_2$, an acceptor. As Figure 7.38 shows, in the poly-AM [TPA/NPDA] film, the absorption band of NPDA film at 400~600 nm disappears and a new band appears at 400~450 nm, indicating

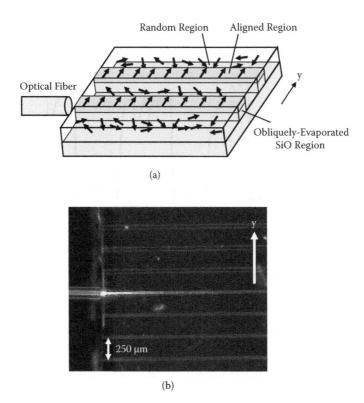

(a)

(b)

FIGURE 7.36 Poly-AM [TPA/PPDA] waveguide fabricated by selectively-aligned growth on surfaces with anisotropic structures. (a) A schematic illustration of the waveguide structure, and (b) photograph of a guided beam. From K. Matsumoto and T. Yoshimura, "Electro-optic waveguides with conjugated polymer films fabricated by the carrier-gas-type organic CVD for chip-scale optical interconnects," *Proc. SPIE* **6899**, 98990E-1-9 (2008).

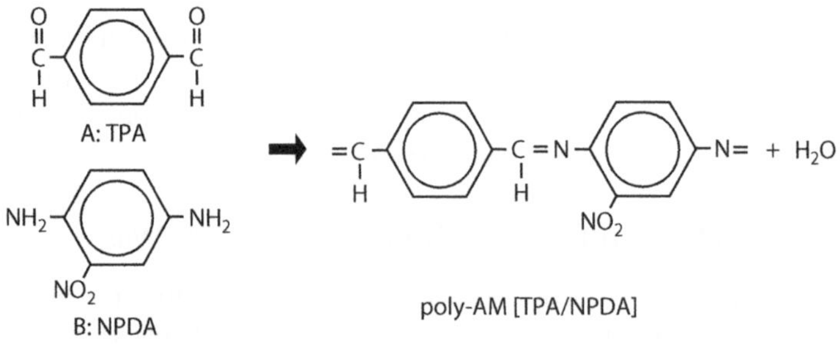

FIGURE 7.37 Molecules and reaction for acceptor-substituted conjugated polymer wire growth.

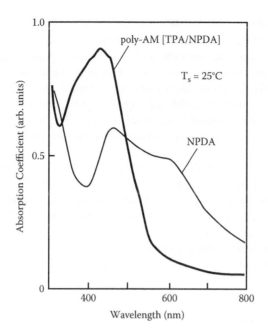

FIGURE 7.38   Absorption spectra of poly-AM [TPA/NPDA] film and NPDA film.

that TPA and NPDA combine to form a polymer. Therefore, in principle, donor–acceptor groups can be introduced into conjugated polymer wires by using molecules that contain groups both for connections and for donor–acceptor.

7.3   NANO-SCALE WAVEGUIDES OF PHOTO-INDUCED REFRACTIVE INDEX INCREASE SOL-GEL MATERIALS

One of the important elements for high-speed and small EO devices is the nano-scale optical circuit. It is usually fabricated using Si waveguides [20] with a large refractive index difference between the core and clad. The fabrication process of Si waveguides is well established in the large-scale integrated circuit (LSIs) industry. However, the process is complicated. In order to simplify the process, the silicon nano composite thin film [21] has been developed for etchingless processes. This material is expected to be useful for intrachip optical interconnects because a large refractive index decrease due to UV photo-oxidation of Si atoms is induced by patterned UV light exposure.

The photoinduced refractive index increase (PRI) sol-gel material, which was developed by Nissan Chemical Industries, Ltd., is another promising material for etchingless processes [22]. The distinguished feature of the PRI sol-gel material is the large refractive index *increase* capability. In spin-coated films of sol-gel materials, the refractive index increases from 1.65 to 1.85 at $\lambda = 633$ nm by UV-light exposure and baking, giving rise to a refractive index contrast of ~13%. Although the contrast of 13% is smaller than ~60% in Si waveguides, it is sufficient to provide nano-scale waveguide structures. Thus, the PRI sol-gel material enables us to fabricate high-index contrast (HIC) waveguides without etching/developing processes, simplifying the fabrication processes of the nano-scale waveguides.

In intrachip optical interconnects, 850 nm is the most popular signal light wavelength. With respect to detector sensitivity, blue light is preferable [23]. The short-wavelength light is also preferable to miniaturize optical circuits. For these reasons, we focused our attention on visible wavelength regions, and thus, a He-Ne laser of 633 nm was used for evaluating the optical circuits. The high transmission capability at short wavelengths is a remarkable advantage of the PRI sol-gel material against Si.

Furthermore, by using the PRI sol-gel material, optical circuits consisting of high-index contrast waveguides are available with structures of ultrathin planar films with a thickness of 100 nm order. This is favorable for stacking optical circuits on LSIs and for constructing 3-D optical circuits with stacked films [24,25]. Therefore, the PRI sol-gel material is promising for nano-scale optical circuits, especially for a constituent of small-size optical switches.

In the present section, nano-scale waveguides made of PRI sol-gel material are reviewed [22].

7.3.1 Fabrication Process

The PRI sol-gel material is a silicon-oxide-based sol-gel material incorporating titanium oxide. The optical waveguide fabrication process is shown in Figure 7.39. On a Si substrate with SiO_2 of 2 μm thickness, the sol-gel material was coated and then prebaked for solvent removal.

The layer was selectively exposed to UV light from a high-pressure mercury lamp through a photomask with light intensity of ~5 mW/cm^2 and energy of 9000 mJ/cm^2. The nominal core pattern width is 1 μm in the photomask. The UV light exposure activates the hydrolytic condensation of the titanium oxide.

Finally, the layer was postbaked at 200°C for 15 min. During postbaking, the hydrolytic condensation proceeds more rapidly in UV light–exposed areas than in

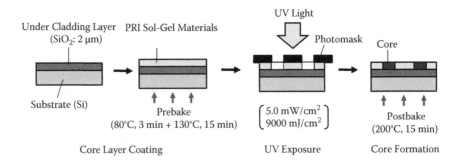

FIGURE 7.39 Fabrication process of optical waveguides using the PRI sol-gel material. From S. Ono, T. Yoshimura, T. Sato, and J. Oshima, "Fabrication and evaluation of nano-scale optical circuits using sol-gel materials with photo-induced refractive index variation characteristics," *J. Lightwave Technol.* **27**, 1229–1235 (2009).

250 μm

S-bending

Y-branching with curvature

Y-branching

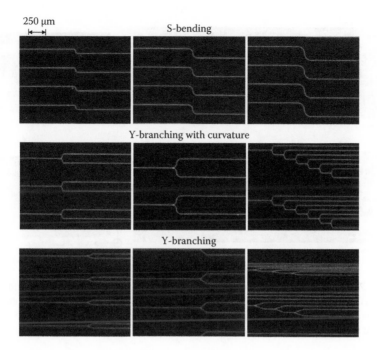

FIGURE 7.40 Photographs of nano-scale optical circuits fabricated using the process shown in Figure 7.39. From S. Ono, T. Yoshimura, T. Sato, and J. Oshima, "Fabrication and evaluation of nano-scale optical circuits using sol-gel materials with photo-induced refractive index variation characteristics," *J. Lightwave Technol.* **27**, 1229–1235 (2009).

unexposed areas, yielding a higher refractive index in the exposed areas than in the unexposed surrounding areas. Thus, the UV light–exposed areas become cores.

The initial refractive index with no UV light dose, which corresponds to the refractive index of the clad, is 1.65. Refractive index differences between the core and clad (Δn) increases with UV light dose. Δn is saturated to be ~0.2 with a UV light dose of 9000 mJ/cm^2. Δn is basically not affected by the prebake and postbake temperature below 200°C. However, high-temperature baking (e.g., over 300°C) induces hydrolytic condensation in unexposed areas, and hence the refractive index contrast between core and clad tends to be reduced with increased baking temperature. Photographs of nano-scale optical circuits fabricated using the process shown in Figure 7.39 are shown in Figure 7.40. The core thickness is ~230 nm and core width is ~1 μm.

7.3.2 Linear Waveguides

Figure 7.41 shows a guided beam and an output beam from a core edge of a linear waveguide. Propagation loss was 1.86 dB/cm for TE mode (polarization // x-axis) and 1.89 dB/cm for TM mode (polarization // y-axis), indicating small polarization dependence.

Far-field patterns (FFPs) of the output beam from the core edge are shown in Figure 7.42. Spread angles of the output beams, which were determined by the width

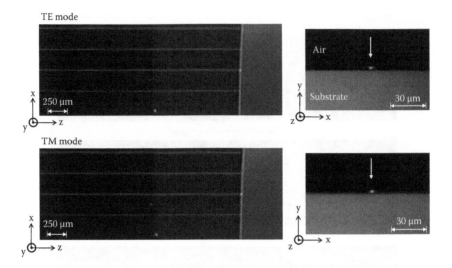

FIGURE 7.41 Photographs of linear waveguides, a guided beam in one of them, and an output beam from the core edge. From S. Ono, T. Yoshimura, T. Sato, and J. Oshima, "Fabrication and evaluation of nano-scale optical circuits using sol-gel materials with photo-induced refractive index variation characteristics," *J. Lightwave Technol.* **27**, 1229–1235 (2009).

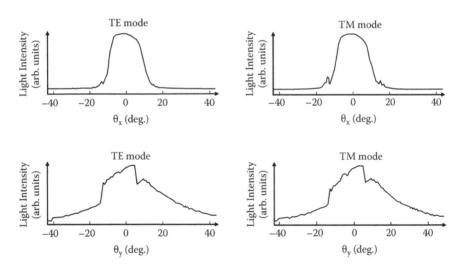

FIGURE 7.42 FFPs of the output beam from the core edge of the linear waveguide. From S. Ono, T. Yoshimura, T. Sato, and J. Oshima, "Fabrication and evaluation of nano-scale optical circuits using sol-gel materials with photo-induced refractive index variation characteristics," *J. Lightwave Technol.* **27**, 1229–1235 (2009).

of the FFPs at a light intensity of 1/e of the maximum height, are 20° and 40° in the core width direction (// x-axis) and in the core thickness direction (// y-axis), respectively, for both TE mode and TM mode. Spot size (spot radius) of the guided beam can be calculated from the spread angle using the following equation.

$$w = \frac{180 \times \lambda}{2\theta \times \pi^2} \qquad (7.12)$$

Here, w [μm] is the spot size, 2θ (°) is the spread angle of the output beam, and λ [μm] is the guided beam wavelength. The refractive index of the free space (air), where beams emitted from the output propagate, is 1. Spot sizes of guided beams along the x-axis and along the y-axis were calculated to be 0.6 μm and 0.3 μm, respectively, for both TE mode and TM mode. From these results, it is concluded that PRI sol-gel materials can realize guided beam confinement within a cross-sectional area less than 1.0 μm² with small polarization dependence. Since, roughly speaking, the criterion for core cross-section areas in intrachip optical interconnects is ~1.0 μm² [26], it is suggested that the PRI sol-gel material can be applied to the internal wiring of chips.

In order to compare the experimental results with theoretical ones, light intensity profiles in the optical waveguides were simulated by the beam propagation method (BPM). The model for the simulation is shown in Figure 7.43. The core thickness is 230 nm, the core width is 1 μm, the refractive index of the core is 1.85, the refractive index of the clad is 1.65, the SiO_2 (under clad) thickness is 2 μm, and the refractive index of the under clad is 1.46. The wavelength of the guided beams is 633 nm. The results of the BPM simulation are shown in Figure 7.44. Spot sizes of the guided beams along the x-axis and along the y-axis are, respectively, 0.5 μm and 0.2 μm for both TE mode and TM mode.

The results for spot sizes of the guided beams are summarized in Table 7.1. It is found that the guided beam confinement is slightly stronger in the BPM simulation than in the experiment. The difference might be attributed to Δn. That is, in the fabricated waveguides, Δn is slightly smaller than Δn in the simulation model, that is, less than ~0.2, resulting in an increase in penetration of evanescent waves into the clad, which arises from the tunneling effect and an increase in the spot sizes.

It should be noted that Si waveguides enable strong guided beam confinement in cores with a width less than 0.5 μm due to its high core refractive index of ~3.5. However, the high refractive index causes a decrease in the propagation speed of optical signals in Si waveguides. The PRI sol-gel material waveguides, on the other hand, exhibit a faster propagation speed of optical signals by a factor of ~2 compared with Si waveguides because the core refractive index is ~1.85. Considering the trade-off relationship between guided beam confinement and propagation speed, it is expected that the sol-gel material waveguides can preferably be applied to global wiring in chips.

It is expected that further reduction of the core width may be achieved in PRI sol-gel materials by optimization of the fabrication processes for increasing Δn to the theoretical value, which enables nano-scale optical circuits with a core width of less than 1 μm.

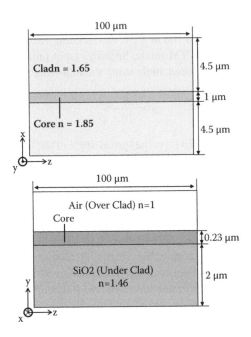

FIGURE 7.43 A model for the BPM simulation. From S. Ono, T. Yoshimura, T. Sato, and J. Oshima, "Fabrication and evaluation of nano-scale optical circuits using sol-gel materials with photo-induced refractive index variation characteristics," *J. Lightwave Technol.* **27**, 1229–1235 (2009).

7.3.3 S-Bending and Y-Branching Waveguides

S-bending waveguides and guided beams of TE mode in them are shown in Figure 7.45. Output beams from the core edges of the Y-branching waveguides are shown in Figure 7.46 for TE mode. It is found that input beams are divided into two branches with branching angles of 20°, 40°, and 80°. In the photographs in Figure 7.46, there is background in addition to the guided beam spots. Because the 3-μm core radius of the single-mode optical fiber used for introducing light beams into the waveguide cores is much larger than the core sizes of the waveguides and the waveguide length is relatively short, say, less than ~5 mm, considerable beams exist in the clad regions.

The S-bending loss was measured by the following process (Figure 7.47).

1. The output power P_1 at waveguide length L_1 is measured.
2. After cutting the waveguide, output power P_2 at waveguide length L_2 is measured.
3. After the waveguide is cut beyond the S-bend, output power P_3 at waveguide length L_3 is measured.
4. The output power in units of dBm at each measurement point is plotted against the waveguide length.

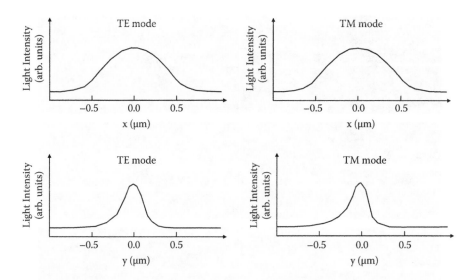

FIGURE 7.44 Light intensity profiles in optical waveguides simulated by BPM. From S. Ono, T. Yoshimura, T. Sato, and J. Oshima, "Fabrication and evaluation of nano-scale optical circuits using sol-gel materials with photo-induced refractive index variation characteristics," *J. Lightwave Technol.* **27**, 1229–1235 (2009).

TABLE 7.1
Estimated Spot Sizes of Guided Beams

	TE Mode	TM Mode
Experiment	0.6×0.3	0.6×0.3
BPM Simulation	0.5×0.2	0.5×0.2

Source: S. Ono, T. Yoshimura, T. Sato, and J. Oshima, "Fabrication and evaluation of nano-scale optical circuits using sol-gel materials with photo-induced refractive index variation characteristics," *J. Lightwave Technol.* **27**, 1229–1235 (2009).

Note: Unit = μm^2

5. (L_1, P_1) and (L_2, P_2) are connected with a straight line of $y = ax + b$. Since slope a gives the propagation loss of the linear waveguide, another straight line of slope a, that is, $y = ax + c$, is traced on (L_3, P_3) to represent power vs. waveguide length characteristics before reaching the S-bend part.

6. Bending loss is obtained from the gap at the S-bend part, $c - b$, because the gap is caused by guided beam leakage at the S-bend part.

Y-branching loss was also measured by the same method. Bending and Y-branching losses are expressed by the following formula. Bending loss for one bend is one half of the S-bending loss.

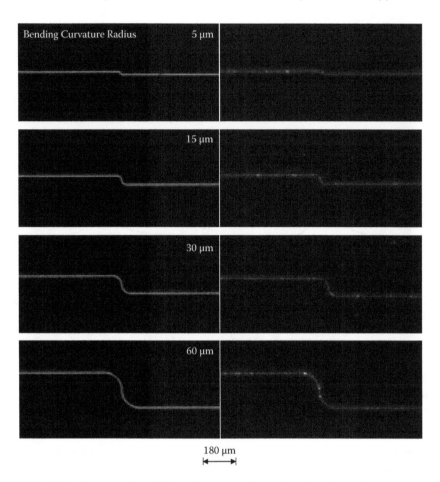

FIGURE 7.45 Photographs of S-bending waveguides and guided beams in them for TE mode. From S. Ono, T. Yoshimura, T. Sato, and J. Oshima, "Fabrication and evaluation of nano-scale optical circuits using sol-gel materials with photo-induced refractive index variation characteristics," *J. Lightwave Technol.* **27**, 1229–1235 (2009).

$$\text{S-Bending loss} = c - b \text{ [dB]}$$

$$\text{Bending loss} = (c - b) / 2 \text{ [dB]} \tag{7.13}$$

$$\text{Y-Branching loss} = c - b - 3 \text{ [dB]}$$

The relationship between bending loss and bending curvature radius is shown in Figure 7.48(a). When the radius is increased from 5 μm to 60 μm, the loss is reduced from 0.44 dB to 0.24 dB for TE mode, and from 0.46 dB to 0.26 dB for TM mode. This suggests that the PRI sol-gel materials can provide miniaturized optical circuits with a bending curvature radius of several tens of μm.

The relationship between Y-branching loss and branching angle is shown in Figure 7.48(b). When the angle is decreased from 80° to 20°, the loss is reduced

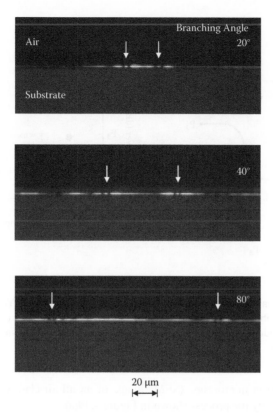

FIGURE 7.46 Photographs of output beams from core edges of Y-branching waveguides for TE mode. From S. Ono, T. Yoshimura, T. Sato, and J. Oshima, "Fabrication and evaluation of nano-scale optical circuits using sol-gel materials with photo-induced refractive index variation characteristics," *J. Lightwave Technol.* **27**, 1229–1235 (2009).

from 1.33 dB to 0.08 dB for TE mode, and from 1.34 dB to 0.12 dB for TM mode. If Y-branching loss of ~ 0.5 dB is permitted, it is expected that PRI sol-gel materials can be used for optical circuits with a branching angle of ~40°.

7.3.4 FINE 3-D STRUCTURES FOR ALL-AIR-CLAD WAVEGUIDES

The PRI sol-gel material becomes insoluble in Al etchant by exposing the material to UV/blue light followed by post-bake. By using this property, fine 3-D structures including all-air-clad waveguides are constructed.

Figure 7.49(a) shows an example of the process for fabricating 3-D waveguide structures. After patterned Al films are deposited on a substrate, a PRI sol-gel film is spin-coated on it. Through a photomask, the sol-gel film is exposed to UV light with a pattern of waveguide cores and is postbaked. By putting the sample in an Al etchant, the unexposed regions of the sol-gel film and the Al film are solved. Thus fine 3-D patterns for 3-D waveguide structures are completed. Figure 7.49(b) shows

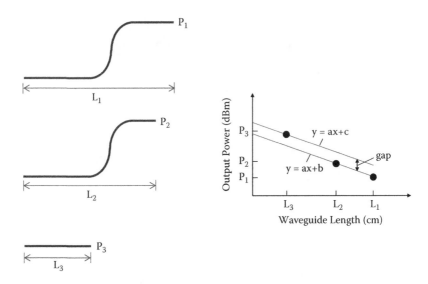

FIGURE 7.47 Evaluation procedure for bending loss. From S. Ono, T. Yoshimura, T. Sato, and J. Oshima, "Fabrication and evaluation of nano-scale optical circuits using sol-gel materials with photo-induced refractive index variation characteristics," *J. Lightwave Technol.* **27**, 1229–1235 (2009).

a scanning electron microscope (SEM) image of an all-air-clad waveguide with a spacer fabricated by the process shown in Figure 7.49(a).

7.4 SELF-ORGANIZED LIGHTWAVE NETWORK (SOLNET) FOR SELF-ALIGNED OPTICAL COUPLINGS AND VERTICAL WAVEGUIDES

When many thin-film photonic devices are integrated in OE systems such as optical interconnect, optical switching, and solar energy conversion systems, there will be an ever-increasing number and density of optical couplings between the photonic devices, requiring enormous effort for assembly with micron or submicron accuracy. To resolve the problem, the self-organized lightwave network (SOLNET) was proposed [24,27–30]. SOLNET enables us to construct self-aligned coupling optical waveguides between misaligned devices.

7.4.1 THE SOLNET CONCEPT

Figure 7.50 shows the SOLNET concept [24,27–30]. Optical devices such as optical waveguides, optical switches, reflectors, vertical cavity surface emitting lasers (VCSELs), photodiodes (PDs), and so on are placed in a photo-induced refractive index increase (PRI) material such as photo-polymers for holography, PRI sol-gel

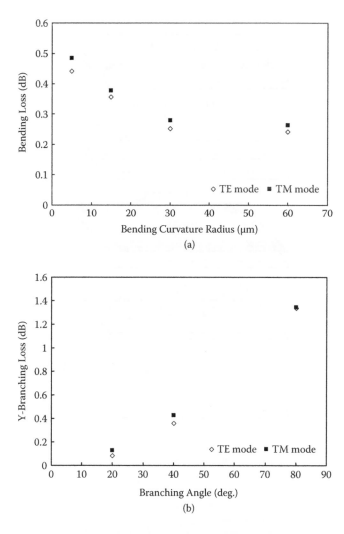

FIGURE 7.48 (a) Relationship between bending loss and bending curvature radius. (b) Relationship between Y-branching loss and branching angle. From S. Ono, T. Yoshimura, T. Sato, and J. Oshima, "Fabrication and evaluation of nano-scale optical circuits using sol-gel materials with photo-induced refractive index variation characteristics," *J. Lightwave Technol.* **27**, 1229–1235 (2009).

materials, photo-definable materials, or photo-refractive crystals, whose refractive index increases with write-beam exposure. By introducing write beams into the PRI material from optical devices, the refractive index distribution of the material continuously changes time by time. The write beam propagation is affected by the varying refractive index distribution, which is generated by the write beams themselves.

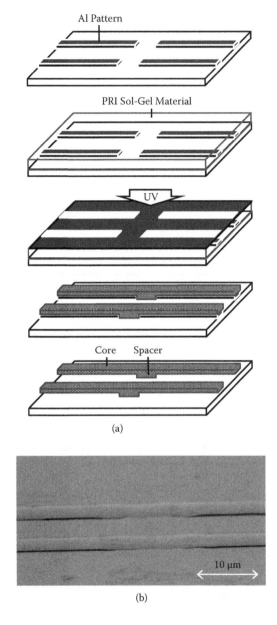

FIGURE 7.49   (a) Process for fabricating 3-D waveguide structures with the PRI sol-gel material. (b) A SEM image of a 3-D waveguide structure for fabricating all-air-clad waveguide with spacers. (This SEM image was provided by T. Sato and J. Oshima.)

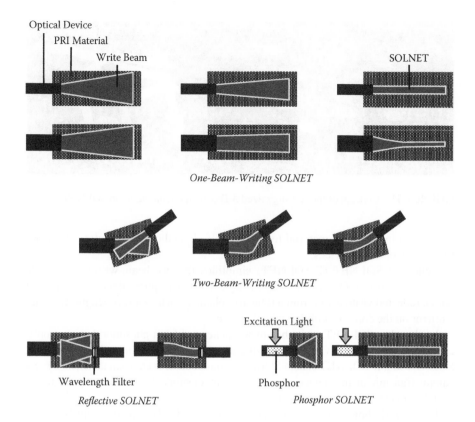

FIGURE 7.50   The SOLNET concept.

In one-beam-writing SOLNET, a write beam is introduced into PRI material from an optical device edge. By the self-focusing effect in the PRI material, the write beam is concentrated along the propagation axis to construct a straight path, forming an optical waveguide with a mode size close to that of the optical device. When a large-mode-size write beam is introduced, the write beam is focused along the propagation axis with a reduced diameter, resulting in a downtaper waveguide that has a smaller mode size at the end than the optical device.

In two-beam-writing SOLNET, two write beams are introduced from two different optical devices. The two beams are attracted each other in PRI material, and merge into one optical waveguide by self-focusing, constructing a coupling optical waveguide between the two optical devices automatically. Even when there is *offset in position*, *offset in angle*, and *mode size mismatching* between the devices, the write beams gradually merge into one, achieving self-aligned coupling between the misaligned devices.

The two-beam-writing SOLNET can be extended to multi-beam-writing SOLNET, in which write beams from two, three, or more optical devices are

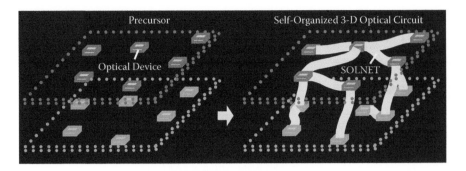

FIGURE 7.51 Concept of the self-organized 3-D optical circuit based on SOLNET.

introduced into the PRI material to achieve self-aligned optical couplings between two, three, or more devices.

Reflective SOLNET (R-SOLNET) simplifies the two-beam-writing SOLNET fabrication process. One of the two write beams from optical devices is replaced with a reflected write beam from a reflective element such as a wavelength filter and a mirror on the core facet of the optical device.

In phosphor SOLNET, write beams are generated from phosphor doped in optical waveguides in the devices by exposing the phosphor to excitation light. Phosphor SOLNET and R-SOLNET enable SOLNET construction even when the write beams cannot transmit in the devices, such as photodetectors, laser diodes, and optical switches made of narrow-band-gap materials.

Figure 7.51 shows a concept of the self-organized 3-D optical circuit based on SOLNET. After a precursor is built with optical devices distributed three-dimensionally, optical waveguides are formed between the devices automatically by write beams emitted from some of the optical devices or write beams from outside to construct self-aligned optical waveguides three-dimensionally.

7.4.2 PROOF OF CONCEPT OF SOLNET

7.4.2.1 One-Beam-Writing SOLNET

For an observation of waveguide construction in a free space filled with PRI materials [27,28], the following setup was made. A single-mode optical fiber for 1.3-μm wavelength with a core diameter of 9.5 μm was put in a photo-polymer with PRI effect. A white light beam (HOYA ML-150) of 80 nW was introduced from the fiber to the photo-polymer. A waveguide core was grown from the optical fiber edge along the light propagation axis, to form a straight waveguide. An interference microscopic photograph of SOLNET constructed by 2-minute exposure is shown in Figure 7.52. The line stretched from the edge of the optical fiber is the waveguide core of SOLNET where the refractive index is higher than the surrounding area. The stretched waveguide has a width that is almost the same as the fiber core diameter, indicating the proof-of-concept for one-beam-writing SOLNET.

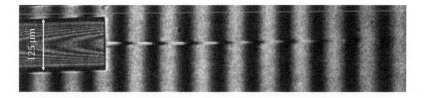

FIGURE 7.52 Experimental demonstration of one-beam-writing SOLNET. From T. Yoshimura, J. Roman, Y. Takahashi, W. V. Wang, M. Inao, T. Ishituka, K. Tsukamoto, K. Motoyoshi, and W. Sotoyama, "Self-organizing lightwave network (SOLNET) and its application to film optical circuit substrates," *IEEE Trans. Comp., Packag. Technol.* **24**, 500–509 (2001).

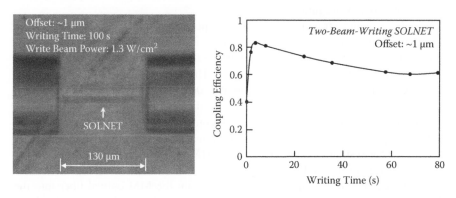

FIGURE 7.53 Experimental demonstration of two-beam-writing SOLNET. From T. Yoshimura, A. Hori, Y. Yoshida, Y. Arai, H. Kurokawa, T. Namiki, and K. Asama, "Coupling efficiencies in reflective self-organized lightwave network (R-SOLNET) simulated by the beam propagation method," *IEEE Photon. Technol. Lett.* **17**, 1653–1655 (2005).

7.4.2.2 Two-Beam-Writing SOLNET

Two-beam-writing SOLNET was constructed between two optical fibers with a 9.5-μm core diameter placed on a glass substrate with a lateral misalignment of ~1 μm [31]. The gap of 130 μm between them was filled with a photo-polymer made of acrylic/epoxy components developed by JSR Corporation [32]. The acrylic component contains an aromatic diacrylate with a high refractive index and a radical photo-initiator. The epoxy component contains an aliphatic diepoxide and a photoacid generator. The refractive index is 1.53 and 1.51 for the acrylic and epoxy components, respectively. The mixing ratio of the two components is 50:50 in wt%. The acrylic component exhibits larger reactivity than the epoxy component upon irradiation of light in the range of wavelength below 400 nm, inducing the PRI phenomena. The photo-polymer gives rise to a maximum refractive index change of around 0.02. The energy density required for saturation is ~6 J/cm^2 [32].

A self-organized optical waveguide formed by 100-s exposure is shown in Figure 7.53. Write beams of 1.3 W/cm^2 from a high-pressure mercury lamp were introduced from the two optical fibers into the photo-polymer. The coupling efficiency measured using 1.3-μm signal beams increases with write beam exposure to reach a maximum over 80%.

7.4.2.3 R-SOLNET

Figure 7.54 shows a schematic illustration of Y-branching R-SOLNET [33]. A write beam from an input optical waveguide and reflected write beams from two wavelength filters on the core facets of two output optical waveguides overlap near the wavelength filters. In the overlap regions, the refractive index of the PRI material increases, pulling the write beams to the wavelength filter locations more and more. We call this effect the *pulling water effect*. Finally, by self-focusing, self-aligned coupling optical waveguides are formed between the input optical waveguide and the two output optical waveguides.

In order to demonstrate the proof of concept of R-SOLNET experimentally, we carried out R-SOLNET formation between a multimode (MM) optical fiber with a core diameter of 50 μm and a mirror deposited on a facet of an optical fiber [33]. Figure 7.55(a) shows the experimental setup. The MM optical fiber and the mirror are placed in a photo-polymer with a gap of ~800 μm and an angular misalignment of 3° between the MM optical fiber direction and the surface-normal direction of the mirror. The photo-polymer of acrylic/epoxy components was the same as that used for the two-beam-writing SOLNET construction.

Figure 7.55(b) shows a photograph of the MM optical fiber and the optical fiber with the Al mirror on the facet in the photo-polymer. When a write beam from a high-pressure mercury lamp was introduced from the MM optical fiber into the photo-polymer, an R-SOLNET was formed. It is found from Figure 7.55(c) that a probe beam of 650 nm in wavelength, which is introduced from the MM optical

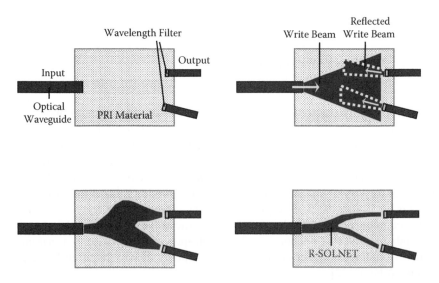

FIGURE 7.54 Schematic illustration of Y-branching R-SOLNET. From T. Yoshimura and H. Kaburagi, "Self-organization of optical waveguides between misaligned devices induced by write-beam reflection," *Appl. Phys. Express* **1**, 062007-1-3 (2008).

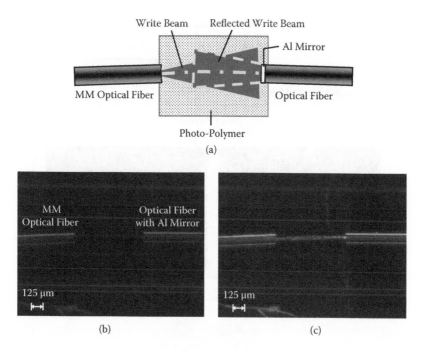

FIGURE 7.55 An experiment of R-SOLNET formation with angular misalignment of 3°. (a) Experimental setup, (b) a photograph before SOLNET formation, and (c) a photograph after SOLNET formation. From T. Yoshimura and H. Kaburagi, "Self-organization of optical waveguides between misaligned devices induced by write-beam reflection," *Appl. Phys. Express* **1**, 062007-1-3 (2008).

fiber, propagates in the optical waveguide of the R-SOLNET. The core width of the R-SOLNET is found to be about 50 μm, which is close to that of the MM optical fiber.

If the pulling water effect does not exist, two optical waveguides of SOLNET with a directional difference of 6° should be formed from the mirror by the incident write beam and the reflected write beam. However, in the photograph shown in Figure 7.55(c), only one bow-shaped optical waveguide of the SOLNET is observed. This indicates that a self-aligned optical waveguide of the R-SOLNET is formed due to the pulling water effect induced by the reflected write beam from the mirror.

Figures 7.56(a) shows the experimental setup for the R-SOLNET with a lateral offset [33]. An MM optical fiber and an optical fiber with an Al mirror are placed with an offset of 30 μm in the photo-polymer (Figure 7.56(b)). When a write beam was introduced from the MM optical fiber into the photo-polymer, the R-SOLNET was formed, connecting the MM optical fiber to the misaligned mirror. It is found from Figure 7.56(c) that a probe beam of 650 nm in wavelength propagates in the S-shaped self-aligned optical waveguide of the R-SOLNET with a core width of about 50 μm.

It is noticed that the R-SOLNET is connected to the upper part of the fiber facet. As can be seen from Figure 7.56(d), the mirror is not deposited uniformly on the fiber

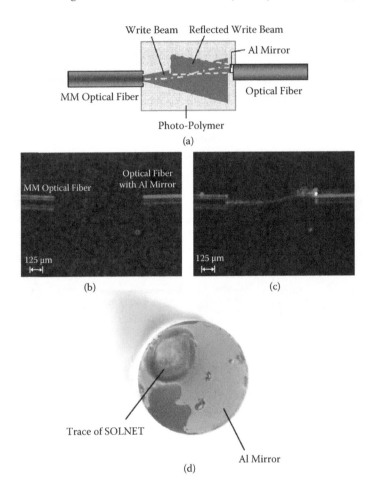

FIGURE 7.56 An experiment of R-SOLNET formation with lateral offset of 30 μm. (a) Experimental setup, (b) a photograph before SOLNET formation, (c) a photograph after SOLNET formation, and (d) a photograph of the Al mirror on the fiber facet. From T. Yoshimura and H. Kaburagi, "Self-organization of optical waveguides between misaligned devices induced by write-beam reflection," *Appl. Phys. Express* **1**, 062007-1-3 (2008).

facet, but is deposited partially. This is the reason why the R-SOLNET is pulled to the upper part. Actually, trace of the SOLNET is found on the upper part of the fiber facet, where the mirror exists.

In addition, we simulated R-SOLNET formation by the finite difference time domain (FDTD) method for a model with two wavelength filters [33,34]. The model and results are shown in Figure 7.57. An input optical waveguide with a core width of 2 μm and two output optical waveguides with a core width of 0.5 μm are placed in a PRI material with a gap of 10 μm. The output optical waveguides are misaligned by 0.5 μm from the input optical waveguide. The refractive index of the core and clad regions are 1.5 and 1, respectively. The refractive index of the PRI material changes from 1.5 to 1.8 with write beam exposure. Dielectric multilayer wavelength filters

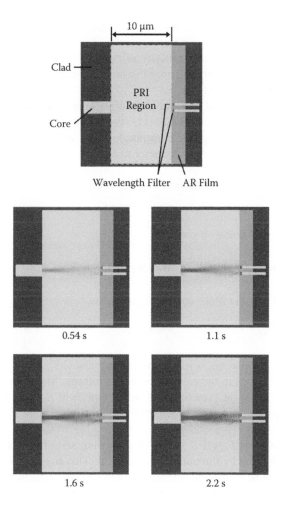

FIGURE 7.57 R-SOLNET formation simulated by the FDTD method for a model with two wavelength filters. From T. Yoshimura and H. Kaburagi, "Self-organization of optical wave-guides between misaligned devices induced by write-beam reflection," *Appl. Phys. Express* **1**, 062007-1-3 (2008).

on the core facets of the output optical waveguides reflect write beams of 650 nm in wavelength. The filters consist of an 80-nm-thick layer with a refractive index of 1.5 and an 80-nm-thick layer with a refractive index of 2.5. The total layer count is four.

The simulation procedure is as follows. First, the initial refractive index distribution for the calculated region is input into a data file. Since the refractive index of the PRI material changes with time, the refractive index distribution is rewritten at each time step with a duration of Δt. Energy density of the write beam introduced into the PRI material over Δt [s] is $(1/2)\varepsilon E^2 v \Delta t$, where, E [V/m] is the electric field of the write beam, ε [J/(mV²)] is the dielectric constant, and v [m/s] is the light velocity. Then, Δn is expressed as

$$\Delta n = \gamma |(1/2)\varepsilon E^2 v \Delta t|, \tag{7.14}$$

Here, γ [m^2/J] is related to the sensitivity of the PRI material. For simplicity, it is assumed that the refractive index changes in proportion to the write beam exposure until it reaches the saturation limit.

In real systems, the typical response time of PRI materials ranges from around a few seconds to a few minutes. In the FDTD method, however, it is difficult to carry out calculations with such long time durations because Δt should be sub-ps. So, in the simulations we rescaled the time parameter to $\Delta t'$ as follows.

$$\Delta n = \gamma' |(1/2)\varepsilon E^2 v \Delta t'|, \tag{7.15}$$

where, $\Delta t' = \Delta t \times 10^{12}$ and $\gamma' = \gamma \times 10^{-12}$.

The electric field of light sources is expressed as follows.

$$E(t) = (1/2)(1 - \cos(\pi t / T_\omega))E_0 \sin \omega_0 t \qquad 0 \leq t \leq T_\omega$$

$$E(t) = E_0 \sin \omega_0 t \qquad t \geq T_\omega$$

Here, ω_0 is the angular frequency of the light wave and $\omega_0 T_\omega = 4\pi$.

The parameters used in the calculation are as follows. Mesh sizes are $\Delta x = \Delta y = 20$ nm and $\Delta t = 0.0134$ ps. Amplitude of the electric field E_0 is 10 V/m, which corresponds to a power density of 13 nW/cm^2 for ε and v in vacuum. γ' is 1 m^2/J with a saturation limit of 0.3.

It can be seen from Figure 7.57 that Y-branching self-aligned coupling optical waveguides are formed by the pulling water effect. This suggests that a plurality of optical devices can be connected automatically in PRI materials by putting reflective materials such as wavelength filters on the core facets. Coupling optical waveguides of R-SOLNET are led to the reflective materials locations. These results lead us to believe that R-SOLNET is effective in forming self-aligned optical couplings in micro/nano-scale optical circuits.

7.5 RESOURCE-SAVING HETEROGENEOUS INTEGRATION

7.5.1 CONCEPT OF PL-PACK WITH SORT

In order to integrate various types of thin-film photonic/electronic devices, a low-cost heterogeneous integration process is required. The photolithographic packaging with SORT (PL-Pack with SORT) shown in Figure 7.58 is a promising candidate [35,36]. The total process flow of PL-Pack with SORT is as follows:

1. Prepare a substrate (LSI, MCM, removable substrate, etc.).
2. Place thin-film devices on the substrate by SORT.

(1) Prepare Substrate

(2) Place Thin-Film Devices by SORT

(3) Embed Thin-Film Devices

Heterogeneous Integration on a Substrate

OE Film

Device-Embedded Optical Waveguide Film

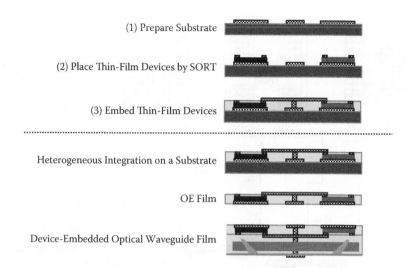

FIGURE 7.58 Total process flow of PL-Pack with SORT. From T. Yoshimura, K. Kumai, T. Mikawa, O. Ibaragi, and M. Bonkohara, "Photolithographic packaging with selectively occupied repeated transfer (PL-Pack with SORT) for scalable optical link multi-chip-module (S-FOLM)," *IEEE Trans. Electron. Packag. Manufact.* **25**, 19–25 (2002).

3. Embed the thin-film devices in an insulator layer followed by pads/electrodes/vias formation.

Thus, heterogeneous integration is completed on a substrate to provide an OE-LSI, OE-MCM, and so on. When the device-embedded film is removed from the substrate, an OE film is obtained. When combining the process (steps 1–3) with optical waveguide formation, a device-embedded optical waveguide film is obtained.

The most critical part is SORT in step (2). Figure 7.59 shows an outline of SORT. For simplicity, two kinds of devices are considered, although in general many kinds of different thin-film devices are integrated. The process flow is as follows. The Device I array and Device II array are respectively grown with a pitch of d on growth wafer I and growth wafer II. The devices are separated by boundary grooves. All the Devices I are picked up on a pick-up substrate at one time by, for example, epitaxial lift-off (ELO) [37,38], followed by pretransfer to a supporting substrate, Sub I. Similarly, Devices II are pretransferred to Sub II. The distribution patterns of the devices on Sub I and Sub II are respectively designed to match the pad distribution patterns for Devices I and Devices II on the final substrate. The devices left on the pick-up substrates are provided to another Sub I and Sub II.

For final transfer, Sub I is attached on a final substrate, where catch-up pads with a pitch of p_p are formed, to selectively place Devices I at sites for Device I on the final substrate. Then, Sub II is attached on the final substrate, to place Devices II at sites for Device II. Sub I and Sub II are further attached to other final substrates, or other places of the final substrate, to provide Devices I and Devices II. By repeating the procedure, all the devices on Sub I and Sub II are transferred to final substrates.

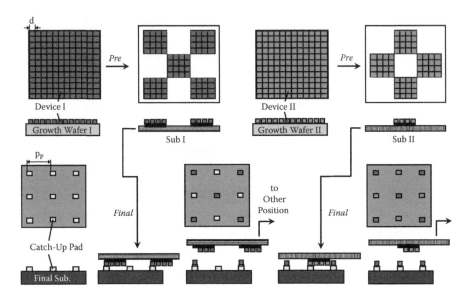

FIGURE 7.59 SORT for integration of two kinds of devices. From T. Yoshimura, M. Ojima, Y. Arai, and K. Asama, "Three-dimensional self-organized micro optoelectronic systems for board-level reconfigurable optical interconnects: performance modeling and simulation," *IEEE J. Select. Topics in Quantum Electron.* **9**, 492–511 (2003).

For the case shown in Figure 7.59, there are $4 \times 4 = 16$ devices corresponding to one catch-up pad. This implies that one supporting substrate can provide devices to 16 final substrates. Two supporting substrates are available for Device I and II, respectively, indicating that 32 final substrates, on which Devices I and II are distributed with a desired configuration, can be obtained from a growth wafer I and a growth wafer II.

A marker alignment procedure can be applied to the device transfer process by using an aligner. SORT can stably be performed by using an adhesive whose adhesion strength is controlled with an *adhesion strength hierarchy* or *dynamic adhesion strength control*. In the former, the adhesion strength is increased in steps. In the latter, adhesion strength is controlled during the process by, for example, using UV light exposure. For fabrication of OE LSIs, OE boards, or 3-D OE LSIs, Devices I and II could be light modulators, VCSELs, and PDs. For fabrication of three-dimensional micro-optical switching system (3D-MOSS), Devices I and II could be optical switches of polymer MQD, lead lanthanum zirconate titanate (PLZT), or III-V epitaxial films, wavelength filters, photonic crystals, waveguide lenses of Ti_xSi_yO thin films, and so on, for example.

7.5.2 Advantages of PL-Pack with SORT

A schematic illustration for structural comparison between PL-Pack with SORT and conventional flip-chip (FC) bonding is shown in Figure 7.60 for the case of devices made of III-V materials, typically, VCSELs and PDs. In FC bonding, a

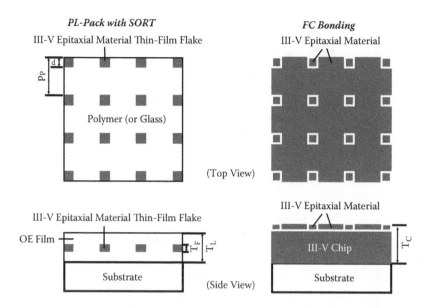

FIGURE 7.60 Schematic illustration of structural comparison between PL-Pack with SORT and conventional FC bonding. From T. Yoshimura, M. Ojima, Y. Arai, and K. Asama, "Three-dimensional self-organized micro optoelectronic systems for board-level reconfigurable optical interconnects: performance modeling and simulation," *IEEE J. Select. Topics in Quantum Electron.* **9**, 492–511 (2003).

bulk chip of III-V material is put on a substrate. In PL-Pack with SORT, an OE film in which III-V epitaxial material thin-film flakes with length and width of d are embedded with a pitch of p_P, is put on a substrate. The difference brings remarkable advantages to the structure built by PL-Pack with SORT over FC bonding [35,36].

7.5.2.1 Resource Saving

The first advantage of PL-Pack with SORT is its resource-saving characteristics. In FC bonding, the whole area is occupied by expensive III-V epitaxial materials. In PL-Pack with SORT, on the other hand, the epitaxial materials exist only at necessary sites. The III-V epitaxial material saving effect can be measured by the material saving factor (MSF) of PL-Pack with SORT defined by

$$MSF = [1/R_{SORT}][1/R_{FC\ Bonding}] = (p_p/d)^2. \tag{7.16}$$

Here, R_{SORT} and $R_{FC\ Bonding}$ represent III-V epitaxial material consumption in PL-Pack with SORT and in FC bonding, respectively. Device length and width d and pitch p_P in the III-V chips/OE films shown in Figure 7.60, respectively, correspond to the device pitch on a growth wafer and the catch-up pad pitch shown in Figure 7.59. Figure 7.61(a) shows MSF as a function of d for various p_P. When p_P is 100 μm, MSF

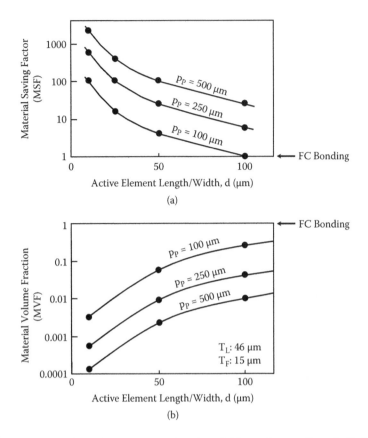

FIGURE 7.61 (a) MSF and (b) MVF as a function of d for various p_P. From T. Yoshimura, M. Ojima, Y. Arai, and K. Asama, "Three-dimensional self-organized micro optoelectronic systems for board-level reconfigurable optical interconnects: performance modeling and simulation," *IEEE J. Select. Topics in Quantum Electron.* **9**, 492–511 (2003).

equals 1 for d of 100 μm. This implies that whole area of the OE film prepared by PL-Pack with SORT is occupied by III-V epitaxial materials since p_P is equal to d. MSF increases with an increase in p_P and with a decrease in d, reaching more than 100~1000.

The resource-saving characteristics of PL-Pack with SORT results in cost reduction of OE systems such as the scalable film optical link module (S-FOLM) described in Section 8.1. Based on the *cost versus active element pitch* characteristics presented by Stirk et al. for an OE module with a VCSEL array [39], the impact of PL-Pack with SORT on cost can be estimated. As the 0th approximation, costs are assumed to be the same for PL-Pack with SORT and FC bonding, except for III-V epitaxial material cost. Due to the epitaxial-material saving effect in PL-Pack with SORT, module cost is reduced with a decrease in d. For example, in the case that p_P and d are, respectively, 200 μm and 50 μm, a cost reduction of ~1/6 is expected. By further decreasing d, a cost reduction of <1/10 might be expected.

7.5.2.2 Process Simplicity

The second advantage of PL-Pack with SORT is process simplicity in heterogeneous integration of thin-film devices, which results in a cost reduction. In FC-bonding-based packaging, generally, different kinds of OE devices should be fabricated on the same chip, which requires a sophisticated film growth method, which raises fabrication cost. In PL-Pack with SORT, as shown in Figure 7.59, heterogeneous integration can be performed by delivering OE devices from a set of wafers. On individual wafers, arrays containing same kind of devices are grown by simple growth method, contributing to cost reduction.

Furthermore, SORT can remove the conventional FC bonding–based packaging to achieve heterogeneous device integration only by semiconductor device fabrication processes, that is, all-photolithographic processes. These features are effective on cost reduction of OE systems.

7.5.2.3 Thermal Stress Reduction

The third advantage of PL-Pack with SORT is the potential to reduce thermal stress. As an example, consider a case where a 2-D VCSEL array of L^2 μm², in which VCSELs are located with a pitch of p_P μm, is put on a plastic substrate. For FC bonding, an L^2-μm² bulk III-V chip of T_C-μm thick is mounted on the substrate. For PL-Pack with SORT, an L^2-μm² OE film of T_L-μm thick, where T_F-μm-thick III-V epitaxial material thin-film flakes with d^2 μm² area are embedded with a pitch of p_P μm, is mounted on a plastic substrate.

In FC bonding, due to the large difference in the coefficient of thermal expansion (CTE) between III-V material and plastics, as well as due to the large thickness of a III-V chip with a large Young's modulus, a large thermally induced stress between the chip and the substrate is induced as is well known. In PL-Pack with SORT, compared with FC bonding, the thickness of the III-V material block is reduced by a factor of T_F/T_C and the area by a factor of d^2/L^2, indicating that the individual III-V material block volume is reduced by a factor of $(T_F/T_C)(d^2/L^2)$. For a case that $T_F = 15$ μm, $T_C = 500$ μm, $d = 20$ μm, and $L = 10,000$ μm, roughly speaking, the power inducing the stress at the boundary is reduced by seven orders of magnitude in PL-Pack with SORT.

Figure 7.61(b) shows a material volume fraction (MVF) of III-V material in an OE film, which is defined by

$$MVF = (d^2 T_F) / (P_P^2 T_L).$$ (7.17)

For a case where $d = 20$ μm, $T_F = 15$ μm, $p_P = 200$ μm, and $T_L = 46$ μm, MVF is calculated to be 0.003, suggesting that the CTE of the OE films is close to that of the host matrix polymer with small Young's modulus. Thus, the stress between the OE film and the plastic substrate is drastically reduced compared with that between the III-V chip and the plastic substrates. Furthermore, basically, no large thermal expansion difference arises between OE films because the CTE of the film is almost defined by the host matrix polymer. Therefore, it is expected that thermally induced displacement may be neglected in S-FOLM where films are pinned by electrical pad connections.

7.5.3 Experimental Demonstrations of SORT

7.5.3.1 SORT of Polymer Waveguide Lenses

The SORT process for polymer waveguide lenses is shown in Figure 7.62 [40]. On a glass substrate, Al patterns of 50 nm thickness with apertures of waveguide lens shapes were formed to make a "built-in mask," followed by spin-coating of a polyvinylalcohol (PVA) film 10 μm thick. After drying the PVA in ambient atmosphere at room temperature for 1 h, an acrylic photo-definable material film was spin-coated on it with a thickness of 40 μm. The film was exposed to UV light from a high-pressure mercury lump of 400 mJ/cm^2 through the built-in mask to form a polymer waveguide lens array by developing. The waveguide lens diameter is 300 μm, and the pitch in the array is 450 μm. For picking up, a supporting substrate, which is a glass substrate covered with an adhesive film (GelPak film-×8 provided by GelPak), was attached on the top of the polymer waveguide lens array. By solving the PVA film in water, the polymer waveguide lens array was picked up to the supporting substrate.

The photograph on the top in Figure 7.63 represents the polymer waveguide lens array on the adhesive film of the supporting substrate. A final substrate was prepared by exposing a glass substrate covered with a UV-curable-material-coated film (UV tape UE-111AJ provided by Nitto Denko) to UV light from a high-pressure mercury lamp of 180 mJ/cm^2 through a photo-mask. The adhesion strength of the exposed

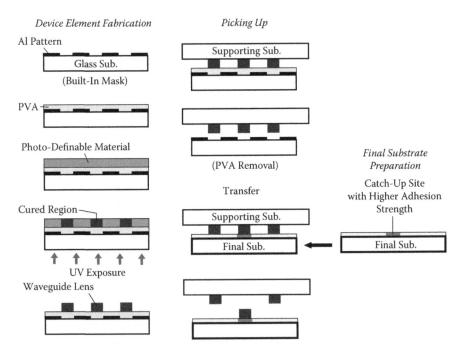

FIGURE 7.62 SORT process for polymer waveguide lenses. From T. Yoshimura, D. Tamaki, S. Kawamura, Y. Arai, and K. Asama, "A device transfer process 'selectively occupied repeated transfer (SORT)' for resource-saving integration of polymer micro optics fabricated by 'built-in mask' method," *IEEE Trans. Comp., Packag. Technol.* **27**, 468–471 (2004).

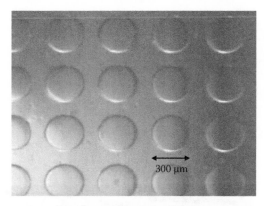

Transfer onto Supporting Substrate

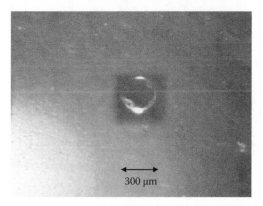

Transfer onto Final Substrate

FIGURE 7.63 Photographs of a polymer waveguide lens array on the adhesive film of the supporting substrate (top), and a polymer waveguide lens transferred selectively onto the catch-up site of the final substrate (bottom). From T. Yoshimura, D. Tamaki, S. Kawamura, Y. Arai, and K. Asama, "A device transfer process 'selectively occupied repeated transfer (SORT)' for resource-saving integration of polymer micro optics fabricated by 'built-in mask' method," *IEEE Trans. Comp., Packag. Technol.* **27**, 468–471 (2004).

region decreased by more than one order of magnitude compared with the unexposed region, providing a final substrate with catch-up sites having stronger adhesion strength than the surrounding area. For transfer, using the SORT equipment shown in Figure 7.64, the polymer waveguide lens array on the supporting substrate was attached to the final substrate, and then the two substrates were detached to each other. As shown in the photograph on the bottom in Figure 7.63, the polymer waveguide lens was transferred selectively onto the catch-up site of the final substrate. This demonstrates the feasibility of the SORT process.

The SORT equipment was proposed by the author and was assembled by Suruga Seiki Co., Ltd. using components for fiber optics with positional accuracy of ~1 μm.

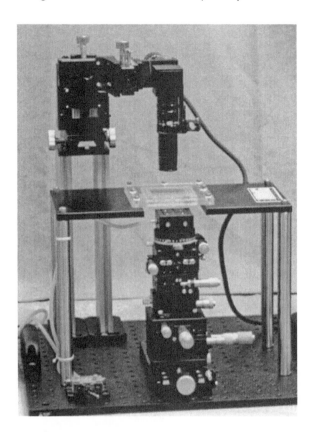

FIGURE 7.64 SORT equipment. From T. Yoshimura, T. Kofudo, T. Kashiwazako, K. Naito, K. Ogushi, and Y. Kitabayashi, "A material-saving optical waveguide fabrication process with selective transfers of cores," *Opt. Eng.* **47**, 014601-1-7 (2008).

Alignment between two substrates was achieved by optical stages, on which a lower substrate was set below the upper substrate holder.

7.5.3.2 SORT of Optical Waveguides

Figure 7.65(a) shows the concept of material-saving optical waveguide fabrication using SORT [41]. On a substrate, denoted by a picking-up substrate, cores for optical waveguides are placed in an array with high density. On final substrates, catch-up sites are formed. By attaching the picking-up substrate to a final substrate and detaching these, the cores are selectively transferred onto the final substrate to construct optical waveguides. In the case shown in Figure 7.65, since core pitch on the final substrate is six times larger than that on the picking-up substrate, one picking-up substrate can provide cores to six final substrates. This implies that six times the core materials are saved. For example, in chip-to-chip optical wiring with a core width of 40 μm and a core pitch of 250 μm, when cores are formed on the picking-up substrate with a 50-μm pitch and catch-up sites are arranged with a 250-μm pitch, it is expected that five times the waveguide materials can be saved.

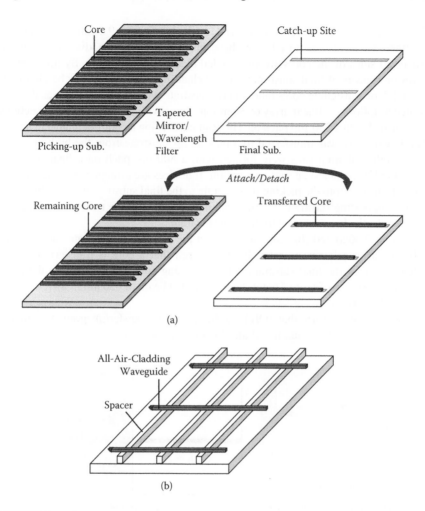

FIGURE 7.65 Concept of (a) material-saving optical waveguide fabrication using SORT, and (b) all-air-cladding waveguides for fine-pattern optical interconnects. From T. Yoshimura, T. Kofudo, T. Kashiwazako, K. Naito, K. Ogushi, and Y. Kitabayashi, "A material-saving optical waveguide fabrication process with selective transfers of cores," *Opt. Eng.* **47**, 014601-1-7 (2008).

In usual optical waveguides for chip-scale optical interconnects, devices such as vertical mirrors and wavelength filters with a metal/dielectric film coating on core end facets coexist. Complicated deposition/photolithography processes are involved in making these devices. The material-saving optical waveguide fabrication process shown in Figure 7.65 can also reduce the process steps for fabricating the coexisting devices by six times.

SORT is applicable to fabrication of all-air-cladding waveguides, as shown in Figure 7.65(b). For intrachip optical interconnects, fine-pattern optical circuits with light beam sizes confined to submicrons are necessary. To realize this, strong confinement of guided beams in cores is required by making the refractive index contrast between core and clad high. In optical waveguide–type light beam collecting films for photo-oltaic devices (which is described in Section 9.2), strong light beam confinement is also required. For these purposes, all-air-cladding waveguides, where all sides of the cores are surrounded by air, are promising. By transferring cores on spacers of final substrates, all-air-cladding waveguides may be obtained.

SORT of waveguide cores onto a glass substrate with catch-up sites was demonstrated as follows. A linear array of 3-cm-long cores with 32×32-μm^2 cross-sections was formed with a 250-μm pitch on a built-in mask using a photo-definable material. The linear array was transferred to a picking-up substrate from the built-in mask. Meanwhile, catch-up sites were formed with a 500-μm pitch on a final substrate using the UV-curable-material-coated film. If devices are arranged in a linear array, direct transfers from the picking-up substrate to the final substrate are possible without the supporting substrate. So, by attaching/detaching of the two substrates, cores can be selectively transferred to the final substrate.

Figures 7.66(a) and (b) show the cores on the picking-up substrate and the final substrate after selective transfer. It can be seen that a core is transferred onto a catch-up site of the final substrate while a missing core site is found on the picking-up substrate. As Figure 7.66(c) shows, a guided beam of 650 nm in wavelength is clearly observed at the core edge.

These results confirm that SORT can be applied to transfer/integration of micro-scale optical circuits, including all-air-cladding waveguides.

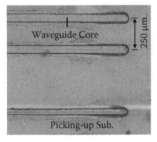
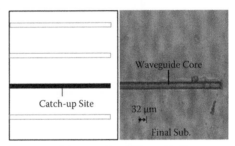

(a) Remained Waveguide Cores on Picking-up Sub.

(b) A Transferred Waveguide Core on Final Sub.

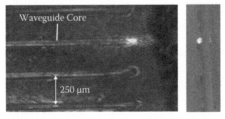

(c) Light Beam Propagation in a Transferred Waveguide Core

FIGURE 7.66 Photographs of cores after selective transfer (a) on the picking-up substrate and (b) on the final substrate. (c) A guided beam in an optical waveguide with a transferred core. From T. Yoshimura, T. Kofudo, T. Kashiwazako, K. Naito, K. Ogushi, and Y. Kitabayashi, "A material-saving optical waveguide fabrication process with selective transfers of cores," *Opt. Eng.* **47**, 014601-1-7 (2008).

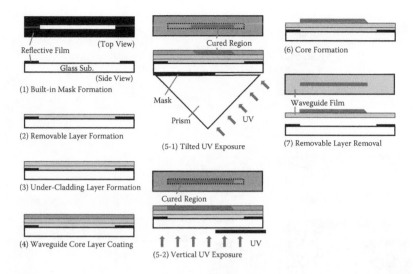

FIGURE 7.67 Duplication process of optical waveguide films with vertical mirrors using the built-in mask method. From T. Yoshimura, M. Miyazaki, Y. Miyamoto, N. Shimoda, A. Hori, and K. Asama, "Three-dimensional optical circuits consisting of waveguide films and optical Z-connections," *J. Lightwave Technol.* **24**, 4345–4352 (2006).

7.6 OPTICAL WAVEGUIDE FILMS WITH VERTICAL MIRRORS AND 3-D OPTICAL CIRCUITS

One of the important elements for thin-film photonic device integration is OE films of optical waveguides with vertical mirrors. The OE films enable 3-D optical systems such as optical interconnect, optical switching, and solar energy conversion systems. In the present section, optical waveguide films with vertical mirrors duplicated by the built-in mask method and 3-D optical circuits constructed by stacking optical waveguide films are described.

7.6.1 OPTICAL WAVEGUIDE FILMS WITH VERTICAL MIRRORS

Figure 7.67 shows a duplication process of optical waveguide films with vertical mirrors using the built-in mask method [24,26,42]. On a glass substrate, Al film was deposited with patterns of cores to make a built-in mask. After a PVA removable layer and an under-cladding layer were successively formed on the mask, a 32-μm thick core layer was coated. For the under-cladding layer and the core, acrylic photo-definable polymers developed by JSR Corporation were used (OPSTAR PJ3045 for the under-cladding layer and OPSTAR PJ3001 for the core) [32]. The core layer was selectively exposed to a tilted UV light with a 45° incident angle through a prism put on the back of the built-in mask. By this process, cured structures with beveled walls with 45° slopes were constructed at the waveguide core ends, providing waveguide end facets for 45° mirrors. Vertical UV exposure was also carried out with a mask that covers the beveled wall regions to construct planar waveguide cores. The energy

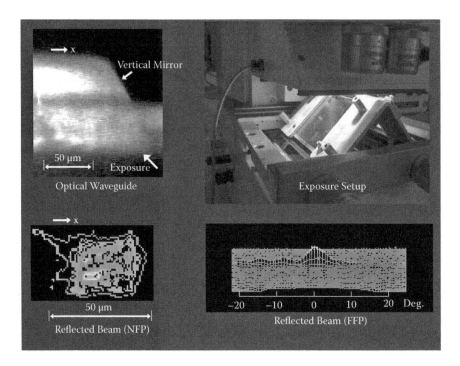

FIGURE 7.68 A photograph of an optical waveguide film with a vertical mirror duplicated by the built-in mask method, a tilted exposure setup, and NFP and FFP of a reflected beam from a vertical mirror.

density of UV exposure using a high-pressure mercury lamp was ~1 J/cm². By developing, core formation was completed. Finally, the removable layer was removed in water to obtain an optical waveguide film with 45° mirrors. The refractive index of the under-cladding layer is 1.56 and that of the core is 1.58 [32].

Figure 7.68 shows a photograph of an optical waveguide film with a vertical mirror duplicated by the built-in mask method. Nominal core width is 32 μm. The waveguide end facet for a vertical mirror can be observed at the right edge of the core. Figure 7.68 also shows an example of a tilted exposure setup, and an near-field pattern (NFP) and an FFP of a reflected beam from a vertical mirror.

7.6.2 3-D Optical Circuits

Figure 7.69 shows two types of 3-D optical circuits for demonstration [26]. Type 1 is a stacked waveguide film with 45° mirrors. Two optical waveguide films (Films A and B) are stacked by contacting their under-cladding layers. No over-cladding layer is coated, meaning that the films have air-cladding waveguides. A probe beam is introduced at the *Input* in Film A. The beam is transferred into Film B through the *Optical Z-Connection* to reach the *Output* in Film B.

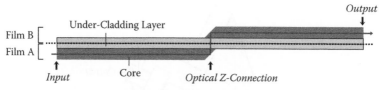

(a) Type 1: Stacked Waveguide Films with 45° Mirrors

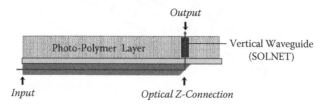

(b) Type 2: Waveguide Films with Vertical Waveguides

FIGURE 7.69 Schematic illustration of two types of 3-D optical circuits for demonstrations. From T. Yoshimura, M. Miyazaki, Y. Miyamoto, N. Shimoda, A. Hori, and K. Asama, "Three-dimensional optical circuits consisting of waveguide films and optical Z-connections," *J. Lightwave Technol.* **24**, 4345–4352 (2006).

Type 2 is a waveguide film with vertical waveguides. A vertical waveguide is constructed above a 45° mirror in an optical waveguide film using SOLNET. When a write beam is introduced at the *Input*, it is reflected into the PRI layer of a photopolymer to grow a vertical waveguide at the *Optical Z-Connection*. The vertical waveguide confines a reflected probe beam to guide it efficiently to the *Output* at the PRI layer surface.

7.6.2.1   Type 1: Stacked Waveguide Films with 45° Mirrors

In Figure 7.70 a fabrication and measurement system for 3-D optical circuits consisting of stacked waveguide films is shown [26]. Film B put on an alignment stage was stacked on Film A with adjusting the positions of the 45° mirrors in Films A and B to construct the *Optical Z-Connection*. A single mode (SM) optical fiber for 1.3 μm was connected to the *Input* of a core in Film A by butt joint. A probe beam was introduced at the *Input* and propagated to the *Output* in Film B through the *Optical Z-Connection*.

Figure 7.71 shows the propagation of a 650-nm-wavelength probe beam in Films A and B. Before alignment, since the 45° mirror in Film B is not located above the 45° mirror in Film A, the probe beam reflected by the 45° mirror in Film A propagates through under-cladding layers to reach a camera set above the 45° mirror. After alignment, as shown in Figure 7.72, the two 45° mirrors are correctly aligned to form the *Optical Z-Connection*. The reflected probe beam can be transmitted into Film B through the *Optical Z-Connection* to reach the *Output*. A ~30 × 30 μm² NFP is observed at the *Output*. Thus, 3-D optical wiring operation in the 3-D optical circuit is demonstrated successfully.

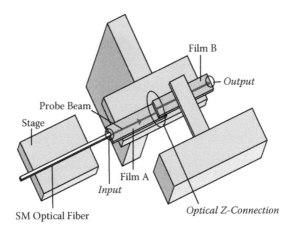

FIGURE 7.70 Fabrication/measurement system for 3-D optical circuits consisting of stacked waveguide films. From T. Yoshimura, M. Miyazaki, Y. Miyamoto, N. Shimoda, A. Hori, and K. Asama, "Three-dimensional optical circuits consisting of waveguide films and optical Z-connections," *J. Lightwave Technol.* **24**, 4345–4352 (2006).

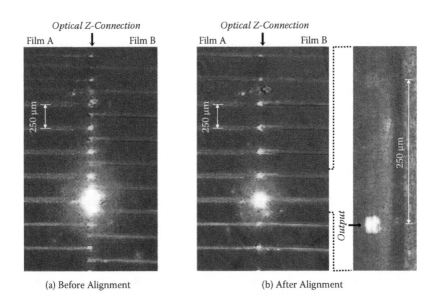

(a) Before Alignment   (b) After Alignment

FIGURE 7.71 Probe beam propagation in the 3-D optical circuit with stacked waveguide films. From T. Yoshimura, M. Miyazaki, Y. Miyamoto, N. Shimoda, A. Hori, and K. Asama, "Three-dimensional optical circuits consisting of waveguide films and optical Z-connections," *J. Lightwave Technol.* **24**, 4345–4352 (2006).

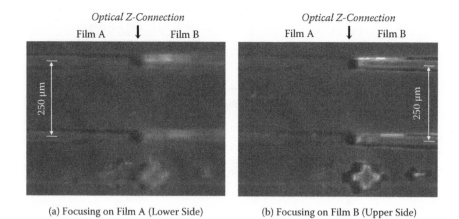

(a) Focusing on Film A (Lower Side)　　　　(b) Focusing on Film B (Upper Side)

FIGURE 7.72 Photographs of the *Optical Z-Connection* part. From T. Yoshimura, M. Miyazaki, Y. Miyamoto, N. Shimoda, A. Hori, and K. Asama, "Three-dimensional optical circuits consisting of waveguide films and optical Z-connections," *J. Lightwave Technol.* **24**, 4345–4352 (2006).

7.6.2.2 Type 2: Waveguide Films with Vertical Waveguides of SOLNET

As one of the ways for loss reduction at the *Optical Z-Connection*, implementation of a 3-D optical circuit consisting of vertical waveguides on an optical waveguide film is effective. The fabrication/measurement system is shown in Figure 7.73 [26]. On the back of an under-cladding layer in an optical waveguide film, a ~500-μm-thick PRI layer of photo-polymer was coated. The photo-polymer developed by JSR Corporation consists of acrylic/epoxy components [32].

A write beam of 405-nm blue laser diode (LD) was introduced from a SM optical fiber at the *Input* of the optical waveguide film with power of ~100 μW or less. The propagated write beam was reflected to the surface-normal direction by a 45° mirror. The write beam passed through the under-cladding layer and was injected into the PRI layer. Then, a vertical waveguide of SOLNET was formed in the PRI layer. A probe beam of 650-nm wavelength was introduced at the *Input* of the optical waveguide film. The beam was guided in the vertical waveguide at the *Optical Z-Connection* to reach the *Output* at the PRI layer surface of the SOLNET.

In Figure 7.74(a), SOLNET grown vertically on the under-cladding layer is shown. It is found that the vertical waveguide is constructed in the photo-polymer layer above the 45° mirror. As shown in Figure 7.74(b), a near-field pattern of the probe beam with an area of ~30 × 30 μm², which is approximately the same as the core size of the optical waveguide film, is observed at the *Output*. These results demonstrate the 3-D optical wiring capability of the vertical waveguide of SOLNET.

It should be noted that SOLNET was formed by a write beam from the under-cladding layer, which is a free space with no beam-confining structures. In other words, SOLNET can be formed even when some spacers are inserted between 45° mirrors and PRI layers. An experimental measurement of SOLNET formed between two optical fibers suggests that loss in SOLNET is around 0.7 dB.

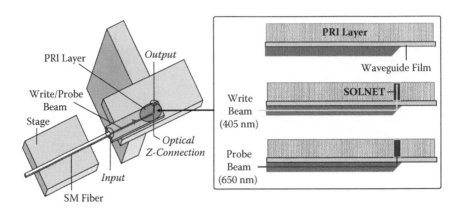

FIGURE 7.73 Fabrication/measurement system for 3-D optical circuits of waveguide films with vertical waveguides of SOLNET. From T. Yoshimura, M. Miyazaki, Y. Miyamoto, N. Shimoda, A. Hori, and K. Asama, "Three-dimensional optical circuits consisting of waveguide films and optical Z-connections," *J. Lightwave Technol.* **24**, 4345–4352 (2006).

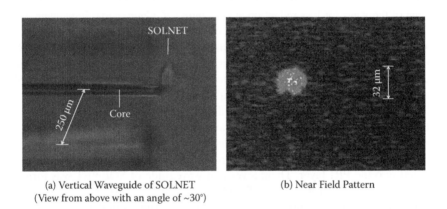

(a) Vertical Waveguide of SOLNET
(View from above with an angle of ~30°)

(b) Near Field Pattern

FIGURE 7.74 (a) Vertical waveguide of SOLNET grown above the 45° mirror. (b) NFP of the probe beam guided in the vertical waveguide. From T. Yoshimura, M. Miyazaki, Y. Miyamoto, N. Shimoda, A. Hori, and K. Asama, "Three-dimensional optical circuits consisting of waveguide films and optical Z-connections," *J. Lightwave Technol.* **24**, 4345–4352 (2006).

By combining 3-D optical circuits of Type 1 and 2, SOLNET can be inserted between two 45° mirrors. By connecting two waveguide films with vertical waveguides of SOLNET, light beams may be guided from one 45° mirror to the other with strong confinement, reducing the leakage of the light beams at the *Optical Z-Connection*.

REFERENCES

1. K. D. Singer, M. G. Kuzyk, W. R. Holland, J. E. Sohn, S. J. Lalama, R. B. Comizzoli, H. E. Katz, and M. L. Schilling, "Electro-optic phase modulation and optical second-harmonic generation in corona-poled polymer films," *Appl. Phys. Lett.* **53**, 1800–1802 (1988).
2. B. F. Levine, C. G. Bethea, C. D. Thurmond, R. T. Rynth, and J. L. Bernstein, "An organic crystal with an exceptionally large optical second-harmonic coefficient: 2-methyl-4-nitroaniline," *J. Appl. Phys.* **50**, 2523–2527 (1979).
3. G. F. Lipscomb, A. F. Garito, and R. S. Narang, "An exceptionally large electro-optic effect in the organic solid MNA," *J. Chem. Phys.* **75**, 1509–1516 (1981).
4. T. Yoshimura, "Enhancing second-order nonlinear optical properties by controlling the wave function in one-dimensional conjugated molecules," *Phys. Rev.* **B40**, 6292–6298 (1989).
5. T. Yoshimura, S. Tatsuura, and W. Sotoyama, "Quantum well formation in one-dimensional polymer films by molecular layer deposition and chemical vapor deposition," *Mat. Res. Soc. Symp. Proc.* **247**, 829–834 (1992).
6. T. Yoshimura, "Characterization of the EO effect in styrylpyridinium cyanine dye thin-film crystals by an ac modulation method," *J. Appl. Phys.* **62**, 2028–2032 (1987).
7. G. R. Meredith, "Design and characterization of molecular and polymeric nonlinear optical materials: Successes and pitfalls," in *Nonlinear optical properties of organic and polymeric materials* (Ed. D. J. Williams), 27–56, American Chemical Society, Washington, DC (1983).
8. J. S. Wei and W. D. Westwood, "A new method for determining thin-film refractive index and thickness using guided optical waves," *Appl. Phys. Lett.* **32**, 819–821 (1978).
9. Y. Kubota and T. Yoshimura, "Endothermic reaction aided crystal growth of 2-methyl-4-nitroaniline and their electro-optic properties," *Appl. Phys. Lett.* **53**, 2579–2580 (1988).
10. T. Yoshimura and Y. Kubota, "Pockels effect in organic thin films," *Springer Proceedings in Physics* **36**, Nonlinear Optics of Organics and Semiconductors, Springer-Verlag Berlin, Heidelberg, 222–226 (1989).
11. T. Yoshimura, S. Tatsuura, and W. Sotoyama, "Organic electro-optic materials and their applications," *Oyo Buturi* **61**, 38–42 (1992) [in Japanese].
12. W. Sotoyama, S. Tatsuura, and T. Yoshimura, "Electro-optic side-chain polyimide system with large optical nonlinearity and high thermal stability," *Appl. Phys. Lett.* **64**, 2197–2199 (1994).
13. C. C. Teng and H. T. Man, "Simple reflection technique for measuring the electro-optic coefficient of poled polymers," *Appl. Phys. Lett.* **56**, 1734–1736 (1990).
14. W. Sotoyama, S. Tatsuura, K. Motoyoshi, and T. Yoshimura, "Directional-coupled optical switch between stacked waveguide layers using electro-optic polymer," *Jpn. J. Appl. Phys.* **31**, L1180–L1181 (1992).
15. W. Sotoyama, S. Tatsuura, and T. Yoshimura, "Fabrication techniques for electro-optic polymer waveguides," *The Review of Laser Engineering*, **21**, 1125–1133 (1993) [in Japanese].
16. S. Tatsuura, W. Sotoyama, and T. Yoshimura, "Epoxy-amine polymer waveguide containing nonlinear optical molecules fabricated by chemical vapor deposition," *Appl. Phys. Lett.* **60**, 1158–1160 (1992).
17. S. Tatsuura, W. Sotoyama, and T. Yoshimura, "Electro-optic polymer waveguide fabricated using electric-field-assisted chemical vapor deposition," *Appl. Phys. Lett.* **60**, 1661–1663 (1992).
18. S. Tatsuura, W. Sotoyama, K. Motoyoshi, A. Matsuura, T. Hayano, and T. Yoshimura, "Polyazomethine conjugated polymer film with second order nonlinear optical properties fabricated by electric-field-assisted chemical vapor deposition," *Appl. Phys. Lett.* **62**, 2182–2184 (1993).

19. K. Matsumoto and T. Yoshimura, "Electro-optic waveguides with conjugated polymer films fabricated by the carrier-gas-type organic CVD for chip-scale optical interconnects," *Proc. SPIE* **6899**, from Photonics West 2008, San Jose, California, 98990E-1-9 (2008).

20. H. Rong, R. Jones, A. Liu, O. Cohen, Y-H. Kuo, and M. Paniccia, "Silicon-based laser, amplifier, and wavelength converter for optoelectronic integration," *Proc. SPIE* **6125**, 612505-1-13 (2006).

21. R. M. Kubacki, "Micro-optic enhancement and fabrication through variable in-plane index of refraction (VIPIR) engineered silicon nanocomposite technology," *Proc. SPIE* **5347**, 233–246 (2004).

22. S. Ono, T. Yoshimura, T. Sato, and J. Oshima, "Fabrication and evaluation of nano-scale optical circuits using sol-gel materials with photo-induced refractive index variation characteristics," *J. Lightwave Technol.* **27**, 1229–1235 (2009).

23. A. Bhatnagar, S. Latif, C. Debaes, and D. A. B. Miller, "Pump-probe measurements of CMOS detector rise time in the blue," *J. Lightwave Technol.* **22**, 2213–2217 (2004).

24. T. Yoshimura, T. Inoguchi, T. Yamamoto, S. Moriya, Y. Teramoto, Y. Arai, T. Namiki, and K. Asama, "Self-organized lightwave network based on waveguide films for three-dimensional optical wiring within boxes," *J. Lightwave Technol.* **22**, 2091–2100 (2004).

25. T. Yoshimura, M. Miyazaki, Y. Miyamoto, N. Shimoda, A. Hori, and K. Asama, "Three-dimensional optical circuits consisting of waveguide films and optical Z-connections," *J. Lightwave Technol.* **24**, 4345–4352 (2006).

26. E. Cassan, S. Laval, S. Lardenois, and A. Koster, "On-chip optical interconnects with compact and low-loss light distribution in silicon-on-insulator rib waveguides," *IEEE J. Select. Topics Quantum Electron.* **9**, 460–464 (2003).

27. T. Yoshimura, J. Roman, Y. Takahashi, W. V. Wang, M. Inao, T. Ishitsuka, K. Tsukamoto, K. Motoyoshi, and W. Sotoyama, "Self-organizing waveguide coupling method 'SOLNET' and its application to film optical circuit substrates," *Proc. 50th Electronic Components & Technology Conference (ECTC)*, Las Vegas, Nevada, 962–969 (2000).

28. T. Yoshimura, J. Roman, Y. Takahashi, W. V. Wang, M. Inao, T. Ishitsuka, K. Tsukamoto, K. Motoyoshi, and W. Sotoyama, "Self-organizing lightwave network (SOLNET) and its application to film optical circuit substrates," *IEEE Trans. Comp., Packag. Technol.* **24**, 500–509 (2001).

29. T. Yoshimura, W. Sotoyama, K. Motoyoshi, T. Ishitsuka, K. Tsukamoto, S. Tatsuura, H. Soda, and T. Yamamoto, Method of producing optical waveguide system, optical device and optical coupler employing the same, optical network and optical circuit board, U.S. Patent 6,081,632 (2000).

30. T. Yoshimura, M. Ojima, Y. Arai, and K. Asama, "Three-dimensional self-organized micro optoelectronic systems for board-level reconfigurable optical interconnects: performance modeling and simulation," *IEEE J. Select. Topics in Quantum Electron.* **9**, 492–511 (2003).

31. T. Yoshimura, A. Hori, Y. Yoshida, Y. Arai, H. Kurokawa, T. Namiki, and K. Asama, "Coupling efficiencies in reflective self-organized lightwave network (R-SOLNET) simulated by the beam propagation method," *IEEE Photon. Technol. Lett.* **17**, 1653–1655 (2005).

32. F. Huang, H. Takase, Y. Eriyama, and T. Ukachi, "Optical fiber interconnection by using self-written waveguides," *Proc. 9th Int. Conf. Rad. Tech. Asia*, Japan, 637–639 (2003).

33. T. Yoshimura and H. Kaburagi, "Self-organization of optical waveguides between misaligned devices induced by write-beam reflection," *Appl. Phys. Express* **1**, 062007-1-3 (2008).

34. H. Kaburagi and T. Yoshimura, "Simulation of micro/nano-scale self-organized light-wave network (SOLNET) using the finite difference time domain method," *Opt. Commun.* **281**, 4019–4022 (2008).

35. T. Yoshimura, J. Roman, Y. Takahashi, M. Lee, B. Chou, S. Beilin, W. Wang, and M. Inao, "Optoelectronic scalable substrates based on film/Z-connection and its application to film optical link module (FOLM)," *Proc. SPIE* **3952**, from Photonics West 2000, San Jose, California, 202–213 (2000).

36. T. Yoshimura, K. Kumai, T. Mikawa, O. Ibaragi, and M. Bonkohara, "Photolithographic packaging with selectively occupied repeated transfer (PL-Pack with SORT) for scalable optical link multi-chip-module (S-FOLM)," *IEEE Trans. Electron. Packag. Manufact.* **25**, 19–25 (2002).

37. N. M. Jokerst, M. A. Brooke, S. Cho, S. Wilkinson, M. Vrazel, S. Fike, J. Tabler, Y. J. Joo, S. Seo, D. S. Wils, and A. Brown, "The heterogeneous integration of optical interconnections into integrated microsystems," *IEEE J. Select. Topics in Quantum Electron.* **9**, 350–360 (2003).

38. E. Yablonovitch, T. Gmitter, J.P. Harbison, and R. Bhat, "Extreme sensitivity in the lift-off of epitaxial GaAs films," *Appl. Phys. Lett.* **51**, 2222–2224 (1987).

39. C. Stirk, Q. Liu, and M. Ball, "Manufacturing cost analysis of integrated photonic packages," *Proc. SPIE* **3631**, 224–233 (1999).

40. T. Yoshimura, D. Tamaki, S. Kawamura, Y. Arai, and K. Asama, "A device transfer process "Selectively occupied repeated transfer (SORT)" for resource-saving integration of polymer micro optics fabricated by "built-in mask" method," *IEEE Trans. Comp., Packag. Technol.* **27**, 468–471 (2004).

41. T. Yoshimura, T. Kofudo, T. Kashiwazako, K. Naito, K. Ogushi, and Y. Kitabayashi, "A material-saving optical waveguide fabrication process with selective transfers of cores," *Opt. Eng.* **47**, 014601- 1-7 (2008).

42. T. Inoguchi, S. Moriya, T. Yamamoto, Y. Arai, T. Yoshimura, and K. Asama, "Film waveguides with micro mirrors for three-dimensional optical circuits fabricated by the built-in mask method," *J. Electronics Jisso* **8**, 237–242, 2005 [in Japanese].

8 Applications to Optical Interconnects and Optical Switching Systems

Chapters 1 through 7 covered artificial materials with atomic/molecular-level tailored structures including polymer wires and molecular wires, and fabrication processes for the artificial materials including molecular layer deposition (MLD). Related thin-film organic photonic/electronic materials and devices, as well as integration processes were also described.

In the present chapter, possible applications of these materials and processes to OE systems such as optical interconnects and optical switching systems are presented, and the impact of the polymer multiple quantum dots (MQDs) built by MLD are discussed.

8.1 3-D OPTOELECTRONIC (OE) PLATFORM BASED ON SCALABLE FILM OPTICAL LINK MODULE (S-FOLM)

The three-dimensional (3-D) optoelectronic (OE) platform is a concept involves thin-film photonic devices embedded in films that are connected by optical waveguides three-dimensionally to provide various kinds of functions including optical wiring, optical switching, and so on.

The 3-D OE platform is based on scalable film optical link modules (S-FOLM) [1–6]. Figure 8.1 shows the concept of S-FOLM. First, a set of thin-film-device-embedded OE films are prepared. The thin-film devices include optical waveguides, vertical cavity surface-emitting lasers (VCSELs), photodetectors (PDs), light modulators, optical switches, wavelength filters, photovoltaic devices, driver/amplifier integrated circuits (ICs), large-scale integrated circuits (LSIs), lenses, mirrors, and so on. By combining the OE films with stacked configurations, various kinds of OE systems are constructed.

By stacking a VCSEL/PD-embedded OE film on an LSI, an OE LSI will be available. When an OE film containing a VCSEL/PD/optical waveguide and an OE film containing IC are stacked, an optical interconnect board is created. If LSI-embedded OE films are added in the stack, a 3-D OE LSI is created. If optical-switch-embedded OE films are used in the stack, a three-dimensional micro-optical switching system (3D-MOSS) will be available [7,8]. When photovoltaic devices are embedded in OE films, solar energy conversion systems will be provided [9–13]. Such scalability

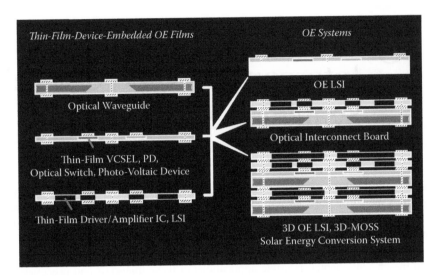

FIGURE 8.1   The S-FOLM concept. From T. Yoshimura, J. Roman, Y. Takahashi, M. Lee, B. Chou, S. Beilin, W. Wang, and M. Inao, "Optoelectronic scalable substrates based on film/Z-connection and its application to film optical link module (FOLM)," *Proc. SPIE* **3952**, 202–213 (2000).

enables S-FOLM to be used for the 3-D OE platform, which contributes to system cost reduction.

8.2   OPTICAL INTERCONNECTS WITHIN BOXES

Within boxes of computers, increases in clock frequency and wiring density cause problems such as RC delay, signal attenuation, power dissipation, and cross-talk. One of the solutions for these problems is the 3-D architecture, which can reduce wiring distances. Another solution is the optical interconnect, which can reduce RC delay, electromagnetic induction and interference, and frequency-dependent signal loss. Due to principled advantages of optical interconnects over electrical interconnects [14], it is expected that optics will be implemented within boxes [15–22].

Optics were first implemented into long-haul communication systems. It has recently been implemented in shorter-distance systems, such as local area network (LAN) and *box-to-box* data transmission. For *within-box* data transmission, the connection channel count is several hundreds to thousands. It is larger than that in conventional OE modules for box-to-box interconnects by one to two orders of magnitude. This means that electrical-to-optical (E-O) and optical-to-electrical (O-E) conversion devices of several hundreds to thousands should be distributed on a board. To achieve such high density, the size and cost of the conversion device should be reduced by one to two orders of magnitude.

We have proposed a low-cost solution for these problems using the 3-D OE platform, where the 3-D architectures and optical interconnects are implemented [1–6].

8.2.1 MULTILAYER OE BOARDS AND 3-D STACKED OE MULTI-CHIP MODULES

Figure 8.2 shows a multilayer OE board and a 3-D stacked OE multi-chip-module (MCM) based on the 3-D OE platform. In the multilayer OE board [1–4], a plurality of optical waveguide films with 45° reflectors are stacked. Thin-film active devices such as light modulators and PDs are embedded for E-O and O-E conversions. In the example shown in Figure 8.2, E-O and O-E conversions are performed in the second layer from the top, where thin-film active devices are embedded. The third and fourth layers are used for multilayer optical signal routing, which avoids wiring interference in high-density interconnect systems.

The first layer is an interface film, where thin-film ICs are embedded. On the interface film, LSIs are mounted. The thin-film ICs in the interface films may perform the functions of MUX/DMUX, error correction, and drivers/amplifiers, providing standardized interface capability.

When we use OE amplifier/driver-less substrate (OE-ADLES) (described in Section 8.2.2), where E-O and O-E conversions are respectively carried out by light modulators that are directly driven by LSI output and by PDs that directly generate LSI input, the interface film is not necessary. OE-ADLES is expected to have *low power dissipation* and *high data rate* capabilities [23–25].

In 3-D stacked OE MCM [1–8], light modulator/PD-embedded waveguide films are stacked together with films containing thin-film LSIs. The structure can be regarded as 3-D stacked OE LSI when the stacked thin-film LSIs are divided chips

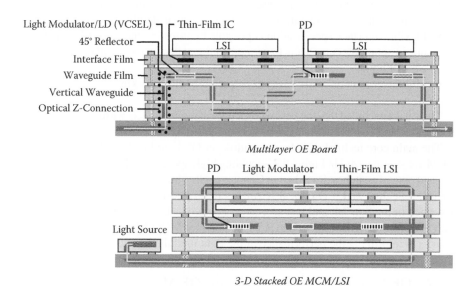

Multilayer OE Board

3-D Stacked OE MCM/LSI

FIGURE 8.2 Schematic illustrations of a multilayer OE board and a 3-D stacked OE MCM/LSI. From T. Yoshimura, T. Inoguchi, T. Yamamoto, S. Moriya, Y. Teramoto, Y. Arai, T. Namiki, and K. Asama, "Self-organized lightwave network based on waveguide films for three-dimensional optical wiring within boxes," *J. Lightwave Technol.* **22**, 2091–2100 (2004).

corresponding to blocks of an LSI chip. This structure is the goal of the 3-D OE platform for computers because it realizes 3-D wiring and optical wiring simultaneously. In the 3-D stacked OE MCM, heat generation is a serious problem. So implementation of OE-ADLES that uses high power external light sources might be mandatory for power dissipation reduction within the 3-D OE platform. By setting light sources with a plurarity of wavelengths, wavelength division multiplexing (WDM) can be performed in the interconnect systems.

Optical Z-connections in the 3-D OE platform are used to achieve optical links between the stacked waveguide films, constructing 3-D optical wiring. The optical Z-connections consist of two 45° reflectors built in the waveguide films. For long-distance connections, vertical waveguides are inserted between the 45° reflectors.

The optical interconnects based on the 3-D OE platform consisting of S-FOLM have the following advantages:

- Excess module spaces for E-O and O-E conversions are not necessary, resulting in size reduction.
- Long electrical lines for connection between LSI pads and E-O and O-E conversion sites are replaced with short vertical electrical lines, resulting in latency and bit error rate reduction.
- Material consumption for the devices can be saved by PL-Pack with SORT, resulting in cost reduction.
- Alignment between optical waveguides and thin-film devices can be performed by photolithography, resulting in low assembly cost.
- Relative positions between optical waveguides and thin-film device flakes are insensitive to temperature, resulting in thermal stability.
- Flexible interface specification is possible by inserting interface films, permitting computer engineers to use conventional design tools without knowledge of optics.
- Multilayer structures are easily constructed by film stacking.

Thus, it is concluded that the key issue for optical interconnects within boxes, that is, drastic size and cost reduction, can be accomplished by S-FOLM implementation.

The main core technologies for the multilayer OE board and the 3-D stacked OE MCM are summarized in Figure 8.3, including high-speed, small-sized light modulators and optical switches described in Chapter 6, OE films of optical waveguides with vertical reflectors and 3-D optical wiring as described in Section 7.6, SOLNET as described in Section 7.4, and resource-saving heterogeneous integration by PL-Pack with SORT as described in Section 7.5. The polymer MQD fabricated by MLD contributes to improving the performance of the light modulators and optical switches.

8.2.2 OE Amplifier/Driver-Less Substrate (OE-ADLES)

To reduce power dissipation and increase the data rate, we proposed OE-ADLES [23–25]. Figure 8.4 shows OE-ADLES for the OE MCM. E-O and O-E conversions are respectively carried out by light modulators and PDs, which are completely embedded in OE films. LSI I/O pads are connected to electrodes of the light

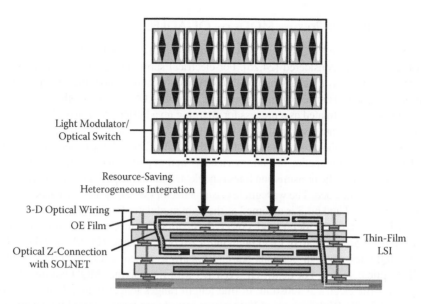

FIGURE 8.3   Main core technologies for the multilayer OE board and the 3-D stacked OE MCM.

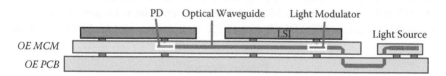

FIGURE 8.4   OE-ADLES for OE MCM. From T. Yoshimura, Y. Suzuki, N. Shimoda, T. Kofudo, K. Okada, Y. Arai, and K. Asama, "Three-dimensional chip-scale optical interconnects and switches with self-organized wiring based on device-embedded waveguide films and molecular nanotechnologies," *Proc. SPIE* **6126**, 612609-1-15 (2006).

modulators and PDs. Strong light beams are introduced from outside into the OE MCM via an OE printed circuit board (PCB). The external light beams are directly modulated by LSI output through the light modulators without drivers. The generated optical signals are transmitted through optical waveguides to the PDs, where the optical signals are converted back to electrical signals for LSI input without amplifiers. OE-ADLES is expected to have low power dissipation and high data rate capabilities for the following reasons:

- Light sources are placed outside, reducing power dissipation within the interconnect systems.
- Light modulators, which are voltage-driven devices, are used for transmitters, reducing driving power and rising/falling time for optical signal generation.
- Strong light is available from external sources (high-power LD, mode-locked LD), achieving fast response of PDs.

8.2.3 IMPACT OF POLYMER MQDs ON OE-ADLES

In order to estimate RC delay and power dissipation in OE-ADLES, the models shown in Figure 8.5 are considered [24,25]. For electrical substrates, data transmissions are treated as a process for charging up a buffer driver, a transmission line, and a receiver through a gate resistance of the buffer driver. For OE-ADLES, on the other hand, data transmissions are treated as a process for charging up a buffer driver and a light modulator through the gate resistance of the buffer driver, and a process for charging up a PD and a receiver by photocurrents generated in the PD. In Table 8.1, the formula for the RC delay and power dissipation are summarized as well as parameters for the calculations. Attenuation along the transmission lines is neglected. External power dissipation at light sources is not included. The waveguide prism deflector (WPD) optical switch described in Sections 6.1.2 and 6.1.3 is assumed as the light modulator for the transmitter.

In Figure 8.6, the delay and power dissipation of an electrical substrate and OE-ADLES are compared. Lead lanthanum zirconate titanate (PLZT) and the polymer MQD described in Sections 5.5 and 5.7 are assumed as EO materials for the WPD optical switch. For delay, OE-ADLES with the polymer MQD has an advantage over the electrical substrate in line length ranges longer than 1.2 cm for light power of 3 mW. With increasing light power introduced from external light sources, the response time becomes shorter, moving the break-even point toward shorter line length ranges of mm. For power dissipation, drastic reduction can be seen in OE-ADLES with the polymer MQD. The break-even point is located at line length of sub mm ranges. In OE-ADLES with PLZT, comparing with OE-ADLES with the

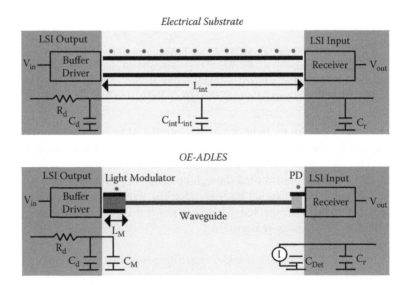

FIGURE 8.5 Models for estimation of RC delay and power dissipation in electrical interconnects and optical interconnects (OE-ADLES). From T. Yoshimura, Y. Suzuki, N. Shimoda, T. Kofudo, K. Okada, Y. Arai, and K. Asama, "Three-dimensional chip-scale optical interconnects and switches with self-organized wiring based on device-embedded waveguide films and molecular nanotechnologies," *Proc. SPIE* **6126**, 612609-1-15 (2006).

TABLE 8.1
Formula and Parameters for Calculations of RC Delay and Power Dissipation

Electrical Substrate	RC Delay	$\tau = R_d (C_{int} L_{int} + C_d + C_r)$
	Power dissipation	$P = f (C_{int} L_{int} + C_d + C_r) V_{out}^2$ $C_{int} = \varepsilon_0 \varepsilon_i$ w/d
OE-ADLES	RC Delay	$\tau = R_d (C_M + C_d) + V_{out} (C_{Det} + C_r)/(\eta_c P_{Det}) + \tau_p$
	Power dissipation	$P = f [(C_M + C_d) V_{in}^2 + (C_{Det} + C_r) V_{out}^2]$
		$C_M = \varepsilon_0 \varepsilon_i A_M / d$

| | | | OE-ADLES | |
| | | | | |
		Electrical Substrate	**PLZT**	**Conjugated Polymer (Calc.)**
dielectric constant	ε_i	3	300	4
refractive index	n		2.48	2
EO coefficient	R, r		$R = 5.7 \times 10^{-16}$ m^2/V^2	$r = 1.35 \times 10^{-9}$ m/V

f	Data rate	d	Gap between electrodes: 1 μm
V_{in}	Power supply voltage: 3 V	A_M	Light modulator area: 16 × 16 μm^2
V_{out}	Received detector voltage: 3 V	C_M	Light modulator capacitance
C_{int}	Transmission line capacitance per cm	C_{Det}	Photodetector capacitance: 0.01 pF
R_d	Driver gate resistance: 50 Ω	η_c	Responsibility of photodetector: 0.8 A/W
C_d	Driver capacitance: 0.02 pF	τ_p	Photodetector internal propagation delay: 25 ps
C_r	Receiver capacitance: 0.02 pF	P_{Det}	Incident power to photodetector
L_{int}	Transmission line length	λ	Wavelength: 1.3 μm
w	Electrode width: 4 μm		

Source: T. Yoshimura, Y. Suzuki, N. Shimoda, T. Kofudo, K. Okada, Y. Arai, and K. Asama, "Three-dimensional chip-scale optical interconnects and switches with self-organized wiring based on device-embedded waveguide films and molecular nanotechnologies," *Proc. SPIE* **6126**, 612609-1-15 (2006).

polymer MQD, the break-even point is located at longer line length ranges for both delay and power dissipation.

The superiority of optics is attributed to the capacitance reduction, that is, in the electrical substrate a large capacitance of the electrical transmission line exists, while in OE-ADLES, only small capacitances exist in the light modulators and PDs. As mentioned above, the polymer MQD exhibits faster responses and much lower power consumption compared with PLZT. The superiority of the polymer MQD comes from its small dielectric constant.

The key issue in OE-ADLES is availability of EO materials with a large EO coefficient and a small dielectric constant like the polymer MQD fabricated by the molecular layer deposition (MLD). Size reduction of the light modulators and PDs is another important issue. When they are miniaturized by one order of magnitude for each dimension, that is, the device thickness is reduced by one order of magnitude

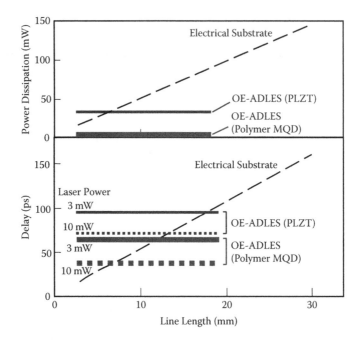

FIGURE 8.6 Comparisons of RC delay and power dissipation between electrical inter-connects and optical interconnects (OE-ADSLES). From T. Yoshimura, Y. Suzuki, N. Shimoda, T. Kofudo, K. Okada, Y. Arai, and K. Asama, "Three-dimensional chip-scale optical interconnects and switches with self-organized wiring based on device-embedded waveguide films and molecular nanotechnologies," *Proc. SPIE* **6126**, 612609-1-15 (2006).

and the device area by two orders of magnitude, their capacitances as well as the internal propagation delay in PDs will be reduced by one order of magnitude in principle. This will move the break-even points to sub mm ranges.

8.3   3-D MICRO OPTICAL SWITCHING SYSTEM (3D-MOSS)

The rapid increase in data rate and volume in networking and processing systems is promoting progress of massive optical switching [26]. This trend results in a key issue—miniaturization of high-speed massive optical switching systems. The miniaturization is useful for high-speed packet switching in optical communication systems and board-level reconfigurable optical interconnects.

In order to provide a solution for the issue, we proposed the three-dimensional micro optical switching system (3D-MOSS) based on S-FOLM, where OE films with embedded optical switches are stacked. 3D-MOSS will overcome speed limits of ~ms in microelectromechanical systems (MEMS) and thermo-optics, to provide high-speed massive switching with large-scale integration of optical switches consisting of EO waveguides.

In the present section, 3D-MOSS structure is designed and the possible size and performance are theoretically predicted [7,27–29] by the beam propagation method (BPM) and the finite difference time domain (FDTD) method. The impact of polymer MQD fabricated by MLD is also discussed.

8.3.1 THE 3D-MOSS CONCEPT

An example of 3D-MOSS configuration is shown in Figure 8.7(a) for a case involving a 32 × 32 Banyan network. It consists of five optical switch arrays, in which sixteen 2 × 2 optical switches are arranged in a 3-D optical network. The optical switch consists of EO waveguides that exhibit high-speed switching characteristics. The

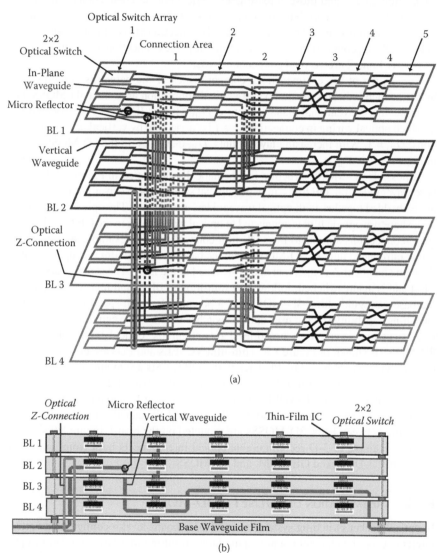

FIGURE 8.7 (a) An example of 3D-MOSS configuration for 32 × 32 switching. (b) A possible 3D-MOSS structure consisting of four layers. From T. Yoshimura, M. Ojima, Y. Arai, and K. Asama, "Three-dimensional self-organized micro optoelectronic systems for board-level reconfigurable optical interconnects: performance modeling and simulation," *IEEE J. Select. Topics in Quantum Electron.* **9**, 492–511 (2003).

optical network is divided into four blocks—BL 1, 2, 3, and 4, which are stacked to make a four-layer structure. Between optical switch arrays, a *connection area* exists. Adjacent optical switch arrays i and $i + 1$ are connected by optical wiring in connection area i. Connection areas consisting of in-plane optical connections with in-plane waveguides and corner-turning micro reflectors are identified as "in-plane connection areas," and those consisting of vertical optical connections with optical Z-connections, which are built from pairs of 45° reflectors and vertical waveguides for interplane links, are identified as "vertical connection areas." In the model shown in Figure 8.7(a), connection areas 3, and 5 are in-plane connection areas, and connection areas 1 and 2 are vertical connection areas.

The degree of three-dimensional characteristics of the networks is represented by D_{3D}, which indicates the number of vertical connection areas in a 3D-MOSS. For example, in a conventional planar structure, $D_{3D} = 0$. For 3D-MOSS with two layers, in which optical wiring in connection area 1 is carried out by vertical optical connections, $D_{3D} = 1$. With increasing layer counts, D_{3D} increases. In the 3D-MOSS shown in Figure 8.7(a), $D_{3D} = 2$.

A possible 3D-MOSS structure is schematically illustrated in Figure 8.7(b). On a base waveguide film, OE films, in which optical switches (and driver ICs if necessary) are embedded together with optical waveguides, are stacked. The optical waveguides in the base waveguide film is connected to optical waveguides in the OE films by optical Z-connections similar to inter-OE-film connections. For example, an input light signal introduced into the 3D-MOSS from the base waveguide film is coupled to BL 2 through an optical Z-connection and is switched in the layer. The signal is coupled to BL 4 and is again switched, followed by transition to BL 3, and finally is output into the base waveguide film. 3D-MOSS has a potential for high-speed massive optical switching capability since it uses optical switches with EO waveguides.

A structural example of board-level reconfigurable optical interconnects with 3D-MOSS is shown in Figure 8.8. Thin-film VCSELs and PDs are embedded in an optical interconnect board by PL-Pack with SORT. LSIs are mounted on the board. For simplicity, interface films that should be inserted between the board and the LSIs are not drawn. When OE-ADLES is adopted, LSIs can be mounted on the board without the interface films. Instead of LSIs, 3D OE LSIs can also be used. For route switching between LSIs, 3D-MOSSs are mounted on the board. Figure 8.9 shows a possible example of 3-D reconfigurable optical interconnects built from multilayer OE boards, 3-D stacked OE MCM, and 3D-MOSSs.

8.3.2 IMPLEMENTATION OF SOLNET IN 3D-MOSSs

In optical packet switching systems, as well as in board-level reconfigurable optical interconnects, the channel count is quite large, resulting in a large number of optical coupling points. This forces us to align a lot of photonic devices precisely with enormous effort. Furthermore, in 3D-MOSS, vertical waveguides are necessary for interlayer connections. In order to achieve drastic cost reduction of the OE system, some technology to construct a 3-D optical network with a self-organized manner is required. One possible candidate is the self-organized 3-D optical circuits with SOLNET described in Section 7.4.

Figure 8.10 shows a concept for SOLNET implementation in 3-D optical circuits in S-FOLM. First, OE devices are arranged in a designed configuration in optical waveguide films to make OE films. The OE films are stacked to construct a 3-D structure. Next, by introducing some excitation to the 3-D structure from outside, self-aligned 3-D optical circuits of SOLNET are automatically constructed, including coupling optical paths between devices and optical Z-connections.

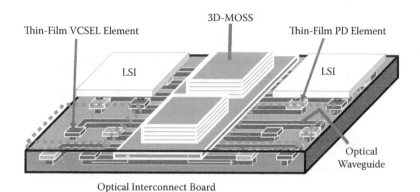

FIGURE 8.8   Structural example of board-level reconfigurable optical interconnects with 3D-MOSS. From T. Yoshimura, M. Ojima, Y. Arai, and K. Asama, "Three-dimensional self-organized micro optoelectronic systems for board-level reconfigurable optical interconnects: performance modeling and simulation," *IEEE J. Select. Topics in Quantum Electron.* **9**, 492–511 (2003).

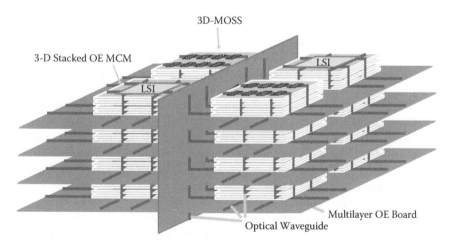

FIGURE 8.9   An example of 3-D reconfigurable optical interconnects built from multilayer OE boards, 3-D stacked OE MCMs, and 3D-MOSSs. From T. Yoshimura, Y. Suzuki, N. Shimoda, T. Kofudo, K. Okada, Y. Arai, and K. Asama, "Three-dimensional chip-scale optical interconnects and switches with self-organized wiring based on device-embedded waveguide films and molecular nanotechnologies," *Proc. SPIE* **6126**, 612609-1-15 (2006).

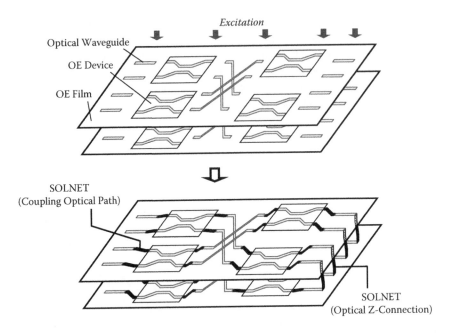

FIGURE 8.10 Concept of SOLNET implementation in 3-D optical circuits in S-FOLM. From T. Yoshimura, M. Ojima, Y. Arai, and K. Asama, "Three-dimensional self-organized micro optoelectronic systems for board-level reconfigurable optical interconnects: performance modeling and simulation," *IEEE J. Select. Topics in Quantum Electron.* **9**, 492–511 (2003).

In self-organized 3-D optical circuits, even if some misalignment of the optical waveguides and the stacked films exist, self-organized optical couplings can be achieved, which will reduce fabrication cost, especially the cost for assembly and optical Z-connection formation.

For SOLNET implementation in S-FOLM, write beam insertion is a serious concern. Because 3D-MOSS has a large number of connection parts and many of them are terminated by two optical switches, it is impossible to introduce write beams from outside. The problem can be resolved, as shown in Figure 8.11, by R-SOLNET and phosphor SOLNET, as described in Section 7.4.1. First, OE films are made by transferring optical switches onto final substrates using SORT. Before the transfer, wavelength filters, which can selectively reflect write beams of wavelengths typically between 0.35 µm and 0.7 µm and pass signal beams of 0.5–1.5 µm, are deposited just on core surfaces at the edge of optical switches using atomic layer deposition (ALD) and MLD with the assistance of photolithography.

Next, as Figure 8.11 shows, a preform is constructed by stacking the OE films. For write beam generation, phosphor is partially doped in optical waveguides. Photoinduced refractive index (PRI) materials are inserted in parts where optical wiring is necessary. Then, by illuminating the phosphor with UV light from outside the 3D-MOSS, write beams generated in the phosphor are introduced into the PRI materials through optical waveguides to make SOLNET. For vertical waveguide construction, the write beams are introduced into the PRI materials from upper and lower

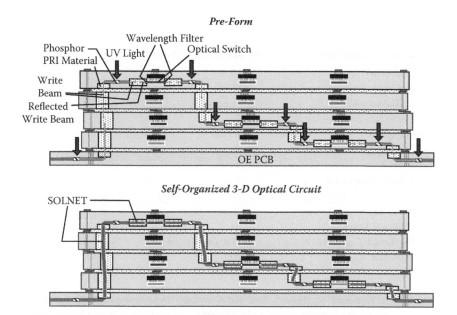

FIGURE 8.11   Fabrication process of self-organized 3-D optical circuits. From T. Yoshimura, M. Ojima, Y. Arai, and K. Asama, "Three-dimensional self-organized micro optoelectronic systems for board-level reconfigurable optical interconnects: performance modeling and simulation," *IEEE J. Select. Topics in Quantum Electron.* **9**, 492–511 (2003).

optical waveguides through 45° reflectors to form SOLNET. For connections between an in-plane waveguide and a waveguide within an optical switch, a write beam introduced into the PRI material from the in-plane waveguide is reflected by a wavelength filter on the waveguide core edge of the optical switch. The two write beams are attracted each other to construct a SOLNET. UV light from outside corresponds to the excitation shown in Figure 8.10. It should be noted that the same method is effective for VCSEL-waveguide and PD-waveguide couplings in optical interconnect boards by depositing the wavelength filters on the active regions of VCSELs and PDs.

8.3.3   STRUCTURAL MODEL OF 3D-MOSS

In Table 8.2, parameters for the 1024 × 1024 3D-MOSS model are summarized. The 3-D optical circuit backbone consists of optical waveguides with a core cross-section of 4 × 4 μm^2, core refractive index n_{Core} of 1.52, and clad refractive index n_{Clad} of 1.50.

Figure 8.12 shows top-view illustrations of Banyan network models for a conventional planar structure and a 3D-MOSS structure. Channel count is N_C, meaning $N_C \times N_C$ switching is considered. In the present model, N_C is 1024. The network consists of 2 × 2 optical switch arrays, which are connected by optical waveguides in connection areas. The optical switch is the WPD optical switch shown in Figure 6.19. Although the PLZT thickness is 400 nm in Figure 6.19, in the present calculation the PLZT thickness is assumed to be 4 μm in order to match the PLZT thickness to

TABLE 8.2

Parameters for the 1024 × 1024 3D-MOSS Model

Channel Count: N_C	1024
Layer Count: N_L	16
Layer Thickness/Gap: T_L/T_G	46 µm/4 µm
Degree of 3D Characteristics: D_{3D}	4
Wavelength: λ	1.3 µm
Optical Waveguide	
Width: w	4 µm
Refractive Index Clad: n_{Clad}	1.50
Core: n_{Core}	1.52
Pitch: P_{WG}	20 µm
2 × 2 Optical Switch	
Size: Length L_{SW}, Width: W_{SW}	1190 µm, 100 µm
Step Count: S_{SW}	10
Pitch in a Array: P_{SW}	200 µm

Source: T. Yoshimura, M. Ojima, Y. Arai, and K. Asama, "Three-dimensional self-organized micro optoelectronic systems for board-level reconfigurable optical interconnects: performance modeling and simulation," *IEEE J. Select. Topics in Quantum Electron.* **9**, 492–511 (2003).

core thickness of the 3-D optical circuit backbone. In this case, switching voltage is 12 V. One array contains $N_C/2$ 2 × 2 optical switches with length L_{SW} of 1190 µm and width W_{SW} of 100 µm. The length of the array equals the individual switch length L_{SW}. The optical switch array count is S_{SW}, which is 10 in the present model.

In 3D-MOSS, the network is divided into N_L blocks, which are stacked to make an N_L-layer 3-D structure. The general expression for 3-D characteristics of networks, D_{3D}, is given as follows.

$$D_{3D} = \log_2 N_L \tag{8.1}$$

Optical wiring is carried out by vertical optical connections in connection areas 1 to D_{3D}, and in-plane optical connections in connection areas $D_{3D} + 1$ to $S_{SW} - 1$.

It is found from Figure 8.12 that the area of the network is drastically reduced in 3D-MOSS, when compared with the conventional planar structure. The sizes inserted in Figure 8.12, as well as in Figures 8.13 through 8.15, are calculated using the general Equations (8.2)–(8.6) described later.

Figure 8.13 shows a side-view illustration of 3D-MOSS. Since (8.1) gives $D_{3D} = 4$ for the present model with $N_L = 16$, vertical optical connections are in connection areas 1–4, and in-plane optical connections in connection areas 5–9. The region that includes connection areas 5–9 corresponds to the 64 × 64 Banyan network with a conventional planar structure shown in Figure 8.14. T_L and T_G are the layer thickness and gap thickness, respectively. In the present model, T_L is 46 µm and T_G is 4 µm. Thickness per one layer is 50 µm.

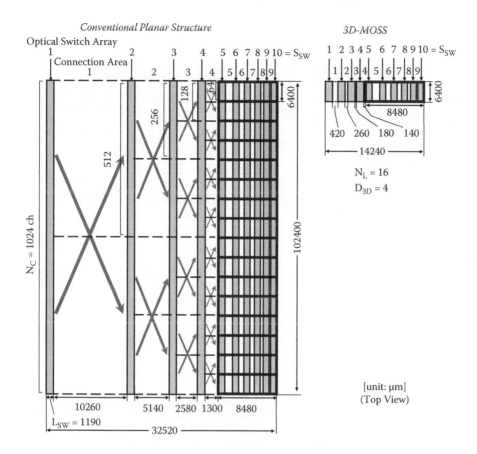

FIGURE 8.12 Top-view illustrations of Banyan network models for a conventional planar structure and a 3D-MOSS structure. From T. Yoshimura, M. Ojima, Y. Arai, and K. Asama, "Three-dimensional self-organized micro optoelectronic systems for board-level reconfigurable optical interconnects: performance modeling and simulation," *IEEE J. Select. Topics in Quantum Electron.* **9**, 492–511 (2003).

As an example of in-plane connection area, detail of connection area 5 is shown in Figure 8.15(a). Connections between optical switch arrays are constructed by crank-shaped waveguides consisting of a pair of corner-turning micro reflectors. P_{SW} and P_{WG} are the pitch of the optical switches and the pitch of the optical waveguides, respectively. Interswitch connection line count is expressed by $N_c/2^5$ for connection area 5. Then, by adding P_{WG} for buffer spaces, the length of connection area 5 is given by $(N_c/2^5 + 1) P_{WG}$. The width is given by $((N_c/2) / 2D_{3D})P_{SW}$. Thus, the expressions for connection area length and width of in-plane connection area i can be generalized as follows.

$$L(IP)_i = (N_C/2^i + 1)P_{WG} \tag{8.2}$$

$$W(IP) = ((N_C/2)/2^{D_{3D}})P_{SW} \tag{8.3}$$

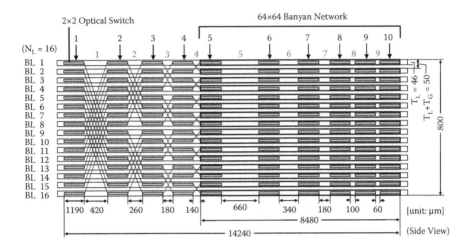

FIGURE 8.13 Side view illustration of Banyan network model for a 3D-MOSS structure with layer count N_L = 16. From T. Yoshimura, M. Ojima, Y. Arai, and K. Asama, "Three-dimensional self-organized micro optoelectronic systems for board-level reconfigurable optical interconnects: performance modeling and simulation," *IEEE J. Select. Topics in Quantum Electron.* **9**, 492–511 (2003).

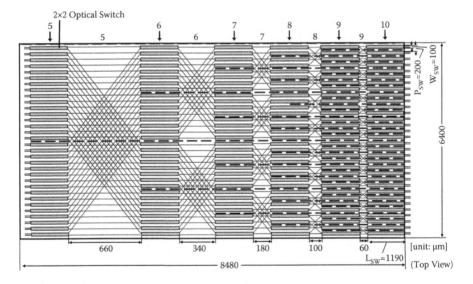

FIGURE 8.14 Top view illustration of 64 × 64 Banyan network with in-pane optical connections. From T. Yoshimura, M. Ojima, Y. Arai, and K. Asama, "Three-dimensional self-organized micro optoelectronic systems for board-level reconfigurable optical interconnects: performance modeling and simulation," *IEEE J. Select. Topics in Quantum Electron.* **9**, 492–511 (2003).

In the present model, $P_{SW} = 200$ μm and $P_{WG} = 20$ μm. P_{WG} of 20 μm is wide enough for parallel wiring with a length of 1 cm to keep cross-talk below −30 dB for the assumed 4-μm-wide waveguides and 1.3-μm wavelength. By substituting the parameters into Equations (8.2) and (8.3), the size of connection area 5 can be calculated to be 660-μm long and 6400-μm wide. Similarly, the sizes for $i = 6$~9 can be calculated as shown in Figures 8.13 and 8.14.

As an example of a vertical connection area, a detailed side view of connection area 1 and a top view of a part of the connection area in BL 12 are shown in Figure 8.15(b). The interswitch connection line count in a cross-section is expressed as $N_L/2^{i-1}$ for connection area 1. Then, adding buffer spaces of $3P_{WG}$ for corner-turning micro reflectors at the front and tail ends of the connection area, the length of connection area 1 is calculated to be $(N_L/2^{i-1} + 5)P_{WG}$. The width is the same as in the in-plane optical connection case. The height is $N_L(T_L + G_L)$. Thus, the expressions for connection area length $L\ (V)$, width $W\ (V)$, and height $H\ (V)$ of vertical connection area i can be generalized as follows.

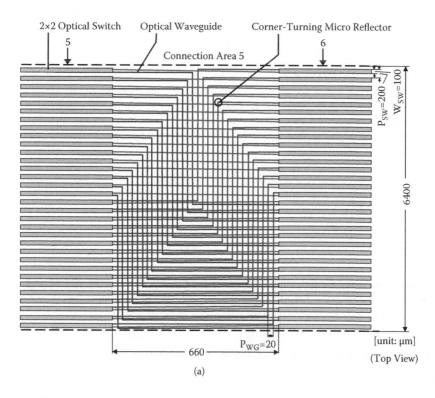

(a)

FIGURE 8.15 Details of (a) top view of connection area 5 and (b) side view of connection area 1 and top view of a part of the connection area in BL 12. From T. Yoshimura, M. Ojima, Y. Arai, and K. Asama, "Three-dimensional self-organized micro optoelectronic systems for board-level reconfigurable optical interconnects: performance modeling and simulation," *IEEE J. Select. Topics in Quantum Electron.* **9**, 492–511 (2003).

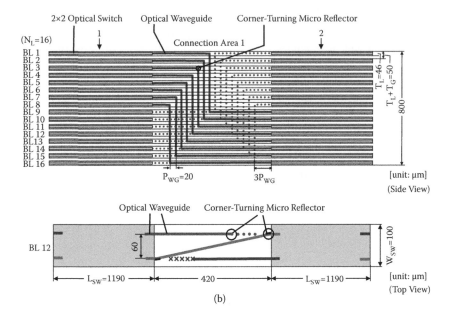

FIGURE 8.15 **(CONTINUED)** Details of (a) top view of connection area 5 and (b) side view of connection area 1 and top view of a part of the connection area in BL 12. From T. Yoshimura, M. Ojima, Y. Arai, and K. Asama, "Three-dimensional self-organized micro optoelectronic systems for board-level reconfigurable optical interconnects: performance modeling and simulation," *IEEE J. Select. Topics in Quantum Electron.* **9**, 492–511 (2003).

$$L(V)_i = (N_L / 2^{i-1} + 5)P_{WG} \qquad (8.4)$$

$$W(V) = ((N_C / 2) / 2^{D_{3D}})P_{SW} \qquad (8.5)$$

$$H(V) = N_L(T_L + G_L) \qquad (8.6)$$

Parameter substitution into Equations (8.4), (8.5), and (8.6) gives the size of connection area 1 of 420 μm in length, 6400 μm in width, and 800 μm in height. Similarly, the sizes for $i = 2{\sim}4$ can be calculated as shown in Figures 8.12 and 8.13.

8.3.4 OPTICAL Z-CONNECTIONS AND OPTICAL SWITCHES

8.3.4.1 Optical Z-Connections

Figure 8.16 shows an optical Z-connection model. Waveguides that are 4 μm wide are in the upper and lower layers. The gap between the waveguides of the adjacent layers is 16 μm. The waveguides are connected by two 45° reflectors and a vertical waveguide that is 4 μm wide. The reflectors can be metal mirrors, dielectric multilayer wavelength filters, gratings, or photonic crystals. In the present work,

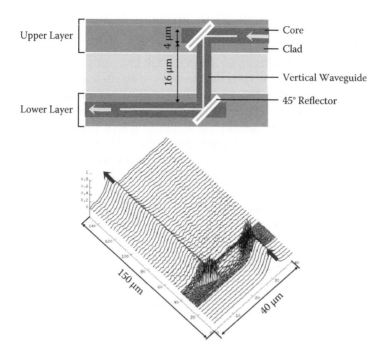

FIGURE 8.16   Optical Z-connection model with 4-μm-wide waveguides. From T. Yoshimura, M. Ojima, Y. Arai, and K. Asama, "Three-dimensional self-organized micro optoelectronic systems for board-level reconfigurable optical interconnects: performance modeling and simulation," *IEEE J. Select. Topics in Quantum Electron.* **9**, 492–511 (2003).

metal mirrors with a refractive index of 0.4 for the real part and −5 for the imaginary part are assumed. Analysis is performed by the BPM/FDTD method coupled simulation. It can be seen in Figure 8.16 that the propagating light beam smoothly propagates from the upper layer to the lower layer through optical Z-connection with little mode disturbance with respect to intensity. Calculated loss of the optical Z-connection (γ_z) is 0.97 dB. The loss is basically induced at the mirror reflection parts.

8.3.4.2   Optical Switches

Optical switches for 3D-MOSS should be small and exhibit high speed and low cross-talk characteristics. At the same time, the switches should be stable against thermal and dimensional fluctuations since a large number of optical switches are involved in 3D-MOSS, say, 5120 2 × 2 switches in 1024 × 1024 Banyan network. This implies that flux-controlled switches are preferable to phase-controlled switches, giving us digital optical switches, internal total reflection switches, variable well optical ICs (VWOICs), and WPD optical switches as candidates. In the present simulation, we selected the WPD optical switches consisting of PLZT as mentioned in Section 8.3.3.

8.3.5 Predicted Performance of 3D-MOSS

Figure 8.17 shows a schematic illustration of a 3-D MOSS calculation model, where only a part of the model is drawn for simplicity. OE films are placed with a pitch of $T_L + T_G$. In OE films, 2 × 2 WPD optical switches of length L_{SW} and width W_{SW} are arranged with a pitch of P_{SW}. Intralayer wiring is carried out by in-plane waveguides with a pitch of P_{WG} while interlayer wiring is by optical Z-connections. Major losses caused in 3D-MOSS are indicated by γ in the figure. γ_P is propagation loss of optical waveguides, γ_{CR} is loss induced at a right-angle waveguide–waveguide cross point, γ_Z is loss of an optical Z-connection, γ_{CT} is loss of a pair of corner turnings with two mirrors, γ_G is loss at a layer gap, γ_{SW} is loss within a 2 × 2 optical switch, and γ_{W-S} is waveguide-switch coupling loss.

Predicted system performance of 1024 × 1024 3D-MOSS is summarized in Table 8.3 as well as those of conventional planar structures for comparison. The process to derive performance is described below.

8.3.5.1 Size and Insertion Loss

For 3D-MOSS with S_{SW} and D_{3D}, connection areas 1 to D_{3D} have vertical optical connections while connection areas $D_{3D} + 1$ to $S_{SW} - 1$ have in-plane optical connections. The connection area length for the in-plane connection areas and the vertical

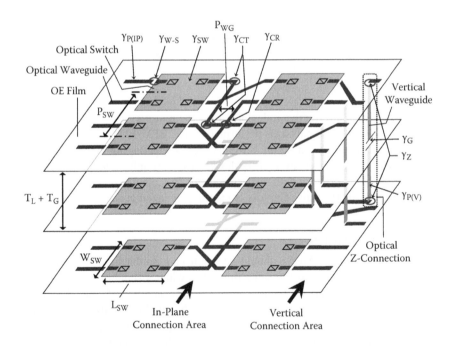

FIGURE 8.17 A schematic illustration of a 3-D MOSS model for insertion loss calculations. From T. Yoshimura, Y. Arai, H. Kurokawa, and K. Asama, "Predicted insertion loss reduction achieved by implementing three-dimensional microoptical network in chip-scale optical interconnects," *IEEE Photon. Technol. Lett.* **16**, 647–649 (2004).

TABLE 8.3
Summary of Predicted System Performance of 1024 × 1024 3D-MOSS with $D_{3D} = 4$

Size	Length	Width	Height
Conventional Planar Structure	3.252 cm	10.24 cm	0.005 cm
3D-MOSS ($D_{3D} = 4$)	1.424 cm	0.64 cm	0.08 cm
Insertion Loss			
Conventional Palanar Structure	43 dB		
3D-MOSS ($D_{3D} = 4$) without Misalignment	29 dB		
with 25% Misalignment	73 dB		
with 25% Misalignment using SOLNET	32 dB		
Electrical Characteristics			
Operation Voltage: V	12 V		
Switching Speed: τ	456 ps		
Power Consumption at a Switching Rate: Power at F	36 W/cm² at 1×10^7/s		

Source: T. Yoshimura, M. Ojima, Y. Arai, and K. Asama, "Three-dimensional self-organized micro opto-electronic systems for board-level reconfigurable optical interconnects: performance modeling and simulation," *IEEE J. Select. Topics in Quantum Electron.* **9**, 492–511 (2003).

connection areas are respectively expressed by Equations (8.2) and (8.4), and the width by Equations (8.3) and (8.5). Therefore, total 3D-MOSS length $L_{3D\text{-MOSS}}$, width $W_{3D\text{-MOSS}}$, and height $H_{3D\text{-MOSS}}$, including switch array length, are given by:

$$L_{3D-MOSS} = L_{SW} S_{SW} + \left(\sum_{i=1}^{D_{3D}} (N_L / 2^{i-1} + 5) \right) P_{WG} + \left(\sum_{i=D_{3D}+1}^{S_{SW}-1} (N_C / 2^i + 1) \right) P_{WG}, \quad (8.7)$$

$$W_{3D-MOSS} = ((N_C / 2) / 2^{D_{3D}}) P_{SW}, \quad (8.8)$$

$$H_{3D-MOSS} = N_L (T_L + G_L). \quad (8.9)$$

Major losses in 3D-MOSS are listed in Table 8.4. Propagation loss γ_P of optical waveguides is assumed to be 0.5 dB/cm, which is typical in fluorinated polyimide. Loss induced at a right-angle waveguide–waveguide cross point γ_{CR} is estimated by the beam propagation method (BPM) simulation to be 0.0088 dB. Loss of an optical Z-connection γ_Z of 0.97 dB was calculated in Section 8.3.4. Since a pair of corner turnings with two mirrors has the same structure as the optical Z-connection, the loss γ_{CT} can also be 0.97 dB. A cross point of a vertical waveguide and a layer gap

TABLE 8.4

Major Losses in 3D-MOSS

Propagation in Optical Waveguide: γ_P		0.5 dB/cm
Waveguide-Waveguide Cross Point: γ_{CR}		0.0088 dB
Pair of Corner Turnings (2 Micro Reflectors): γ_{CT}		0.97 dB
Optical Z-Connection: γ_Z		0.97 dB
Layer Gap: γ_G	with No Misalignment	0.0088 dB
	with 25% Misalignment	0.89 dB
	with 25% Misalignment using SOLNET	0.14 dB
2×2 Optical Switch: γ_{SW}		1.7 dB
Waveguide-Switch Coupling: γ_{W-S}	with No Misalignment	0 dB
	with 25% Misalignment	0.83 dB
	with 25% Misalignment using SOLNET	0.14 dB

Source: T. Yoshimura, M. Ojima, Y. Arai, and K. Asama, "Three-dimensional self-organized micro optoelectronic systems for board-level reconfigurable optical interconnects: performance modeling and simulation," *IEEE J. Select. Topics in Quantum Electron.* **9**, 492–511 (2003).

of 4 μm can be regarded as a kind of right-angle waveguide–waveguide cross point. So, loss at a layer gap γ_G is estimated to be 0.0088 dB. BPM calculation reveals that γ_G with 25% misalignment is 0.89 dB, and γ_G with 25% misalignment is 0.14 dB when SOLNET is used for the coupling. Loss within a 2×2 optical switch γ_{SW} was estimated to be 1.7 dB in Section 6.1.3. Waveguide-switch coupling loss γ_{W-S} is basically 0 dB for no misalignment. For the case with 25% misalignment, the BPM calculation reveals that γ_{W-S} is 0.83 dB. For SOLNET coupling with 25% misalignment, similar to the case of γ_G, γ_{W-S} is 0.14 dB. In the present simulation, boundary reflection losses are neglected, assuming an antireflective (AR) coating on the edge surfaces of individual parts.

General expressions for eight kinds of losses in 3D-MOSS are derived in terms of the major losses listed in Table 8.4. The results are summarized in Table 8.5. In-plane optical connections include in-plane propagation, waveguide–waveguide cross points, pairs of corner turnings, optical switches, and waveguide–switch couplings. Vertical optical connections include vertical propagation, optical Z-connections, and layer gaps.

For total loss due to in-plane propagation, the first and the second terms are caused by the horizontal line length given by the second and the third terms of Equation (8.7). The third term is caused from line length perpendicular to the horizontal direction. Since lines in vertical connection areas have no contribution to the line length perpendicular to the horizontal direction basically, the summation is carried out only for the in-plane connection areas. For total loss due to waveguide–waveguide cross points, as shown in Figure 8.15(a), an optical waveguide has a cross-point count of $N_C/2^i$ in in-plane connection area i. In vertical connection areas, no waveguide cross points exist. Thus, the summation for the total loss calculation is carried out for the in-plane connection areas. Total loss due to corner turnings is determined by the number of

TABLE 8.5

General Expressions for Losses in 3D-MOSS

In-Plane Optical Connections

In-Plane Propagation

$$\left[\left(\sum_{i=1}^{D_{3D}}(N_L/2^{i-1}+5)\right)P_{WG}+\left(\sum_{i=D_{3D}+1}^{S_{SW}-1}(N_C/2^i+1)\right)P_{WG}+\left(\sum_{i=D_{3D}+1}^{S_{SW}-1}(N_C/2)/2^i\right)P_{SW}\right]\gamma_P$$

Waveguide–Waveguide Cross Point   **Pair of Corner Turnings**

$$\left[\sum_{i=D_{3D}+1}^{S_{SW}-1}(N_C/2^i)\right]\gamma_{CR}$$

$$[S_{SW}-1-D_{3D}]\gamma_{CT}$$

Optical Switch   **Waveguide–Switch Coupling**

$[S_{SW}]\gamma_{SW}$ $[2S_{SW}]\gamma_{W-S}$

Vertical Optical Connections

Vertical Propagation

$[2N_L(T_L+T_G)]\gamma_P$

Optical Z-Connection   **Layer Gap**

$[2+D_{3D}]\gamma_Z$ $[2N_L]\gamma_G$ $[2+D_{3D}]\gamma_G$ for SOLNET

Source: T. Yoshimura, M. Ojima, Y. Arai, and K. Asama, "Three-dimensional self-organized micro
optoelectronic systems for board-level reconfigurable optical interconnects: performance modeling and simulation," *IEEE J. Select. Topics in Quantum Electron.* **9**, 492–511 (2003).

pair of corner turnings that equals the in-plane connection area count since optical
signals experience one pair of corner turnings in one in-plane connection area. Total
loss within optical switches and total loss at waveguide–switch coupling points are
respectively induced by S_{SW} optical switches and 2 S_{SW} waveguide–switch coupling
points. For vertical propagation, as can be seen from Figure 8.13 and Figure 8.15(b),
the total loss is determined by the optical path length that corresponds to two times
the 3D-MOSS thickness. Since optical signals experience one optical Z-connection
in one vertical connection area as shown in Figure 8.15(b), total loss caused by optical Z-connections, with adding two optical Z-connections for input and output, is
calculated to be $[2+D_{3D}]\gamma_Z$. Total loss due to layer gaps is determined by two times
the layer gap count contained in a 3D-MOSS. When a SOLNET is implemented in
vertical optical connections, total loss due to layer gaps is replaced with a product of
total Z-connection count times γ_G for the SOLNET.

By substituting the parameters in Table 8.2 into Equations (8.7), (8.8), and (8.9),
length, width, and height of 3D-MOSS can be respectively calculated to be 1.424
cm, 0.64 cm, and 0.08 cm as shown in Table 8.3. For conventional planar structures,
the length is 3.252 cm, the width is 10.24 cm, and the height is 0.005 cm. Therefore,
it is found that drastic occupation area reduction of 3/100 is achieved in 3D-MOSS

with $D_{3D} = 4$. Insertion losses of 3D-MOSS, shown in Table 8.3, can be obtained by summation of the eight losses listed in Table 8.5. While insertion loss of the conventional planar structure is 43 dB, that of the 3D-MOSS is 29 dB, indicating that a 14-dB loss reduction is achieved in 3D-MOSS. In Figure 8.18, occupation areas and insertion losses of 3D-MOSS are shown as a function of D_{3D} [30]. $D_{3D} = 0$ implies conventional structure, and $D_{3D} = 9$ implies that all the connection areas consist of vertical optical connections. It is found that the occupation area decreases drastically with D_{3D}. Insertion loss, on the other hand, has a peak in the region of $D_{3D} = 4~6$. For small D_{3D}, the main contribution to the loss is waveguide–waveguide cross points while layer gaps for large D_{3D}. The results indicate that size and loss of 3D-MOSS can be optimized by selecting D_{3D}.

The misalignment issue is a serious concern in 3D-MOSS. As shown in Table 8.4, misalignment between adjacent layers causes a layer gap loss γ_G of 0.89 dB. At a waveguide–switch coupling, which is equivalent to a coupling between 4-μm-wide waveguides, 25% misalignment induces a loss γ_{W-S} of 0.83 dB. By substituting these loss values into the summation of the eight losses shown in Table 8.5, insertion loss of 3D-MOSS with 25% misalignment is estimated to be 73 dB as shown in Table 8.3. In this case, in order to transmit optical signals with reasonable power, three times optical amplification might be required in 3D-MOSS as a

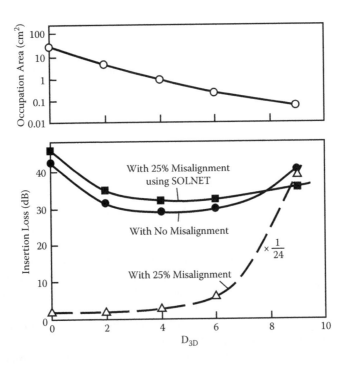

FIGURE 8.18 D_{3D} dependence of occupation areas and insertion losses of 3D-MOSS. From T. Yoshimura, Y. Arai, H. Kurokawa, and K. Asama, "Predicted insertion loss reduction achieved by implementing three-dimensional microoptical network in chip-scale optical interconnects," *IEEE Photon. Technol. Lett.* **16**, 647–649 (2004).

result of inserting amplifier layers. As can be seen in Figure 8.18, SOLNET implementation is effective to reduce the insertion loss. By using γ_G and γ_{W-S} of 0.14 dB listed in Table 8.4 for 25% misalignment using SOLNET, the insertion loss of the SOLNET-implemented 3D-MOSS is calculated to be 32 dB, which is close to the loss in 3D-MOSS without misalignment.

It was reported that high-accuracy bonding equipment enables layer stacking with positional accuracy of 1 μm for 1-cm² chips [31]. Therefore, for 3D-MOSS of 1.424 cm × 0.64 cm, it is consistent to assume maximum positional misalignment of ~1 μm, which corresponds to 25% misalignment for optical circuit backbones consisting of optical waveguides with 4 × 4-μm² core cross-sections. The insertion loss of 32 dB for 1024 × 1024 SOLNET-implemented 3D-MOSS is still high, which requires an optical amplifier layer insertion or 10 mW-order input light signals. For further simplification of the system, each loss listed in Table 8.4 should be decreased by optimizing individual structures and materials.

8.3.5.2 Electrical Characteristics

Switching speed and power consumption can be estimated based on the charge-discharge model for the prism-shaped capacitor in the waveguide prism deflector. The WPD optical switch model used in the present calculation is the same as the model shown in Figure 6.19 except that the PLZT thickness is increased from 400 nm to 4 μm and the applied voltage is increased from 1.2 V to 12V. Therefore, the capacitance C, switching speed τ, and power consumption $Power$ that are given by Equations (6.9), (6.8), and (6.10) for one prism-shaped capacitor are respectively calculated to be 0.57 pF, 57 ps for driver on-resistivity $R_{ON} = 100\ \Omega$, and 41 mW at $F = 5 \times 10^8$ 1/s.

When the number of the waveguide prism deflectors in a WPD optical switch is N_{Prism}, switching speed τ for R_{ON} is given by

$$\tau = N_{Prism}\ R_{ON}C. \tag{8.10}$$

For power consumption, total WPD optical switch count in a 3D-MOSS is $(N_C/2)S_{SW}$, giving $N_{Prism}\ (N_C/2)S_{SW}$ for the total count of the prism-shaped capacitors contained in a 3D-MOSS. Then, total capacitance arising from the deflectors is given by $C\ N_{Prism}\ (N_C/2)S_{SW}$. As a result, power consumption at a switching rate of F in a 3D-MOSS is expressed as follows:

$$Power = F\ CV^2\ N_{Prism}\ (N_C/2)S_{SW} \tag{8.11}$$

By substituting $C = 0.57$ pF, $V = 12$ V, $N_{Prism} = 8$, $N_c = 1024$, and $S_{SW} = 10$ into Equations (8.10) and (8.11), $\tau = 456$ ps for $R_{ON} = 100\ \Omega$, and $Power = 33$ W for $F = 1 \times 10^7$ 1/s are obtained. The power consumption of 33 W corresponds to a power density of 36 W/cm² since the area of the 3D-MOSS model is 1.42 × 0.64 cm². This value is comparable to acceptable heat generation of 30 W/cm² reported by K. Takahashi [31] for stacked LSI of 1-cm² area. These results are included in Table 8.3.

8.3.6 IMPACT OF NANO-SCALE WAVEGUIDES AND POLYMER MQDs ON 3D-MOSS PERFORMANCE

When high-index contrast (HIC) nano-scale waveguides having 400×400 nm² core cross-sections are used for the 3-D optical circuit backbone in the 3D-MOSS, the PLZT thickness, that is, the electrode gap in the WPD optical switches, can be reduced to 400 nm. In this case, switching voltage is reduced to 1.2 V to achieve low-voltage switching, which is favorable for low-power and high-speed operation. The HIC-waveguide-based optical circuits can be fabricated by PRI sol-gel materials as described in Section 7.3.

Figure 8.19 shows switching rate dependence of power density generated in 3D-MOSS for various EO materials of a WPD optical switch. Power density increases with switching rate. Possible switching rate is determined by a balance of heat generation and releasing rates in 3D-MOSS. For a given heat releasing rate, the lower the heat generation rate becomes, the higher the available switching rate becomes as far as switching speed of WPD optical switch τ is sufficiently fast. The power density versus switching rate relationship varies depending on EO materials. In PLZT, the switching rate is limited below 1×10^7 1/s when the limit of the heat releasing capability is 30 W/cm². In the styrylpyridinium cyanine dye (SPCD) organic crystal, which has an EO coefficient of 430 pm/V, an expected dielectric constant of $\sim 4\varepsilon_0$ and a refractive index of 1.55 as mentioned in Section 7.1.2, the line moves to the right-hand side, indicating that higher switching rates are expected than in PLZT.

Drastic switching rate improvement is expected in polymer MQDs [32–34], which has a potential EO coefficient of ~ 1000 pm/V, dielectric constant of $\sim 4\varepsilon_0$, and refractive index of ~ 2, resulting in a possible switching rate around 10^{10} 1/s.

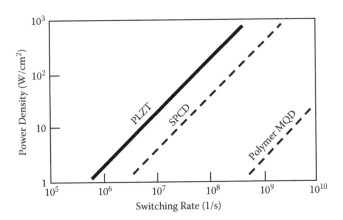

FIGURE 8.19 Switching rate dependence of power density generated in 3D-MOSS for various EO materials of a WPD optical switch. From T. Yoshimura, M. Ojima, Y. Arai, and K. Asama, "Three-dimensional self-organized micro optoelectronic systems for board-level reconfigurable optical interconnects: performance modeling and simulation," *IEEE J. Select. Topics in Quantum Electron.* **9**, 492–511 (2003).

These results of performance simulation indicate that 1024 × 1024 3D-MOSS is viable for massive optical packet switching systems and board-level reconfigurable optical interconnects from the perspective of size, insertion loss, switching speed, and thermal management. It is also suggested that the polymer MQD fabricated by MLD as well as the nano-scale waveguides will have a significant impact on high-speed massive optical switching systems.

REFERENCES

1. T. Yoshimura, J. Roman, Y. Takahashi, M. Lee, B. Chou, S. Beilin, W. Wang, and M. Inao, "Optoelectronic scalable substrates based on film/Z-connection and its application to film optical link module (FOLM)," *Proc. SPIE* **3952**, from Photonics West 2000, San Jose, California, 202–213 (2000).
2. T. Yoshimura, J. Roman, Y. Takahashi, M. Lee, B. Chou, S. I. Beilin, W. V. Wang, and M. Inao, "Proposal of optoelectronic substrate with Film/Z-connection based on OE-film," *Proc. 3rd IEMT/IMC Symposium*, 140–145 (1999).
3. T. Yoshimura, Y. Takahashi, M. Inao, M. Lee, W. Chou, S. Beilin, W-C. Wang, J. Roman, and T. Massingill, Three-dimensional opto-electronic modules with electrical and optical interconnections and methods for making, U.S. Patent 6,690,845 (2004).
4. T. Yoshimura, Y. Takahashi, M. Inao, M. Lee, W. Chou, S. Beilin, W-C. Wang, J. Roman, and T. Massingill, Multi-layer opto-electronic substrates with electrical and optical interconnections and methods for making, U.S. Patent 6,845,184 (2005).
5. T. Yoshimura, J. Roman, Y. Takahashi, W. V. Wang, M. Inao, T. Ishitsuka, K. Tsukamoto, K. Motoyoshi, and W. Sotoyama, "Self-organizing waveguide coupling method "SOLNET" and its application to film optical circuit substrates," *Proc. 50th Electronic Components & Technology Conference (ECTC)*, Las Vegas, Nevada, 962–969 (2000).
6. T. Yoshimura, T. Inoguchi, T. Yamamoto, S. Moriya, Y. Teramoto, Y. Arai, T. Namiki, and K. Asama, "Self-organized lightwave network based on waveguide films for three-dimensional optical wiring within boxes," *J. Lightwave Technol.* **22**, 2091–2100 (2004).
7. T. Yoshimura, M. Ojima, Y. Arai, and K. Asama, "Three-dimensional self-organized micro optoelectronic systems for board-level reconfigurable optical interconnects: performance modeling and simulation," *IEEE J. Select. Topics in Quantum Electron.* **9**, 492–511 (2003).
8. T. Yoshimura and Y. Arai, 3D optoelectronic micro system, U.S. Patent 2006/0261432 (2006).
9. T. Yoshimura, "Proposed applications of 3-D optical interconnect technologies to integrated chemical systems," *2007 Digest of the LEOS Summer Topical Meetings* [Bio-Inspired Sensors and Application/Imprinting on Photonic Integrated Circuits], Portland, Oregon, 129–130 (2007).
10. T. Yoshimura, K. Ogushi, Y. Kitabayashi, K Naito, Y. Miyamoto, and M. Miyazaki, "Optical waveguide films with two-layer skirt-type core end facets for beam leakage reduction at 45° mirrors," *Proc. SPIE* **6899**, from Photonics West 2008, San Jose, California, 689913-1-11 (2008).
11. T. Yoshimura, Integrated photonic/electronic/chemical systems, solar energy conversion systems, light collectors, and optical waveguides, Japanese Patent Tokukai 2009-4717 (2009) [in Japanese].
12. T. Yoshimura and K. Asama, Integrated chemical systems and integrated light energy conversion systems, Japanese Patent Tokukai 2007-107085 (2007) [in Japanese].
13. T. Yoshimura, *Molecular nano systems: Applications to optoelectronic computers and solar energy conversion*, Corona Publishing Co., Ltd., Tokyo (2007) [in Japanese].

14. D. A. B. Miller, "How large a system can we build without optics?" *Workshop Notes, 8th Annual Workshop on Interconnections Within High Speed Digital Systems*, Santa Fe, New Mexico, Lecture 1.2 (1997).

15. N. M. Jokerst, M. A. Brooke, S. Cho, S. ilkinson, M. Vrazel, S. Fike, J. Tabler, Y. J. Joo, S. Seo, D. S. Wils, and A. Brown, "The heterogeneous integration of optical interconnections into integrated microsystems," *IEEE J. Select. Topics in Quantum Electron.* **9**, 350–360 (2003).

16. C. Choi, L. Lin, Y. Liu, J. Choi, L. Wang, D. Haas, J. Magera, and R. T. Chen, "Flexible optical waveguide film fabrications and optoelectronic devices integration for fully embedded board-level optical interconnects," *J. Lightwave Technol.* **22**, 2168–2176 (2004).

17. H. Takahara, "Optoelectronic multichip module packaging technologies and optical input/output interface chip-level packages for the next generation of hardware systems," *IEEE J. Select. Topics in Quantum Electron.* **9**, 443–451 (2003).

18. T. Mikawa, M. Kinoshita, K. Hiruma, T. Ishitsuka, M. Okabe, S. Hiramatsu, H. Furuyama, T. Matsui, K. Kumai, O. Ibaragi, and M. Bonkohara, "Implementation of active interposer for high-speed and low-cost chip level optical interconnects," *IEEE J. Select. Topics in Quantum Electron.* **9**, 452–459 (2003).

19. B. S. Rho, S. Kang, H. S. Cho, H.-H. Park, S.-W. Ha, and B.-H. Rhee, "PCB-Compatible optical interconnection using 45°-ended connection rods and via-holed waveguides," *J. Lightwave Technol.* **22**, 2128–2134 (2004).

20. C. Berger, B. Offrein, and M. Schmatz, "Challenges for the introduction of board-level optical interconnect technology into product development roadmaps," *Proc. SPIE* **6124**, 61240J-1-12 (2006).

21. E.-H. Lee, S. G. Lee, B. H. O, S. G. Park, and K. H. Kim, "Fabrication of a hybrid electrical-optical printed circuit board (EO-PCB) by lamination of an optical printed circuit board (O-PCB) and an electrical printed circuit board (E-PCB)," *Proc. SPIE* **6126**, 61260L-1-10 (2006).

22. A. Glebov, M. G. Lee, D. Kudzuma, J. Roman, M. Peters, L. Huang, and S. Zhou, "Integrated waveguide microoptic elements for 3D routing in board-level optical interconnects," *Proc. SPIE* **6126**, 61260L-1-10 (2006).

23. T. Yoshimura, J. Roman, Y. Takahashi, S. Beilin, W. Wang, and M. Inao, 4th International Conference on Organic Nonlinear Optics (ICONO'4), Extended Abstracts, Chitose, Japan, 278 (1998).

24. T. Yoshimura, J. Roman, Y. Takahashi, S. I. Beilin, W. V. Wang, and M. Inao, "Optoelectronic amplifier/driver-less substrate <OE-ADLES> for polymer-waveguide-based board-level interconnection: calculation of delay and power dissipation," *Nonlinear Optics* **22**, 453–456 (1999).

25. T. Yoshimura, Y. Suzuki, N. Shimoda, T. Kofudo, K. Okada, Y. Arai, and K. Asama, "Three-dimensional chip-scale optical interconnects and switches with self-organized wiring based on device-embedded waveguide films and molecular nanotechnologies," *Proc. SPIE* **6126**, from Photonics West 2006, San Jose, California, 612609-1-15 (2006).

26. M. W. Haney and M. P. Christensen, "Sliding banyan network," *J. Lightwave Technol.* **14**, 703–710 (1996).

27. T. Yoshimura, A. Hori, Y. Yoshida, Y. Arai, H. Kurokawa, T. Namiki, and K. Asama, "Coupling efficiencies in reflective self-organized lightwave network (R-SOLNET) simulated by the beam propagation method," *IEEE Photon. Technol. Lett.* **17**, 1653–1655 (2005).

28. T. Yoshimura, S. Tsukada, S. Kawakami, M. Ninomiya, Y. Arai, H. Kurokawa, and K. Asama, "Three-dimensional micro-optical switching system architecture using slab-waveguide-based micro-optical switches," *Opt. Eng.* **42**, 439–446 (2003).

29. T. Yoshimura, S. Tsukada, S. Kawakami, Y. Arai, H. Kurokawa, and K. Asama,, "3D micro optical switching system (3D-MOSS) architecture," *Proc. SPIE* **4653**, from Photonics West 2002, San Jose, California, 62–70 (2002).

30. T. Yoshimura, Y. Arai, H. Kurokawa, and K. Asama, "Predicted insertion loss reduction achieved by implementing three-dimensional microoptical network in chip-scale optical interconnects," IEEE *Photon. Technol. Lett.* **16**, 647–649 (2004).

31. K. Takahashi, "Research and development on ultra-high-density 3-dimensional LSI-chip-stack packaging technologies," *The 3rd Annual Meeting on Electronics System Integration Technologies Digest,* edited and published by the Electronic System Integration Technology Research Department, Association of Super-Advanced Electronics Technologies (ASET), 43–94 (2002).

32. T. Yoshimura, "Enhancing second-order nonlinear optical properties by controlling the wave function in one-dimensional conjugated molecules," *Phys. Rev.* **B40**, 6292–6298 (1989).

33. T. Yoshimura, "Theoretically predicted influence of donors and acceptors on quadratic hyperpolarizabilities in conjugated long-chain molecules," *Appl. Phys. Lett.* **55**, 534–536 (1989).

34. T. Yoshimura, "Design of organic nonlinear optical materials for electro-optic and all-optical devices by computer simulation," *FUJITSU Sc. Tech. J.* **27**, 115–131 (1991).

9 Applications to Solar Energy Conversion Systems

This chapter presents expected applications of molecular layer deposition (MLD) and related thin-film organic photonic/electronic materials to solar energy conversion systems. Applications to photosynthesis devices are also discussed.

9.1 SENSITIZED PHOTOVOLTAIC DEVICES

We propose multidye sensitization and polymer-MQD (multiple quantum dot) sensitization for photovoltaic devices. In the former, the sensitizing layer consists of molecular wires in which a plurality of kinds of dye molecules are stacked with designated sequences by liquid-phase MLD. In the latter, the sensitizing layer consists of polymer MQDs grown by gas-phase MLD. We also propose waveguide-type photovoltaic devices, which enhance light absorption of the sensitizing layers.

In the present section, after details of the proposed concepts are described, several experimental proofs of concept are demonstrated.

9.1.1 CONCEPT OF MULTIDYE SENSITIZATION AND POLYMER-MQD SENSITIZATION

Structures for multidye sensitization and polymer-MQD sensitization are shown in Figure 9.1. In multidye sensitization [1–3], molecular wires consisting of dye molecules are grown on a semiconductor surface. In the example shown in Figure 9.1(a), four different kinds of molecules are stacked on a surface. In polymer-MQD sensitization [1–4], polymer wires having MQDs with different dot lengths are grown on a semiconductor surface. In the example shown in Figure 9.1(b), four MQDs of different lengths are contained in a polymer wire. The polymer wire directions can be vertical or horizontal.

When Si is used for the semiconductor, as illustrated in Figure 9.2(a), the band gap of Si is relatively narrow, and therefore the photon energy of visible ultraviolet light is much larger than the band gap. In such a case, excess energy of photons is lost as heat, which reduces the light energy conversion efficiency of the Si solar cells.

When multidye-sensitized ZnO with molecular wires consisting of p-type and n-type dye molecules is used for the semiconductor, the energy diagrams and absorption spectra can be drawn as in Figure 9.2(b) [5]. In this example, four kinds of dye molecules are aligned in a p/n/p/n sequence on the n-type ZnO. Since dye molecules have narrow absorption bands as compared with ordinary semiconductors, by appropriate selection of dye molecules, the absorption wavelength region from near-infrared

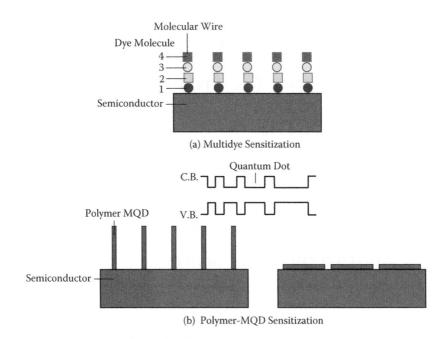

FIGURE 9.1 Multidye sensitization and polymer-MQD sensitization.

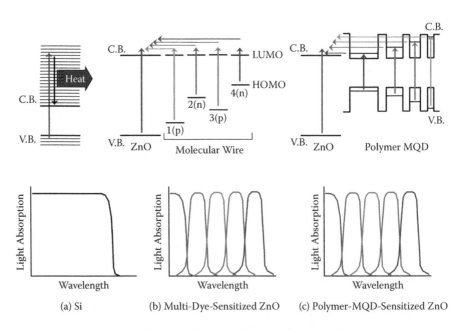

FIGURE 9.2 Schematic illustrations of energy levels and absorption spectra in (a) Si, (b) multidye-sensitized ZnO, and (c) polymer-MQD-sensitized ZnO.

to visible can be divided using several kinds of dye molecules. Each kind of dye molecule absorbs light with wavelengths matched to its own energy gap and injects the excited electrons into the ZnO. This might suppress the energy loss arising from the excess photon energy. The structure is regarded as a molecular tandem structure. Such molecular wire structures are grown by MLD described in Sections 3.1.1 and 3.1.2. In some cases, it is also possible to use "MLD with molecule groups" described in Section 3.1.3. For example, Molecule Group I containing p-type molecules 1 and 3, and Molecule Group II containing n-type molecules 2 and 4, are prepared. By supplying Molecule Group I and Molecule Group II sequentially, structures of molecular wires 1/2, 1/4, 3/2, 3/4 are grown simultaneously. This enables us to realize the four-divided light absorption scheme shown in Fig. 9.2(b) by two-dye molecular wires grown by two-step MLD instead of four-dye molecular wires grown by four-step MLD.

When the polymer-MQD-sensitized ZnO is used for the semiconductor, the same effect might be obtained as in the multidye-sensitized ZnO. The energy diagrams and absorption spectra are drawn in Figure 9.2(c). Absorption bands of quantum dots are narrow as compared with ordinary semiconductors and the absorption peak energy is controlled by the quantum dot length. Therefore, by adjusting the quantum dot lengths in the polymer wire appropriately, the absorption wavelength region can be divided using quantum dots with different lengths. Each quantum dot absorbs light with wavelengths matched to its own energy gap and injects the excited electrons into the ZnO.

We propose two more multidye sensitization processes, which utilize processes similar to the photosynthesis process in plants. In the process shown in Figure 9.3(a), gradual energy steps are formed in the molecular wire, which is similar to the electron transporting system in the z-scheme of the photosynthesis process, to perform smooth electron transfer to the ZnO. In the process drawn on the left, the electron transfer from the redox system occurs at the top of the molecular wire. In the process drawn on the right, it occurs at the inner part of the molecular wires.

In the process shown on the left in Fig. 9.3(b), the z-scheme-like molecular arrangements are formed. In this example, an electron excited from a highest occupied molecular orbital (HOMO) of molecule 1 by a photon with a wavelength of λ_1 is transferred into the ZnO. An electron excited by a photon with λ_2 in molecule 2 is transferred to the HOMO of molecule 1 to compensate the hole left in the HOMO. The hole left in the HOMO of molecule 2 is compensated by a redox system. In the process shown on the right in Fig. 9.3(b), an electron excited by a photon with a wavelength of λ_2 in molecule 2 is transferred into the ZnO via molecule 1. An electron excited by a photon with λ_4 in molecule 4 is transferred to the highest occupied molecular orbital (HOMO) of molecule 2 via molecule 3. The hole left in molecule 4 is compensated by a redox system. A similar same electron flow can be generated when an electron is excited by a photon with λ_1 in molecule 1 instead of in molecule 2, and when an electron is excited by a photon with λ_3 in molecule 3 instead of in molecule 4.

The sensitization mechanism with z-scheme-like molecular arrangements has following two advantages.

- It increases the difference in energy between the Fermi level of ZnO and the electrode potential of the redox system to increase the generated voltage in the photovoltaic device.

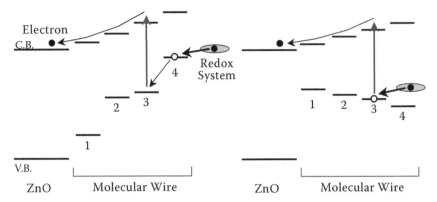

(a) Sensitization by Electron Transfers via Gradual Energy Steps

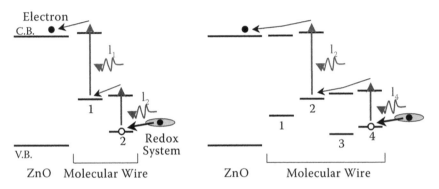

(b) Sensitization by Z-Scheme-Like Electron Transfers with Multi-Photons

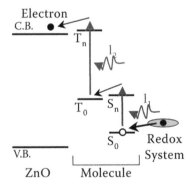

(c) Sensitization by Electron Transfers from Multi-Photon Molecules

FIGURE 9.3   Proposed molecular arrangements of molecular wires on ZnO for sensitization utilizing the photosynthesis process. Sensitization by (a) electron transfers via gradual energy steps, and (b) z-scheme-like electron transfers with multiphotons, and (c) by electron transfers from molecules with four-level two-photon absorption capability.

- It suppresses the energy loss arising from the excess photon energy, which occurs in Si photovoltaic devices, to improve the light energy conversion efficiency.

For the z-scheme-like molecular arrangements, a molecule with HOMO of a high energy level and a molecule with LUMO of a low energy level should be combined. For example, the former could be crystal violet, piacyanol etc., and the latter could be phthalocyanine, C_{60} etc.

The above-mentioned molecular wire structures are grown by MLD described in sections 3.1.1, 3.1.2, or "MLD with molecule groups" described in section 3.1.3.

The similar sensitization as the z-scheme-like electron transfer might arise from molecules with multi-photon absorption characteristics. The concept of the four-level two-photon absorption is shown in Fig. 9.3(c). A photon with a wavelength of λ_1 excites an electron from singlet S_0 state to S_n state in a molecule, and the excited electron in S_n state transfers to triplet T_0 state. Then, a photon with another wavelength of λ_2 further excites the electron to T_n state to complete the four-level two-photon absorption [6,7]. The transfer of the excited electron to the conduction band of ZnO generates photocurrents. Protoporphyrin, tetraphenylporphyrin, biacetyl, comphorquinone, benzyl, etc. are known as two-photon-absorption materials [6,7].

9.1.2 Waveguide-Type Photovoltaic Device Concept

Dye-sensitized photovoltaic devices [8] have attracted much attention as cost-effective solar energy conversion modules. In Figure 9.4, the conventional dye-sensitized solar cell and the waveguide-type dye-sensitized photovoltaic device are compared. In the conventional device shown in Figure 9.4(a), in order to increase the number of adsorbed dye molecules on semiconductor surfaces, porous semiconductors are used. In this case, although sufficient light absorption is accomplished for normally incident light into the semiconductor surfaces, the crystalline quality of semiconductors tends to become poor, suppressing electron mobility. Furthermore, the porous structure narrows the electron transport channels. Consequently, the internal resistivity increases to degrade the photovoltaic performance in the conventional dye-sensitized solar cell.

In the waveguide-type dye-sensitized photovoltaic device shown in Figure 9.4(b) [2,4], on the other hand, thin film semiconductors with flat surfaces and high crystalline quality, that is, high electron mobility, are used. So, the internal resistivity decreases to improve photovoltaic performance. However, in this case, the number of adsorbed dye molecules on the semiconductor surfaces is insufficient. Because normally incident light into the devices passes through only a mono layer of dye molecules, very little light absorption is available. To resolve this problem, instead of normally incident light, guided light is used in semiconductor thin films. In this configuration, light beams pass through a lot of dye molecules to enhance light absorption. Thus, high-performance dye-sensitized photovoltaic devices will be realized. If necessary, transparent electrodes can be replaced with metal electrodes, which will reduce material costs and series internal resistance of electrodes.

Figure 9.4(c) shows a waveguide-type photovoltaic device with multidye sensitization and polymer-MQD sensitization. To form the sensitizing layer, molecular wires containing dye molecules or polymer MQDs are grown on a thin-film semiconductor

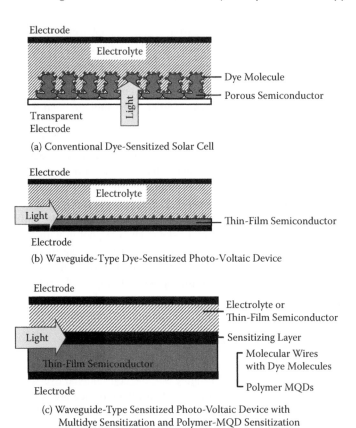

FIGURE 9.4   Comparison of (a) the conventional dye-sensitized solar cell and (b) the waveguide-type dye-sensitized photovoltaic device. (c) The waveguide-type sensitized photovoltaic device with multidye sensitization and polymer-MQD sensitization. From T. Yoshimura, A. Oshima, D. Kim, and Y. Morita, "Quantum dot formation in polymer wires by three-molecule molecular layer deposition (MLD) and applications to electro-optic/photovoltaic devices," *ECS Transactions* **25**, No. 4 "Atomic Layer Deposition Applications 5," 15–25 (2009).

surface. The electrolyte is located between the semiconductor and the counter electrode. A solid-state electrolyte is preferable for film-shaped devices. The electrolyte can also be replaced with a thin-film semiconductor to form this structure: n-type semiconductor/sensitizing layer/p-type semiconductor. The light beam is guided in the layered structures.

A serious issue for waveguide-type photovoltaic devices is optical coupling. Some of the candidates for coupling methods are shown in Figure 9.5. One of the promising ways is to embed the photovoltaic device in waveguide-type light beam collecting films (described in Section 9.2). For the optical coupling between the photovoltaic devices and optical waveguides in the light beam collecting film, reflective SOLNET (R-SOLNET; described in Section 7.4) might be useful because a self-organized lightwave network (SOLNET) can connect optical waveguides with different mode sizes.

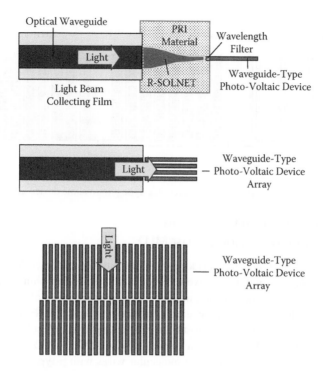

FIGURE 9.5 Possible optical coupling methods to introduce light beams into waveguide-type photovoltaic devices.

Another way utilizes arrays of waveguide-type photovoltaic devices. In this case, the light beams from the optical waveguides in the light beam collecting film can be coupled into the photovoltaic devices without SOLNET. The waveguide-type photovoltaic device arrays can also be used in the normally incident light configuration by placing the array vertically. By stacking the arrays to form multilayered structures, the light beams passing through the arrays in the upper layers can be caught by the array in the lower layers to improve light energy conversion efficiency.

An example of the fabrication process of waveguide-type photovaoltaic device arrays is illustrated in Figure 9.6. A thin film of a semiconductor such as ZnO is grown on a transparent electrode layer having deep trenches with a high aspect ratio. As reported by H. Shin et al. [9], and as mentioned in Section 2.4, such a 3-D structure can be formed by atomic layer deposition (ALD). The sensitizing layer growth can be carried out by MLD [10,11]. Liquid-phase MLD [12,13] and gas-phase MLD [4,13] are respectively applied to the molecular wires of dye molecules and polymer MQDs. By supplying electrolytes to the thin film surface, a waveguide-type photovaoltaic device array with the structure in Figure 9.4(c) can be obtained.

It should be noted that the light beams propagating in both the electrolyte and semiconductor thin films can be used for waveguide-type photovoltaic devices. In this case, although light intensity in the semiconductor region is reduced, the tolerance of optical coupling to the waveguide-type photovoltaic device is considerably widened.

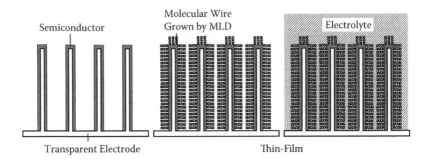

FIGURE 9.6 Fabrication process of a waveguide-type photovoltaic device array.

9.1.3 PROOF OF CONCEPT OF MULTIDYE SENSITIZATION BY LIQUID-PHASE MLD

As mentioned in Section 3.3.2, surface potential measurements revealed that, when p-type dye molecules and n-type dye molecules are coadsorbed simultaneously on n-type ZnO, dye molecules are self-assembled into the structure of n-type ZnO/p-type dye molecule/n-type dye molecule to form an npn structure (Figure 3.17) [14]. The structure is formed by the electrostatic force between dye molecules and between a dye molecule and ZnO. This indicates a possibility that MLD utilizing electrostatic force can be applied to growth of the molecular wires with p-type and n-type dye molecules in the multidye-sensitized photovoltaic device shown in Figure 9.1(a).

The spectral sensitization effect of the p/n-stacked structures on ZnO surfaces was first observed by Kiyota et al. in dyed ZnO for photoreceptors in electrophotographic systems [15]. The sensitization mechanism in photovoltaic devices is similar to that in photoreceptors. Therefore, the p/n-stacked structures on ZnO are expected to be applied to photovoltaic devices.

This section presents sensitization of ZnO by stacked structures of p-type and n-type dye molecules, which are regarded as molecular wires consisting of two dye molecules. The section then describes widening of photocurrent spectra in ZnO sensitized by the p/n-stacked structures, which are formed by the liquid-phase MLD.

9.1.3.1 Spectral Sensitization of ZnO by p/n-Stacked Structures

Investigation of spectral sensitization by p/n-stacked structures was carried out by surface potential measurements [15] and photocurrent measurements [16].

The samples for measuring surface potential and photocurrent were prepared as follows. A ZnO layer (thickness: 20 μm, ZnO powder: Sakai Chem. Co. Sazex 2000 #) was first formed on a glass substrate with a slit-type SnO_2 electrode (slit length: 10 mm, slit gap: 0.5 mm). For the adsorption of dye molecules, 0.01 mL of an alcohol solution (dye concentration: 1.6×10^{-3} mol/l) was dropped on the ZnO layer (area: 3 cm^2). For the coadsorption of the two dye molecules, 0.01 mL of an alcohol solution containing the two kinds of dye in equal concentration (1.6×10^{-3} mol/l) was dropped on the ZnO layer.

Surface potential of the dyed ZnO surfaces were measured by a surface potentiometer. For photocurrent, 30 V was applied at the slit gap of the sample, which was illuminated by monochromatic light, and the current was measured by an electrometer. In the experiments, the following six dyes were used: fluorescein (FL), eosine (EO), and rose bengal (RB) as the p-type dye molecules, and crystal violet (CV), malachite green (MG), and brilliant green (BG) as the n-type dye molecules. The structures of the dye molecules are shown in Figure 3.14. Figure 9.7 shows the reflection spectra of the dyed ZnO. It was found that p-type dye molecules exhibit an absorption band in shorter wavelength regions as compared to the n-type dye molecules.

9.1.3.1.1   Surface Potential Variation Induced by Illumination [16]

In ZnO, zinc atoms in interstitial sites donate their electrons to oxygen adsorbed on the ZnO surface to form an electric double layer, causing a negative surface potential for the ZnO layer as shown in Figure 3.15. The surface potential of a ZnO layer that has adsorbed p-type dye molecules is more negative than that of an undyed ZnO layer [14] since the electrons in the ZnO layer are transferred to the p-type dye layer. Conversely, in the case of n-type dye molecules, the surface potential is less negative than that of an undyed ZnO layer [14] since the electrons in the n-type dye layer are transferred to the ZnO layer. When visible light falls on a dyed ZnO layer, the surface potential of the layer varies as shown in Figure 9.8(a). Here, ZnO/RB, ZnO/CV, and ZnO/MG indicate that the ZnO layer adsorbs RB, CV, and MG, respectively. The variation of the surface potential arises as follows.

1. The dye molecules absorb light and electron-hole pairs are generated in the dye molecules.
2. The electrons in the dye molecules are injected into the ZnO layer and the holes left in the dye molecules are compensated by the ionized oxygen on the ZnO surface.
3. Consequently, a decrease in the strength of the electric double layer at the ZnO surface is observed as the variation of the surface potential.

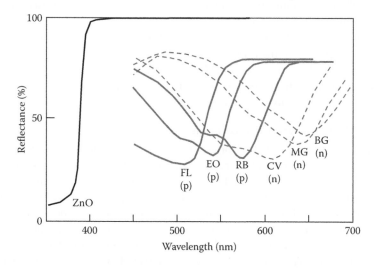

FIGURE 9.7   Reflection spectra of dyed ZnO.

As Figure 9.8(b) shows, when UV light illuminates the samples, the initial variation rates of the surface potentials are not dependent on the adsorbed dye molecules. This is because most of the UV light is transmitted through the dye layers and electron-hole pairs are generated in the ZnO.

As Figure 9.8(a) shows, variations of the surface potential of the samples illuminated by visible light depend heavily on the adsorbed dye molecules. In this case, the saturated values of the surface potential are determined by both the injection rate of the electrons from the dye molecules into the ZnO and the trapping rate of the electrons in the ZnO by adsorbed oxygen. Because the trapping rate does not depend on the adsorbed dye molecules, the results shown in Figure 9.8(a) indicate that the injection rates for RB, CV, and MG become, in that order, successively lower. We found that the saturated values become larger as the light intensity increases, which suggests that the larger the injection rate of the electrons becomes, the larger the saturated value becomes.

The initial variation rate of the surface potential, defined by $(dV/dt)_{t=0} = (1/C)(dQ/dt)_{t=0}$, is proportional to the injection rate of the electrons from the dye molecule to the ZnO layer. Here Q and C are the density of electric charge and the capacitance of the electric double layer at the ZnO surface, respectively.

Figure 9.9 shows the wavelength dependence of the initial slope of the surface potential variation, which is proportional to the injection rates. Here, ZnO/RB/CV indicates the sample of ZnO layer adsorbing the RB molecule and the CV molecule, in that order. Figure 9.10 shows the reflection spectra of ZnO/RB, ZnO/CV, and ZnO/RB/CV. It can be seen from Figures 9.9 and 9.10 that the maximum value of the spectrum of the electron injection rate in ZnO/CV is much smaller than that in ZnO/RB, while the peak height of the reflection spectrum of ZnO/CV is nearly equal to that of ZnO/RB. This indicates that the electron injection efficiency in sample ZnO/CV is lower than that in ZnO/RB.

In the case of ZnO/RB/CV, while the reflection spectrum is almost the same as that of ZnO/CV, the electron injection rate in ZnO/RB/CV is much larger than that in ZnO/CV. The electron injection efficiency from the CV molecule is raised by inserting the RB molecule between the ZnO and the CV molecule. Similar observations were obtained when MG was used instead of CV. This indicates that the electron injection efficiency from the n-type dye molecule to ZnO is enhanced by inserting a p-type dye molecule between the ZnO layer and the n-type dye molecule. This effect makes the multidye sensitization, in which p/n/p/n/... molecular wire structures are utilized as shown in Figures 9.1 and 9.2, viable.

9.1.3.1.2 Sensitization of Photocurrents [17]

Photocurrents in the dyed ZnO layers at the maximum absorption wavelengths of the adsorbed dye molecules are listed in Table 9.1 for dyes whose reflection spectra are shown in Figure 9.7. It can be seen that the photocurrents are much larger for the p-type dye molecules than for the n-type dye molecules, while the peak values of light absorption are almost the same for both p-type dye and n-type dye molecules. This confirms that the p-type dye molecule injects photocarriers into the ZnO layer more efficiently than the n-type dye molecule does, as H. Meier demonstrated [18]. The results are parallel to those obtained by surface potential measurements.

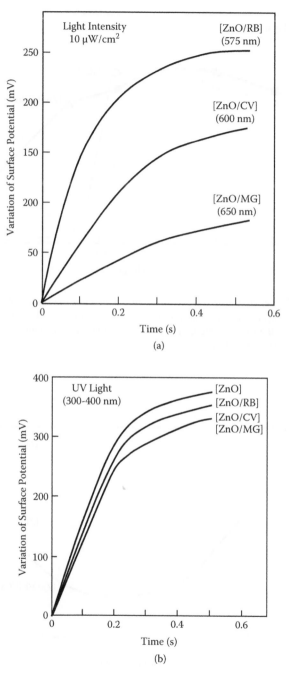

FIGURE 9.8 Variations of the surface potential of the dyed ZnO layer illuminated by (a) visible light, (b) UV light. From T. Yoshimura, K. Kiyota, H. Ueda, and M. Tanaka, "Influence of illumination on the surface potential of ZnO adsorbing p-type dye and n-type dye," *Jpn. J. Appl. Phys.* **19**, 1007–1008 (1980).

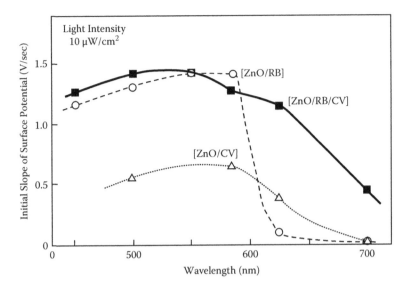

FIGURE 9.9 Wavelength dependence of the initial slope of the surface potential variation of the dyed ZnO layer. From T. Yoshimura, K. Kiyota, H. Ueda, and M. Tanaka, "Influence of illumination on the surface potential of ZnO adsorbing p-type dye and n-type dye," *Jpn. J. Appl. Phys.* **19**, 1007–1008 (1980).

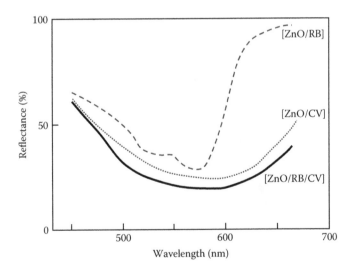

FIGURE 9.10 Reflection spectra of ZnO/RB, ZnO/CV, and ZnO/RB/CV. From T. Yoshimura, K. Kiyota, H. Ueda, and M. Tanaka, "Influence of illumination on the surface potential of ZnO adsorbing p-type dye and n-type dye," *Jpn. J. Appl. Phys.* **19**, 1007–1008 (1980).

TABLE 9.1

Photocurrents in Dyed ZnO Layers at the Maximum Absorption Wavelengths of the Dyes

Dye	Type	Photocurrent (nA)
FL	p	120
EO	p	190
RB	p	180
CV	n	3
MG	n	0.02
BG	n	0.01

Source: T. Yoshimura, K. Kiyota, H. Ueda, and M. Tanaka, "Mechanism of spectral sensitaization of ZnO coadsorbing p-type and n-type dyes," *Jpn. J. Appl. Phys.* **20**, 1671–1674 (1981).

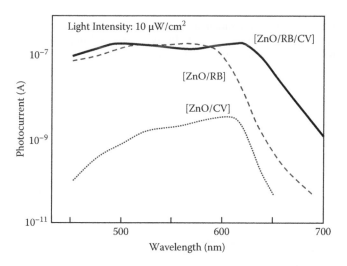

FIGURE 9.11 Spectral sensitization of ZnO/RB, ZnO/CV, and ZnO/RB/CV. From T. Yoshimura, K. Kiyota, H. Ueda, and M. Tanaka, "Mechanism of spectral sensitization of ZnO coadsorbing p-type and n-type dyes," *Jpn. J. Appl. Phys.* **20**, 1671–1674 (1981).

A typical spectral sensitization by the p/n-stacked structures is shown in Figure 9.11. Here, the p-type dye is RB and the n-type dye is CV. The photocurrents are for a light intensity of 10 µW/cm². It is found that photocurrents in ZnO/RB/CV are much larger than in ZnO/RB and ZnO/CV in the 600 to 700 nm wavelength region. As Figures 9.12 and 9.13 show, the same kind of sensitization with two dyes is observed in ZnO/RB/MG and ZnO/EO/MG. In the 600 to 700 nm wavelength region, profiles of the photocurrent spectra for the ZnO layers with p/n-stacked structures are similar to the profiles of the reflection spectra of the n-type dyes adsorbed on the top.

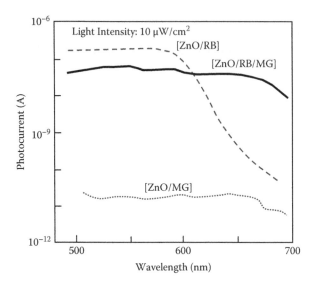

FIGURE 9.12 Spectral sensitization of ZnO/RB, ZnO/MG, and ZnO/RB/MG. From T. Yoshimura, K. Kiyota, H. Ueda, and M. Tanaka, "Mechanism of spectral sensitization of ZnO coadsorbing p-type and n-type dyes," *Jpn. J. Appl. Phys.* **20**, 1671–1674 (1981).

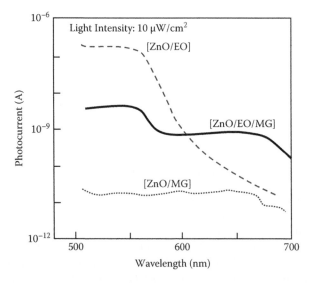

FIGURE 9.13 Spectral sensitization of ZnO/EO, ZnO/MG, and ZnO/EO/MG. From T. Yoshimura, K. Kiyota, H. Ueda, and M. Tanaka, "Mechanism of spectral sensitization of ZnO coadsorbing p-type and n-type dyes," *Jpn. J. Appl. Phys.* **20**, 1671–1674 (1981).

It should be noted that the photocurrent in ZnO/RB/MG is much larger than that in ZnO/EO/MG in the red light region, indicating that the degree of sensitization with p/n-stacked structures greatly depends on the combination of the two dyes.

Pioneering study on super spectral sensitization by two-molecule coadsorption was reported by Kokado et al. [19] and Hauffe [20]. The sensitization with the p/n-stacked structure reported here has a distinct advantage in that it will be able to extend toward multistacked structures consisting of molecules of more than three kinds because the structures might be fabricated by MLD utilizing electrostatic force.

9.1.3.1.3 *Mechanism of Sensitization with p/n-Stacked Structures [17]*

We have already clarified the following phenomena on sensitization with p/n-stacked structures:

1. Dramatic sensitization with dye-molecule-stacked structures arises only from the combination of a p-type dye with an n-type dye.
2. When p-type and n-type dye moleculaes are coadsorbed on ZnO, p-type dye molecules are inserted between ZnO and n-type dye molecules.

A schematic model of sensitization with p/n-stacked structures is shown in Figure 9.14(a). K_p is the efficiency with which the electrons are injected from the p-type dye molecule into the ZnO. K_{pn} is the efficiency with which the electrons excited in the n-type dye molecule are injected into the p-type dye molecule. Then, the efficiency with which the photocarriers generated in the n-type dye molecule are injected into the ZnO is given by the following relationship:

$$\eta = K_p K_{pn} \tag{9.1}$$

K_{pn} depends to a large extent on the combination of the p-type dye molecules and n-type dye molecules. Figure 9.14(b) shows the photocurrents caused by the various combinations of p-type dyes (FL, EO, and RB) and n-type dyes (CV, MG, and BG) at the maximum absorption wavelengths of the n-type dye molecules. The horizontal axis of this figure, ΔE_g, represents the difference in the energy gaps between the p-type dye molecules and the n-type dye molecules combined, that is, $\Delta E_g = E_{gp} - E_{gn}$. Here, E_{gp} and E_{gn} are, respectively, energy gaps of the p-type dye molecules and the n-type dye molecules, which were obtained from the absorption peaks of the dye molecules.

As can be seen in Figure 9.14(b), each point is approximately on a line with a slope of −15. If it is assumed that the p-type dye molecule and n-type dye molecule form an energy structure as shown in Figure 9.15, K_{pn} is proportional to $e^{-\Delta E/k_B T}$, where ΔE is the difference in energy between the excited states of the p-type dye molecule and the n-type dye molecule, and k_B is Boltzmann's constant. When K_p is assumed not to depend on dye molecules, the following relationship is obtained:

$$\log I_p \propto -(\log e / k_B T)\Delta E + a \tag{9.2}$$

Here, I_p is the photocurrent and a is a constant. At room temperature, $-\log e / k_B T \approx -17$. This value is close to the slope of −15 for the line in Figure 9.14(b).

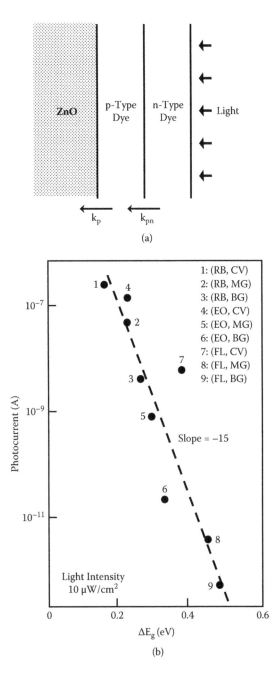

FIGURE 9.14   (a) A schematic model of sensitization with p/n-stacked structures. (b) Photocurrents caused by the various combinations of p-type dyes and n-type dyes at the maximum absorption wavelengths for n-type dye molecules. From T. Yoshimura, K. Kiyota, H. Ueda, and M. Tanaka, "Mechanism of spectral sensitization of ZnO coadsorbing p-type and n-type dyes," *Jpn. J. Appl. Phys.* **20**, 1671–1674 (1981).

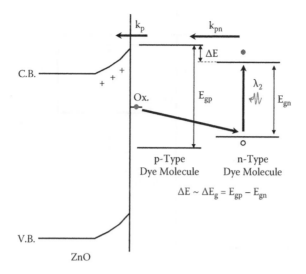

FIGURE 9.15 Mechanism of sensitization with p/n-stacked structures. From T. Yoshimura, K. Kiyota, H. Ueda, and M. Tanaka, "Mechanism of spectral sensitization of ZnO coadsorbing p-type and n-type dyes," *Jpn. J. Appl. Phys.* **20**, 1671–1674 (1981).

Assuming that ΔE corresponds approximately to ΔE_g, the coincidence of the line slope in Figure 9.14(b) with the value of $-\log e \,/\, kT$ confirms the consistency of the model shown in Figure 9.15 for sensitization with p/n-stacked structures. The slight deviation in the line slope from -17 might arise from the fact that ΔE_g contributes not only to ΔE but also to the difference in energy between the ground states of the two dye molecules.

Thus, it is concluded that sensitization with p/n-stacked structures arises from the following process: electrons excited in the n-type dye molecule are transferred to the p-type dye molecule over the barrier of ΔE, and then injected into the ZnO layer. With decreasing ΔE, K_{pn} approaches unity and the sensitization with two dyes becomes large. Thus, the ideal energy structure for sensitization with p/n-stacked structures can be drawn as seen in Figure 9.16. Molecular design to realize such ideal energy structures is a future challenge.

A typical model for sensitization with p/n-stacked structures is summarized in Figure 9.17 for RB and CV. Since the efficiency of the electron injection into ZnO is much higher for the p-type dye molecule than for the n-type dye molecule, the injection efficiency of the electrons from the n-type dye molecule into the ZnO increases as a result of the insertion of the p-type dye molecule between the n-type dye molecule and the ZnO. This is the reason why the dramatic sensitization with p/n-stacked structures occurs for the combination of p-type dye molecules and n-type dye molecules. The electrons excited in the p-type dye moleculae are directly injected into the ZnO. The sensitization effect enables us to widen the photocurrent spectrum, arising from the superposition of the spectra for the p-type molecule and the n-type molecule.

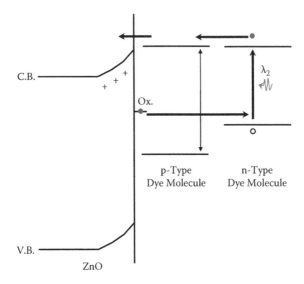

FIGURE 9.16   The ideal energy structure for the sensitization with p/n-stacked structures.

9.1.3.2   Sensitization by p/n-Stacked Structures Constructed by Liquid-Phase MLD

We fabricated the p/n stacked structure on ZnO, that is, the n/p/n structure of ZnO (n-type)/p-type dye molecule/n-type dye molecule, by liquid-phase MLD [12,13] utilizing ZnO thin films with flat surfaces. Photocurrents in the the structure were measured by exposing it to light beams in two configurations—the *guided light configuration* and the *normally incident light configuration*—in order to demonstrate (1) multidye sensitization and (2) the superiority of the waveguide-type dye-sensitized photovoltaic device. The details for issue number 2 are described in Section 9.1.5.

Figure 9.18 shows the liquid-phase MLD process for growth of a stacked structure of p-type and n-type dye molecules on a ZnO surface using p-type dye molecules, RB for molecule A, and n-type dye molecules, CV, for molecule B. When RB in solvent is provided onto the ZnO surface, a monolayer of RB is formed. Once the surface is covered with RB, the deposition of RB is automatically terminated due to the repulsive electrostatic force between the RB molecules. After rinsing the surface with pure solvent, CV in solvent is provided onto the surface. Then, CV is attached to RB due to the attractive force between the RB molecule and the CV molecule, exhibiting monomolecular-step growth. In the present experiment, liquid-phase MLD was performed in two steps to fabricate an n/p/n structure of ZnO/RB/CV, which is regarded as the shortest molecular wire.

Figure 9.19 shows a ZnO sample for photocurrent measurements. A 100-μm-wide slit-type Al electrode with a gap of 60 μm was formed on a glass substrate. A ZnO thin film was deposited on it by vacuum evaporation followed by annealing at 400°C for 1 h in air. To measure photocurrents, 8.5 V was applied to the slit-type electrode.

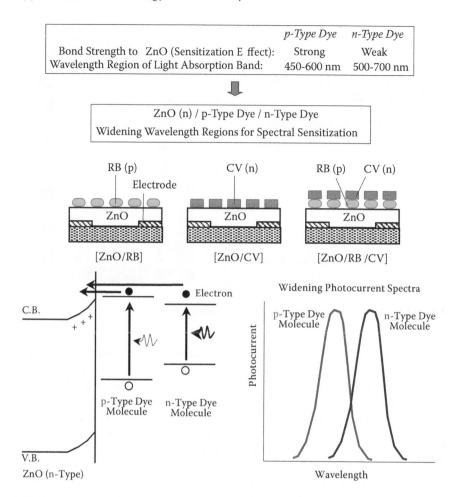

	p-Type Dye	n-Type Dye
Bond Strength to ZnO (Sensitization E ffect):	Strong	Weak
Wavelength Region of Light Absorption Band:	450-600 nm	500-700 nm

ZnO (n) / p-Type Dye / n-Type Dye
Widening Wavelength Regions for Spectral Sensitization

RB (p)   CV (n)   RB (p)   CV (n)

Electrode

ZnO   ZnO   ZnO

[ZnO/RB]   [ZnO/CV]   [ZnO/RB /CV]

C.B.

Electron

Widening Photocurrent Spectra

p-Type Dye Molecule   n-Type Dye Molecule

p-Type Dye Molecule   n-Type Dye Molecule

Photocurrent

V.B.
ZnO (n-Type)

Wavelength

FIGURE 9.17   A typical model for the sensitization with p/n-stacked structures.

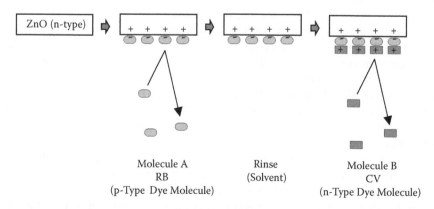

ZnO (n-type)

Molecule A
RB
(p-Type Dye Molecule)

Rinse
(Solvent)

Molecule B
CV
(n-Type Dye Molecule)

FIGURE 9.18   Liquid-phase MLD process for growth of a stacked structure of p-type and n-type dye molecules.

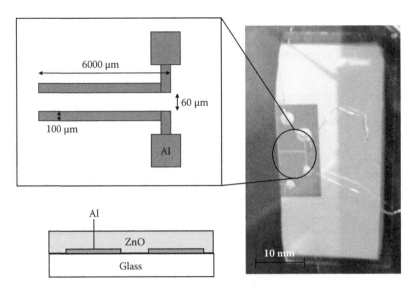

FIGURE 9.19 A ZnO sample for photocurrent measurements.

Figure 9.20 shows setups of photocurrent measurements in the normally incident light configuration and the guided light configuration using light beams from a 405-nm laser diode (LD). For the normally incident light configuration, a light beam was introduced onto the ZnO surface from an optical fiber. In Figure 9.20(a), the light beam spot is put on a place apart from the slit-type electrode in order to observe the spot clearly. For the guided light configuration, a light beam was introduced into the ZnO thin film from the edge of the film. In Figure 9.20(b), a guided light beam is clearly observed. Since 405 nm in wavelength corresponds to photon energy below the band gap of ZnO, small light absorption occurs so that the guided light beam propagates over 5 mm of distance. Photocurrent measurements of ZnO with dye molecules on the surface were carried out using the same configurations described above using lasers of 405 nm, 532 nm, and 633 nm.

To estimate the per-unit light power of photocurrents, we assumed that light power in the gap region of the slit-type electrode is the effective power for photocurrent generation. In our experiments, the effective power for the normally incident light configuration was estimated to be about 25% of incident light power. The effective power for the guided light configuration depends slightly on the wavelength since the fiber core diameters and diffraction angles are different for different wavelengths. In the present experiments it was estimated to be in the range of approximately 5~7% of incident light power.

In Figure 9.21(a), the absorption spectra of RB and CV in alcohol solution are shown. The absorption spectrum of CV is located in a longer wavelength region compared to that of RB. Figure 9.21(b) shows wavelength dependence of photocurrents measured by the guided light configuration in ZnO, ZnO/RB, ZnO/CV, and ZnO/RB/CV. In ZnO, large photocurrents are generated at 405 nm, and no photocurrents

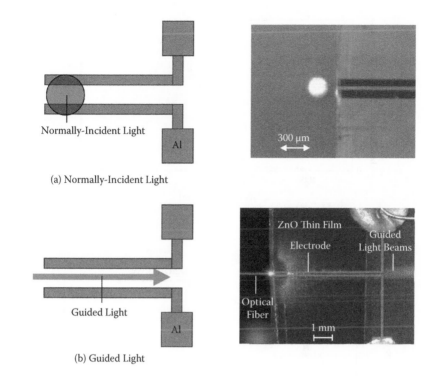

FIGURE 9.20 Setups of photocurrent measurements in (a) the normally incident light configuration and (b) the guided light configuration. From T. Yoshimura, A. Oshima, D. Kim, and Y. Morita, "Quantum dot formation in polymer wires by three-molecule molecular layer deposition (MLD) and applications to electro-optic/photovoltaic devices," *ECS Transactions* **25**, No. 4 "Atomic Layer Deposition Applications 5," 15–25 (2009).

at 532 nm and 633 nm. In ZnO/RB, the photocurrent spectrum extends to 532 nm. In ZnO/RB/CV, the photocurrent spectrum further extends to 633 nm. This is attributed to the fact that the absorption band of CV is located in a longer-wavelength reagion comapred with that of RB.

Note that in ZnO/CV, only small photocurrents are generated at 633 nm in spite of the fact that it has large light absorption at 633 nm. As mentioned in Section 9.1.3.1, electron injection efficiency from an n-type dye molecule like CV to ZnO is low, so only small photocurrents are generated. By inserting RB, which is a p-type dye molecule, between CV and ZnO, the electrons can be injected from CV to ZnO through RB with high efficiency.

As described above, in the structure of ZnO/RB/CV fabricated by liquid-phase MLD, widening of the photocurrent spectra was observed, providing the proof of concept for multidye sensitization shown in Figure 9.2(b). To extend the sensitized wavelength regions, longer molecular wires with more than three kinds of dye molecules should be formed on ZnO thin films by increasing the MLD step counts. The experimental trial is underway.

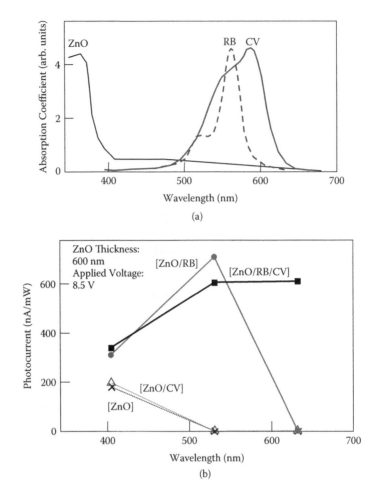

FIGURE 9.21   (a) Absorption spectra of ZnO, RB and CV, and (b) wavelength dependence of photocurrents measured by the guided light configuration in ZnO, ZnO/RB, ZnO/CV, and ZnO/RB/CV.

9.1.4   PROOF OF CONCEPT OF POLYMER-MQD SENSITIZATION

As mentioned in Section 4.2.2, by controlling molecular arrangements in conjugated poly-AM wires with designated sequences of three kinds of molecules using MLD, polymer wires with MQDs can be fabricated [4]. Figure 9.22 shows an example of a poly-AM-based polymer MQD by MLD with terephthalaldehyde (TPA), *p*-phenylenediamine (PPDA), and oxalic dihydrazide (ODH). Quantum dots of OT, OTPT, and OTPTPT are formed in a polymer wire to provide quantum dots with different lengths. The structure can be used for the polymer-MQD sensitization shown in Figure 9.2(c) [3,4].

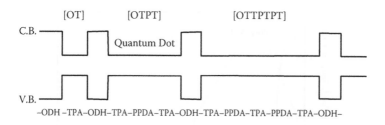

[OT]   [OTPT]   [OTTPTPT]

C.B. ──

Quantum Dot

V.B. ──

−ODH −TPA−ODH−TPA−PPDA−TPA−ODH−TPA−PPDA−TPA−PPDA−TPA−ODH−

FIGURE 9.22 An example of poly-AM-based polymer MQD by MLD with TPA, PPDA, and ODH.

In order to clarify the feasibility of the polymer MQD as a sensitizer for ZnO thin films, we examined the sensitizing ability of poly-AM wires as a preliminary trial.

In poly-AM, since the absorption band is located in wavelength regions of 350–500 nm as Figure 9.23(a) shows, sensitization in the blue-green regions is expected. Figure 9.23(b) shows wavelength dependence of photocurrents measured by the guided light configuration in ZnO and ZnO/poly-AM, which is a ZnO thin film with poly-AM grown on the surface by the carrier gas–type MLD. In ZnO/poly-AM, the photocurrents at 405 nm are enhanced, and the photocurrent spectrum extends to 532 nm, indicating that poly-AM wires have sensitizing ability in the blue-green regions.

It is preferable to grow poly-AM wires vertically from the ZnO surface to construct the polymer MQD structures shown in Figure 9.2(c). However, the sensitization effect is available even when the poly-AM wires are grown horizontally on the ZnO surface, as shown on the right-hand illustration of Figure 9.1(b). The control of the wire orientation can be done by SAM or other surface treatment [21–23].

9.1.5 Proof of Concept of Waveguide-Type Photovoltaic Devices

As a measure of the photocurrent enhancement induced by the guided light configuration, photocurrent enhancement ratio (PCER) is defined as follows.

$$PCER = \frac{[\text{Photocurrent/Incident Light Power in Guided Light Configuration}]}{[\text{Photocurrent/Incident Light Power in Normally Incident Light Configuration}]}$$

As Figure 9.24 shows, large PCERs of ~5, ~15, and ~8 are observed in ZnO at 405 nm, in ZnO/RB at 532 nm, and in ZnO/RB/CV at 633 nm, demonstrating the possibility of waveguide-type photovoltaic devices.

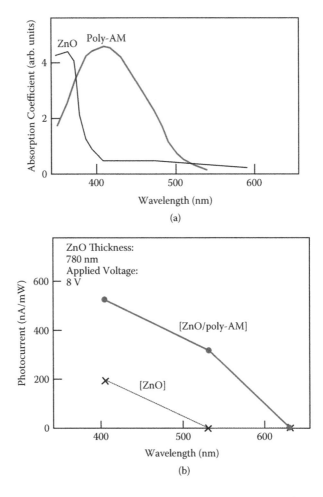

FIGURE 9.23   (a) Absorption spectra of ZnO and poly-AM, and (b) wavelength dependence of photocurrents measured by the guided light configuration in ZnO and ZnO/poly-AM.

Figure 9.25 shows wavelength dependence of the photocurrents in ZnO/poly-AM. Here, GL and NIL, respectively, represent the guided light configuration and the normally incident light configuration. PCERs of ~3 and ~12 are observed at 405 nm and at 532 nm, respectively. PCER is larger at 532 nm, where the light absorption of poly-AM is small, than PCER at 405 nm, where the light absorption of poly-AM is relatively large. It is found from the result that the photocurrent enhancement in the guided light configuration is more remarkable in wavelength regions with smaller light absorption. This is because in a wavelength region where the thin films have large light absorption, the thin films can absorb light to some extent even in the normally incident light configuration to generate photocurrents, and consequently, the PCER is reduced.

Thus, it is clarified that guided light beams can greatly enhance light absorption efficiency. This suggests that waveguide-type photovoltaic devices are promising to

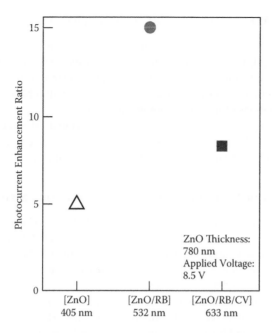

FIGURE 9.24 Photocurrent enhancement in ZnO, ZnO/RB, and ZnO/RB/CV.

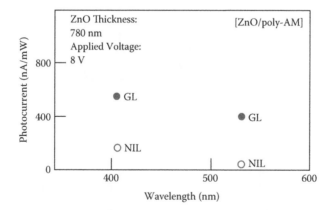

FIGURE 9.25 Wavelength dependence of the photocurrents in ZnO/poly-AM for the guided light configuration (GL) and the normally incident light configuration (NIL).

improve the efficiency of sensitized photovoltaic devices with flat-surface thin-film semiconductors. As future work, practical waveguide-type photovoltaic devices with multidye sensitization and polymer-MQD sensitization will be fabricated.

In waveguide-type photovoltaic devices, the optical coupling method for introducing light beams into the optical waveguides of the photovoltaic device is an important issue, which is discussed in Section 9.1.2.

9.2 INTEGRATED SOLAR ENERGY CONVERSION SYSTEMS

The integrated film systems shown in Figure 9.26 will provide various types of ubiquitous functions [1], including solar energy conversion films, wearable computers, system glasses, thin-film speakers, and so on. The integrated film systems whose structures are similar to those of optical interconnects within boxes, are one of the important applications of thin-film photonic devices.

In the present section, concepts of film-based integrated solar energy conversion systems and related devices such as the integrated photonic/electronic/chemical system (IPECS) and light beam collecting films are proposed [1,2,24–27].

9.2.1 CONCEPT OF INTEGRATED SOLAR ENERGY CONVERSION SYSTEMS

Solar energy conversion systems have received much attention because of their ability to resolve the problems of the environment and serious energy shortages. In conventional energy conversion systems, semiconductors such as Si are placed all over the solar energy conversion modules, so it is foreseen that the cost of the systems will increase because of a shortage of semiconductor materials. In systems with Fresnel lens/mirror concentrators, semiconductor materials can be saved because solar beams are focused on small semiconductor cells. However, the system size becomes very large, and furthermore, they need solar-beam tracking mechanisms to keep the focusing point of the solar beams on the small-sized semiconductor cells.

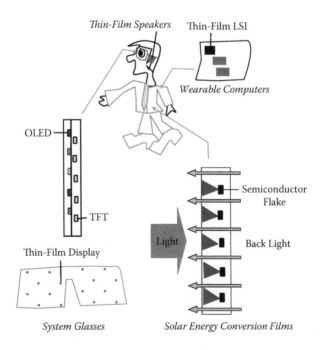

FIGURE 9.26 Integrated film systems.

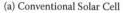

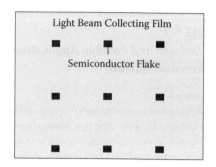

(a) Conventional Solar Cell

(b) Film-Based Integrated
Solar Energy Conversion System

FIGURE 9.27 Concept of (a) the conventional solar cell and (b) the film-based integrated solar energy conversion system. From R. Shioya and T. Yoshimura, "Design of solar beam collectors consisting of multi-layer optical waveguide films for integrated solar energy conversion systems" *Journal of Renewable and Sustainable Energy* **1**, 033106-1-15 (2009).

To reduce the consumption of semiconductor materials, to provide wide-angle light beam collecting capability to these systems, and to make the systems flexible and compact, we proposed film-based integrated solar energy conversion systems with optical waveguides [1,2,24–27].

Figure 9.27 shows schematic illustrations of the conventional solar cell and the film-based integrated solar energy conversion system using optical waveguides for light beam collecting. In integrated solar energy conversion systems, semiconductor flakes are placed partially in a light beam collecting film by a heterogeneous integration process such as PL-Pack with SORT described in Section 7.5. In some cases, the semiconductor flakes are placed outside of the film with solar beams guided by optical fibers.

The advantages and possible applications of the film-based integrated solar energy conversion systems are summarized in Table 9.2. The proposed structures can reduce semiconductor material consumption. This enables us to use expensive III-V compound semiconductors, which have higher photovoltaic conversion efficiency than Si. Conversion efficiency could be improved by delivering collected light beams of different wavelength regions to semiconductors having matched sensitivity spectra using wavelength filters in the optical circuits of the light beam collecting films. Since the integrated solar energy conversion system has waveguide structures, the waveguide-type photovoltaic devices described in Section 9.1 are easily integrated into the system.

9.2.2 THE INTEGRATED PHOTONIC/ELECTRONIC/CHEMICAL SYSTEM (IPECS)

A schematic model of IPECS is shown in Figure 9.28(a) [1,2,24–26]. IPECS has a structure similar to the integrated optical interconnects shown in Figure 9.28(b), and is built by following core technologies: 3-D optical circuits for constructing the

TABLE 9.2

Advantages and Possible Applications of Film-Based Integrated Solar Energy Conversion Systems

Advantages

- Reduction of semiconductor/ITO consumption, cost, size, weight
- Flexibility and wide-angle light beam collection capability
- Function availability by implementation of optical circuits (demultiplexing for delivering appropriate wavelength light beams to semiconductors having matched sensitivity spectra)
- Function availability of embedding thin-film batteries and ICs (providing charge storage and current regulation capability to the films).
- Preferable for integration of the waveguide-type photovoltaic devices

Applications

- Cars, robots, laptop computers, cellular phones, consumer electronics, desk, wall, cloths, hand-carried equipment, sheet, umbrella, buildings, fences, roads, etc.

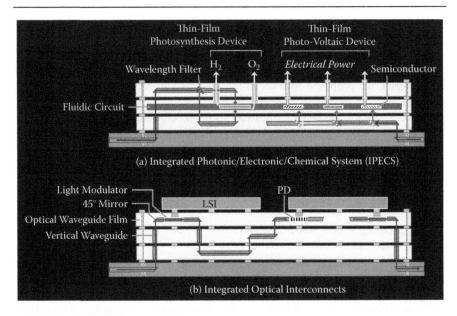

FIGURE 9.28   (a) Integrated photonic/electronic/chemical system (IPECS) and (b) integrated optical interconnects. From T. Yoshimura, "Proposed Applications of 3-D optical interconnect technologies to integrated chemical systems," *Digest of the LEOS Summer Topical Meetings* (Bio-Inspired Sensors and Application/Imprinting on Photonic Integrated Circuits), Portland, Oregon, 129–130 (2007).

backbone of light beam collecting films, PL-Pack with SORT for embedding thin-film photovoltaic devices, SOLNET for self-aligned optical couplings between optical waveguides and thin-film photovoltaic devices, and MLD for high-performance thin-film photovoltaic devices. In some cases, IPECS contains fluidic circuits for supplying electrolytes to the system.

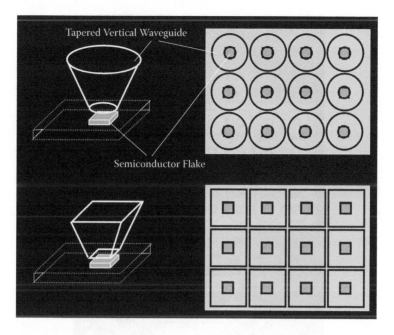

FIGURE 9.29   Integrated solar energy conversion system with a tapered vertical waveguide-type light beam collecting film.

9.2.3 Structures of Light Beam Collecting Films

We proposed three types of light beam collecting films: The tapered vertical waveguide type, the tapered vertical/horizontal waveguide type, and the multilayer waveguide type.

Figure 9.29 shows an example of the integrated solar energy conversion system with a tapered vertical waveguide-type light beam collecting film. Thin-film photovoltaic devices or photosynthesis devices containing semiconductor flakes are embedded in the film. The incident light beams received by the tapered vertical waveguides are guided to the semiconductor flakes, whose areas are smaller than the areas of the light beam collecting aperture of the vertical waveguide. This enables a reduction of semiconductor consumption, that is, cost for solar energy conversion. In this example, the material consumption can be reduced to ~1/9 of the consumption in conventional solar cells.

Figure 9.30 shows an example of the system with a tapered vertical/horizontal waveguide-type light beam collecting film. A film with tapered vertical waveguides and a film with horizontal waveguides having vertical mirrors are stacked. The incident light beams are introduced into the horizontal waveguides through the vertical mirrors and guided to embedded semiconductor flakes. This enables further reduction of semiconductor consumption. In this example, the material consumption can be reduced to ~(1/9) × (1/16) = ~1/144 of the consumption in conventional solar cells. By embedding wavelength filters into the waveguide films, solar beams can branch

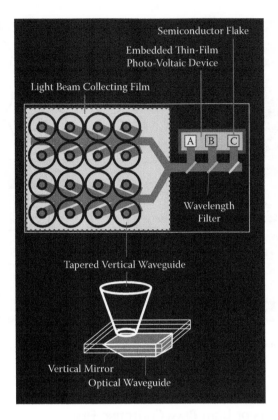

FIGURE 9.30 Integrated solar energy conversion system with a tapered vertical/horizontal waveguide-type light beam collecting film. From T. Yoshimura, "Proposed Applications of 3-D optical interconnect technologies to integrated chemical systems," *Digest of the LEOS Summer Topical Meetings* (Bio-Inspired Sensors and Application/Imprinting on Photonic Integrated Circuits), 129–130 (2007).

to semiconductors having matched spectral responses. For example, UV-blue beams to ZnO, visible beams to Si, and infrared beams to InP.

Figure 9.31 shows an example of the system with a multilayer waveguide-type light beam collecting film. Many films with horizontal waveguides having vertical mirrors are stacked. The incident light beams are introduced through the vertical mirrors into the horizontal waveguides and are guided to semiconductor flakes.

An example of the integration of a waveguide-type photovoltaic device and a light beam collecting film is shown in Figure 9.32. The incident light beams received by the tapered vertical waveguides are guided to the bottom, where cone-shaped 45° mirrors are placed. The light beams are introduced into waveguide-type photovoltaic devices through the 45° mirrors.

Figure 9.33 shows an example of coloring of the integrated solar energy conversion systems by back lights. In conventional solar cells, since semiconductors are placed on the whole substrate, the cells look black. To avoid the monochromatic characteristics of solar cells, in some dye-sensitized solar cells, colored appearance

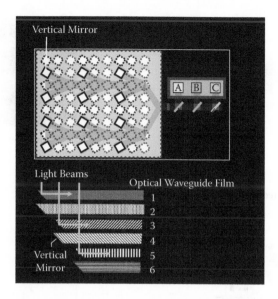

FIGURE 9.31 Integrated solar energy conversion system with a multilayer waveguide-type light beam collecting film.

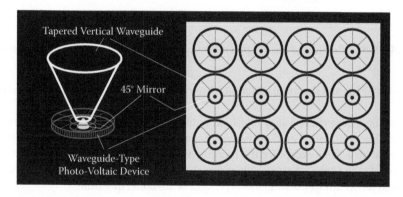

FIGURE 9.32 An example of integration of a waveguide-type photovoltaic device and a light beam collecting film.

is available by suppressing light absorption in some wavelength regions. This causes reduction of energy conversion efficiency. In integrated solar energy conversion systems, back light can pass through the light beam collecting film. By adjusting the color of the back lighting, specific coloration of the film may be achievable, which improves the appearance of solar cells.

9.2.4 Design of Light Beam Collecting Films

In the present subsection, the results of the design of the light beam collecting films and estimation of their light beam collecting ability by the beam propagation

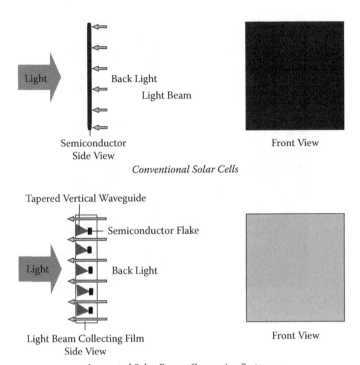

FIGURE 9.33   Coloring of the integrated solar energy conversion systems by back lights.

method (BPM) and the finite difference time domain (FDTD) method are presented [27].

The tapered vertical waveguide-type light beam collecting film and the tapered vertical/horizontal waveguide-type light beam collecting film are basically regarded as similar structures except for the merging waveguides. So the simulation was carried out for the tapered vertical/horizontal waveguide-type light beam collecting film and the multilayer waveguide-type light beam collecting film.

9.2.4.1   Simulation Procedure

An example of a simulation model for light beam collection in the waveguide-type light beam collecting film is shown in Figure 9.34. An incident light beam from a metal mirror is introduced into a horizontal waveguide via a tapered vertical waveguide and a vertical mirror. The refractive index of the optical waveguides is 1.6, and the refractive index of the background is 1.0, assuming air. The width of the horizontal waveguide is 4 μm. An antireflective (AR) coating with a 100-nm thickness and a refractive index of 1.3 is applied to the top and the bottom of the optical waveguide. An incident light beam with a large width is preferable to reproduce the condition of plane waves of sunlight. However, when the width is broadened, the area of the FDTD calculation must be large, making computation time extremely long. So in the present

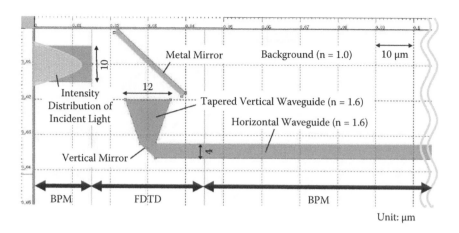

FIGURE 9.34 An example of a simulation model for light beam collection in waveguide-type light beam collecting films.

work, the simulations were performed using a fundamental guided beam in an optical waveguide with a width of 10 μm for the incident light beam. Incident angles of the light beam are adjusted by the metal mirror. Wavelengths of the incident light beams are 400, 600, and 800 nm. The length of the model from the left boundary to the right boundary is 200 μm. Light intensity at the output (the right boundary of the model) is calculated to estimate the light beam collecting efficiency for various models.

9.2.4.2 Tapered Vertical/Horizontal Waveguide-Type Light Beam Collecting Films

In Figure 9.35, models and the light intensity for light beam propagation in wave-guide-type light beam collecting films are shown. Figure 9.35(a) shows a model without the tapered vertical waveguide. Considerable leakage is observed at the vertical mirror. The light beam collecting efficiency was 61%.

Figures 9.35(b) and (c) show models of tapered vertical/horizontal waveguide-type light beam collecting films with a tapered vertical waveguide having a high-refractive-index upper part (Model UH) and a high-refractive-index lower part (Model LH), respectively. The tapered angle of the tapered vertical waveguide is $\lambda_T = 17°$. While the light beam collecting efficiency was 61% in Model UH, the efficiency increased up to 77% in Model LH. This can be explained as follows. Since the light beam spot size becomes small with propagation of the light beams to the lower region of the tapered vertical waveguide, stronger light beam confinement is necessary in the lower region. By putting the high-refractive-index material in the lower regions, strong light beam confinement can be effectively achieved, resulting in an increase in the light beam collecting efficiency.

In order to increase the collecting efficiency further, the tapered angles and refractive index distribution profiles in the tapered vertical waveguides should be optimized.

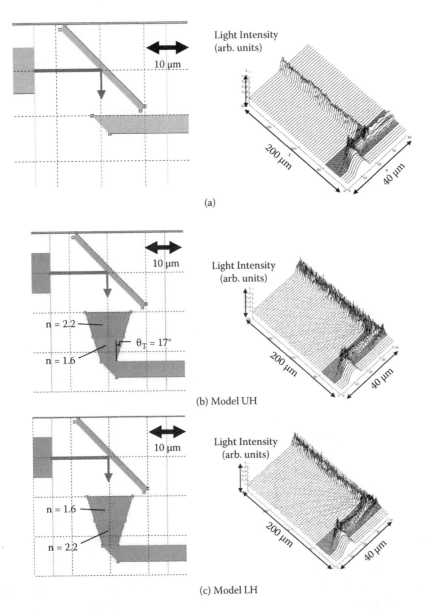

FIGURE 9.35 Models and the propagating light intensity for tapered vertical/horizontal waveguide-type light beam collecting films. (a) Without the tapered vertical waveguide, (b) with a tapered vertical waveguide having high-refractive-index upper part (Model UH), and (c) with a tapered vertical waveguide having high-refractive-index lower part (Model LH).

9.2.4.3 Multilayer Waveguide-Type Light Beam Collecting Films

A model for the multilayer waveguide-type light beam collecting film is shown in Figure 9.36. The model consists of stacked waveguide films. In this example, as shown in the side view, 18 optical waveguide films are stacked. Numbers in the top view indicate vertical mirror apertures, that is, the light beam collecting apertures, of a film denoted by the numbers. An example of optical circuits in a film is illustrated in Figure 9.37. Merging waveguides for light beam collection from the vertical mirrors through the branching waveguides to the main waveguide are integrated. By arranging the films regularly with one-aperture displacement, the surface can be occupied by the light beam collecting apertures in high density, achieving aperture occupation of 100%, theoretically.

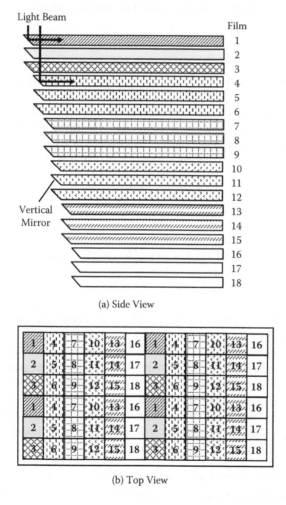

(a) Side View

(b) Top View

FIGURE 9.36 Model for the multilayer waveguide-type light beam collecting film. From R. Shioya and T. Yoshimura, "Design of solar beam collectors consisting of multi-layer optical waveguide films for integrated solar energy conversion systems" *Journal of Renewable and Sustainable Energy* **1**, 033106-1-15 (2009).

Vertical Mirror   Branching Waveguide   Joint   Main Waveguide

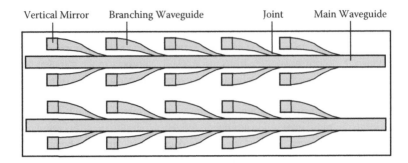

FIGURE 9.37   An example of optical circuits of merging waveguides.

9.2.4.3.1 Effects of Incident Angles on Collecting Efficiency for Single-Layer Optical Waveguide Films

Figure 9.38 shows models with 45° mirrors and the propagating light intensity for incident angles of 0°, 45°, and −45°. The wavelength of incident light beams is 600 nm. Here, in the model of −45°, the waveguide direction is flipped horizontally, which effectively reproduces a configuration for −45° incidence with the same incident light beam direction as for the 45° incidence. For an incident angle of 0°, a light beam is introduced into the optical waveguide through the 45° mirror, and then guided in the optical waveguide smoothly, indicating that the fundamental mode is dominant. For incident angles of 45° and −45° a lot of leakage of light beams passing through the 45° mirror are observed. At the same time considerable higher modes are excited in the guided beams.

Figure 9.39 shows models with 30° mirrors and the propagating light intensity for incident angles of 0° and 45°. For an incident angle of 0°, a light beam guided in the optical waveguide contains considerable higher modes. For an incident angle of 45°, on the other hand, the guided beam propagates smoothly with a small amount of higher modes. Figure 9.40 shows models with 60° mirrors and the propagating light intensity for incident angles of 0° and 45°. The guided beam contains a lot of higher modes for both incident angles.

The light beam collecting efficiency is summarized in Figure 9.41. For an incident angle of 0°, the 45° mirror exhibits the largest efficiency of 62%, and for an incident angle of 45°, the 30° mirror exhibits the largest efficiency of 47%. The 60° mirror does not exhibit great efficiency for incident angles of 45° and 0°. From these results, it is concluded that optical waveguides with 45° mirrors and those with 30° mirrors are respectively useful for the incident light beam with a vertical direction and for that with a tilted direction.

9.2.4.3.2 Effects of Layer Counts on Collecting Efficiency

Figure 9.42 shows a model of a three-layer waveguide structure with 45° mirrors and the propagating light intensity for an incident light beam with a wavelength of 600 nm. The distance between optical waveguides is 1 μm. It is found that the incident light beam, which goes through the first layer, is received by the second layer and the third layer to increase the light beam collecting efficiency. As Table 9.3 shows,

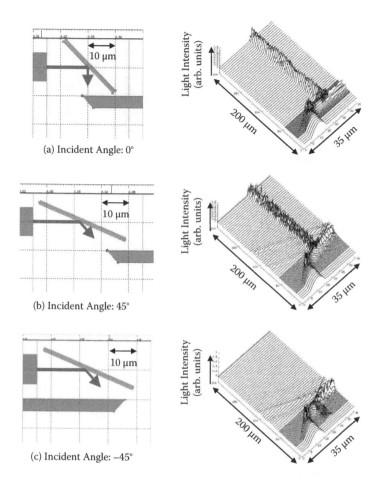

(a) Incident Angle: 0°

(b) Incident Angle: 45°

(c) Incident Angle: −45°

FIGURE 9.38 Models with 45° mirrors and propagating light intensity for incident angles of (a) 0°, (b) 45°, and (c) −45°. From R. Shioya and T. Yoshimura, "Design of solar beam collectors consisting of multi-layer optical waveguide films for integrated solar energy conversion systems" *Journal of Renewable and Sustainable Energy* **1**, 033106-1-15 (2009).

light beam collecting efficiency increases from 62% to 87% with increasing the layer counts from one to three.

9.2.4.3.3   Collecting Efficiency in Multi-Layer Waveguides with 45° and 30° Mirrors

A schematic structure of a light beam collecting film consisting of multilayer optical waveguides with 45° and 30° mirrors is illustrated in Figure 9.43. The first layer is an optical waveguide with a 45° mirror, and the second and third layers are optical waveguides with a 30° mirror. The two optical waveguides with 30° mirrors are arranged in opposite directions. The first layer is for incident light beams with vertical directions. The second layer is for beams from the left side direction, and the third layer is

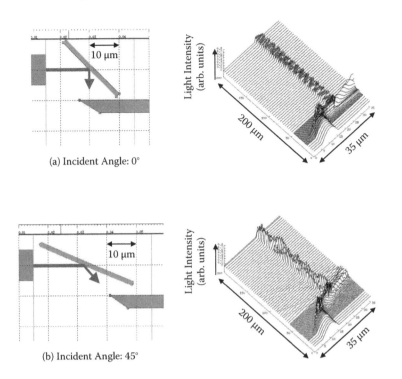

(a) Incident Angle: 0°

(b) Incident Angle: 45°

FIGURE 9.39   Models with 30° mirrors and propagating light intensity for incident angles of (a) 0° and (b) 45°. From R. Shioya and T. Yoshimura, "Design of solar beam collectors consisting of multi-layer optical waveguide films for integrated solar energy conversion systems" *Journal of Renewable and Sustainable Energy* **1**, 033106-1-15 (2009).

for beams from the right side direction. This structure is expected to provide wide-angle light beam collecting capability without solar beam tracking mechanisms.

Figure 9.44 shows models of the three-layer structure and the propagating light intensity. The wavelength of the incident light beams is 600 nm. It is found that the light beam is mainly collected by the first layer for vertical incidence, by the second layer for tilted incidence from the left, and by the third layer for tilted incidence from the right.

For comparison, the light beam collecting characteristics of a light beam collecting film consisting of a three-layer structure with only 45° mirrors are examined. The structure is shown in Figure 9.45. It is the same as the model shown in Figure 9.44 except for the mirror angles. While the collected light beam propagates in the optical waveguides smoothly for vertical incidence, for tilted incidence the collected light beam is reduced and propagates in considerably higher modes.

Figure 9.46 shows dependencies of light beam collecting efficiency on incident angles for the three-layer structure having optical waveguides with 45° and 30° mirrors (denoted by 45°/30°/30°) and the structure having optical waveguides with only 45° mirrors (denoted by 45°/45°/45°). The 45°/30°/30° version exhibits higher efficiency than 45°/45°/45°. Especially for an incident angle of 45°, the efficiency is 1.5 times larger in 45°/30°/30° than in 45°/45°/45°. This indicates that the light beam

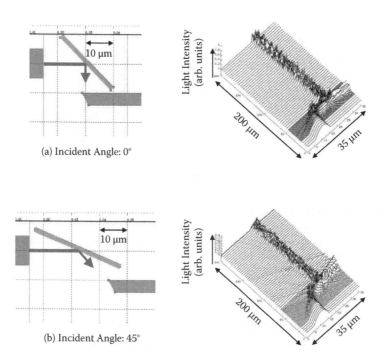

(a) Incident Angle: 0°

(b) Incident Angle: 45°

FIGURE 9.40 Models with 60° mirrors and propagating light intensity for incident angles of (a) 0° and (b) 45°. R. Shioya and T. Yoshimura, "Design of solar beam collectors consisting of multi-layer optical waveguide films for integrated solar energy conversion systems" *Journal of Renewable and Sustainable Energy* **1**, 033106-1-15 (2009).

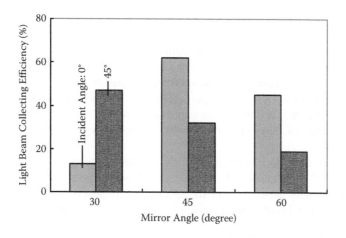

FIGURE 9.41 Light beam collecting efficiency in models with 30°, 45°, and 60° mirrors for incident angles of 0° and 45°. From R. Shioya and T. Yoshimura, "Design of solar beam collectors consisting of multi-layer optical waveguide films for integrated solar energy conversion systems" *Journal of Renewable and Sustainable Energy* **1**, 033106-1-15 (2009).

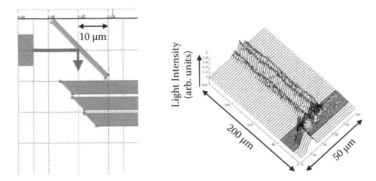

FIGURE 9.42 Model of a three-layer waveguide structure with 45° mirrors and propagating light intensity. From R. Shioya and T. Yoshimura, "Design of solar beam collectors consisting of multi-layer optical waveguide films for integrated solar energy conversion systems" *Journal of Renewable and Sustainable Energy* **1**, 033106-1-15 (2009).

TABLE 9.3
Light Beam Collecting Efficiency for
Layer Counts of One, Two, and Three

Layer Count	Light Beam Collecting Efficiency (%)
1	62
2	83
3	87

collecting film consisting of multilayer optical waveguides with mirrors of different angles provides wide-angle light beam collecting capability.

It should be noted that the model shown in Figure 9.44 has 4 μm × 4 μm aperture in top views. This corresponds to the aperture occupation for a layer count of one in Table 9.3. So, when a 3 × (4 μm × 4 μm) aperture, which corresponds to a layer count of three in Table 9.3, is built by stacking two more film units of 45°/30°/30°, light beam collecting efficiency of more than 80% might be expected for incident angles of 45° and 0°.

The light beam collecting efficiency for an incident angle of 45° were calculated to be 60% and 65% for wavelengths of 400 and 800 nm, respectively. Thus, it is found that the wavelength dependence of the efficiency is relatively small, indicating that the proposed model could operate in wavelength regions from 400 nm to 800 nm.

9.2.4.3.4 Overall Consideration

Overall light beam collecting efficiency $\eta_{\text{Over All}}$ is given as follows:

$$\eta_{\text{Over All}} = \eta_{\text{Mirror}} \times \eta_{\text{Merging}} \times \eta_{\text{Semiconductor}} \qquad (9.3)$$

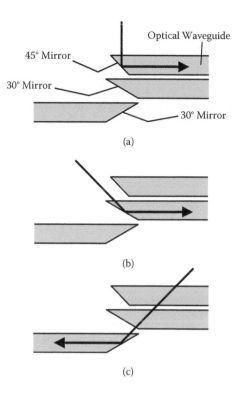

45° Mirror

Optical Waveguide

30° Mirror

30° Mirror

(a)

(b)

(c)

FIGURE 9.43 Schematic structure of a light beam collecting film consisting of multilayer optical waveguides with 45° and 30° mirrors. (a) Incident light beams with vertical directions, (b) incident light beams from the left side directions, and (c) incident light beams from the right side directions. From R. Shioya and T. Yoshimura, "Design of solar beam collectors consisting of multi-layer optical waveguide films for integrated solar energy conversion systems" *Journal of Renewable and Sustainable Energy* **1**, 033106-1-15 (2009).

Here, η_{Mirror}, $\eta_{Merging}$, and $\eta_{Semiconductor}$ are, respectively, the light beam collecting efficiency via a vertical mirror, the light beam propagation efficiency via merging waveguides, and the coupling efficiency between an optical waveguide and a semiconductor flake.

As mentioned in Sections 9.2.4.3.2 and 9.2.4.3.3, η_{Mirror} of the tapered vertical/horizontal waveguide-type light beam collecting film (Model LH) and η_{Mirror} of the multilayer waveguide-type light beam collecting film (three-layer structure of 45°/30°/30°) are expected to be around 80%. For $\eta_{Semiconductor}$, when ZnO (refractive index n: 2.1) is used for the semiconductor and polymers (n: 1.52) for optical waveguides, reflection loss at the interfaces is ~16%. The coupling efficiency may increase to ~100% by inserting an appropriate AR coating.

$\eta_{Merging}$ was calculated by BPM. The model is similar to that shown in Figure 9.37. The branching angle is 10°. The branching waveguides are tapered from a width of 4 μm at the vertical mirror to 0.1 μm at the joint, which is effective to increase the light beam propagation efficiency. Figure 9.47(a) shows the dependence of total propagation efficiency on the number of joints when light beams are introduced into all the

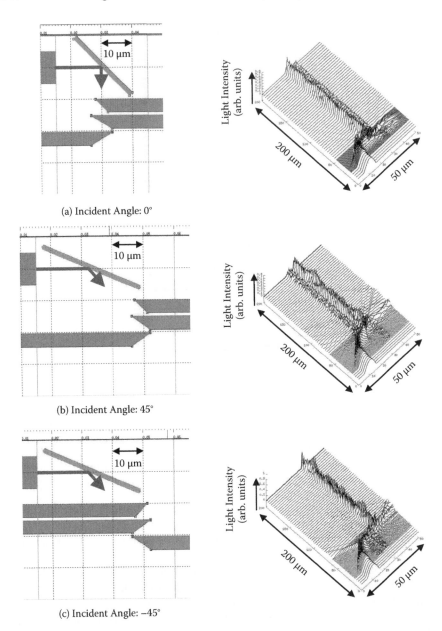

(a) Incident Angle: 0°

(b) Incident Angle: 45°

(c) Incident Angle: −45°

FIGURE 9.44 Models of the three-layer structure with 45° and 30° mirrors and propagating light intensity for incident angles of (a) 0°, (b) 45°, and (c) −45°. From R. Shioya and T. Yoshimura, "Design of solar beam collectors consisting of multi-layer optical waveguide films for integrated solar energy conversion systems" *Journal of Renewable and Sustainable Energy* **1**, 033106-1-15 (2009).

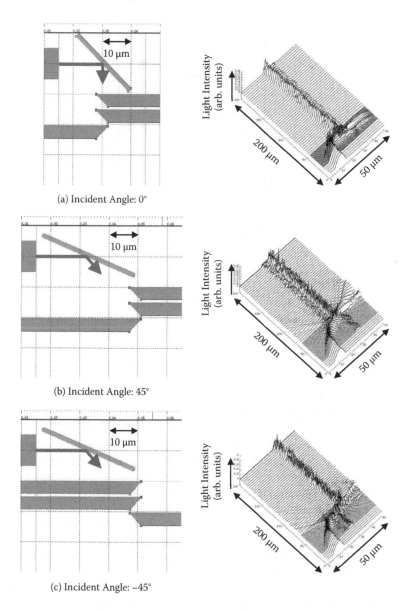

(a) Incident Angle: 0°

(b) Incident Angle: 45°

(c) Incident Angle: −45°

FIGURE 9.45 Models of the three-layer structure with only 45° mirrors and propagating light intensity for incident angles of (a) 0°, (b) 45°, and (c) −45°. From R. Shioya and T. Yoshimura, "Design of solar beam collectors consisting of multi-layer optical waveguide films for integrated solar energy conversion systems" *Journal of Renewable and Sustainable Energy* **1**, 033106-1-15 (2009).

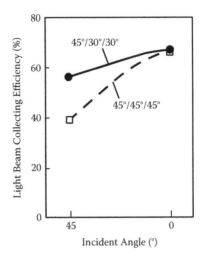

FIGURE 9.46 Dependence of light beam collecting efficiency on incident angles for 45°/30°/30° and 45°/45°/45°. From R. Shioya and T. Yoshimura, "Design of solar beam collectors consisting of multi-layer optical waveguide films for integrated solar energy conversion systems" *Journal of Renewable and Sustainable Energy* **1**, 033106-1-15 (2009).

vertical mirrors. The efficiency decreases with increases in the number of joints. For a joint count of 8, the efficiency is 52%.

By substituting $\eta_{Mirror} \sim 80\%$, $\eta_{Merging} \sim 50\%$, and $\eta_{Semiconductor} \sim 100\%$ into Equation (9.3), the expected overall light beam collecting efficiency is calculated to be ~40% for the joint count of 8.

In the case of the eight-step merging, light beams from 16 vertical mirrors are collected into one semiconductor flake. Therefore, the semiconductor consumption reduction factor, denoted by F_{SCR}, is 1/16 in an integrated solar energy conversion system with the multilayer waveguide-type light beam collecting film. For the systems with the tapered vertical/horizontal waveguide-type light beam collecting film, as discussed in Section 9.2.3, semiconductor consumption reduction of 1/9 is obtained at the tapered waveguide part. Therefore, the eight-step merging gives us F_{SCR} of $(1/9) \times (1/16) = 1/144$. In the conventional solar cell, F_{SCR} is 1 because the semiconductor is placed over the entire module.

As a measure of resource-saving ability, a ratio of $\eta_{Over\ All} /F_{SCR}$ is defined. This ratio is proportional to the amount of energy conversion achieved by semiconductors of unit volume. In the conventional solar cell, since $\eta_{Over\ All}$ is nearly equal to 100%, $\eta_{Over\ All} /F_{SCR}$ ratio is 100/1 = 100. For integrated solar energy conversion systems with multilayer waveguide-type light beam collecting film, influence of the number of joints in the merging waveguides on the $\eta_{Over\ All} /F_{SCR}$ ratio is shown in Figure 9.47(b). The ratio increases with the joint counts, followed by saturation. In the case of eight-step merging, the ratio is 40/(1/16) = 640, which is 6.4 times larger than the ratio in the conventional solar cell. For the systems with the tapered vertical/horizontal waveguide-type light beam collecting film, the $\eta_{Over\ All} /F_{SCR}$ ratio might be 40/(1/144) = 5760, which is 58 times larger than the ratio in the conventional solar

Branching Angle: 10°
Taper of the Branching Waveguide:
4 μm – 0.6 μm – 0.1 μm

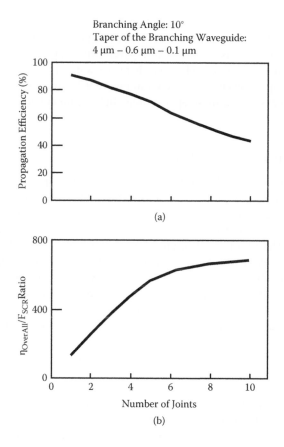

(a)

(b)

FIGURE 9.47 Dependence of (a) total propagation efficiency and (b) ratio of $\eta_{Over\ All}/F_{SCR}$ on the number of joints in the merging waveguides.

cell. Thus, it is concluded that the proposed integrated solar energy conversion systems are expected to realize savings in materials.

In the multilayer waveguide-type light beam collecting film, in order to achieve aperture occupation of 100%, 39 layers are necessary for one angle of the vertical mirrors when the branching angle and the joint counts are, respectively, 10° and 8 in the merging waveguide. This 39-layer structure forms one module unit. To construct a multilayer structure consisting of optical waveguides with 45° and 30° mirrors, one module unit for 45° mirrors and four module units for 30° mirrors are necessary to be stacked in order to collect light beams with vertical incident angles and tilted incident angles (from the left side, right side, front side, and back side). When the waveguide core thickness is 4 μm and the distance between waveguides is 1 μm, the module unit thickness is around 195 μm. Then, total thickness of the stack of five module units is 975 μm.

The estimated efficiency of the waveguide-type light beam collecting films is lower than that of Fresnel lens/mirror concentrators. So, at present, the light beam collecting films do not seem to be suitable for large-scale solar energy conversion centers. However, they have an outstanding feature of flexible light-weight films, enabling us to carry them as sheets everywhere and put them on cars, robots, buildings, roofs of fragile houses, surfaces of pavements/roads, walls, tables, various kinds of equipment, clothes, and everywhere we want. They also have a unique feature of color appearance as illustrated in Figure 9.33. Since very high efficiency is not always necessary in these applications, light beam collecting film may still be useful.

9.2.5 POSSIBLE FABRICATION PROCESS

The built-in mask method and PL-Pack with SORT respectively described in Sections 7.6 and 7.5 can be applied to fabrication of light beam collecting films. For the tapered vertical waveguide-type light beam collecting film, a film with an array of tapered vertical waveguides is formed by tilted exposure using the built-in mask method. The film is stacked on a film, into which semiconductor flakes are embedded by PL-Pack with SORT. For the tapered vertical/horizontal waveguide-type light beam collecting film, a film with an array of tapered vertical waveguides is stacked on an optical waveguide film containing optical circuits of merging horizontal waveguides with embedded semiconductor flakes.

For the multilayer waveguide-type light beam collecting film, optical waveguide films containing optical circuits with merging horizontal waveguides are stacked on each other with sliding with one-vertical mirror-aperture displacement. The film separation can be kept by inserting spacers between films. With respect to light beam confinement, all-air-cladding waveguides are preferable. The all-air-cladding structures can be built by core transfers from one substrate to another substrate using PL-Pack with SORT, which enables us to place waveguide cores and spacers onto films at one time as demonstrated in Section 7.5.3. For waveguide fabrication, the built-in mask method as well as imprinting and stamping can be used.

9.2.6 IMPACT OF POLYMER MQDS ON INTEGRATED SOLAR ENERGY CONVERSION SYSTEMS

For the photovoltaic device embedded in the waveguide-type light beam collecting films, the waveguide-type photovoltaic device described in Section 9.1 is a promising candidate. The waveguide-type photovoltaic device with multidye sensitization and polymer-MQD sensitization achieved by MLD might be especially promising for future high-performance energy conversion systems.

9.3 NOVEL STRUCTURES OF PHOTOVOLTAIC AND PHOTOSYNTHESIS DEVICES

In addition to the waveguide-type photo-voltaic devices described in Section 9.1, other kinds of novel photo-voltaic and photo-synthesis devices might be used in the integrated solar energy conversion systems. Figure 9.48 shows examples of a thin-film photo-voltaic device and a photo-synthesis device, which might be integrated in the light beam

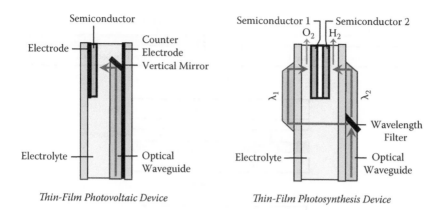

Thin-Film Photovoltaic Device   *Thin-Film Photosynthesis Device*

FIGURE 9.48 Examples of thin-film photovoltaic and photosynthesis devices. From T. Yoshimura, "Proposed applications of 3-D optical interconnect technologies to integrated chemical systems," *Digest of the LEOS Summer Topical Meetings* (Bio-Inspired Sensors and Application/Imprinting on Photonic Integrated Circuits), Portland, Oregon, 129–130 (2007).

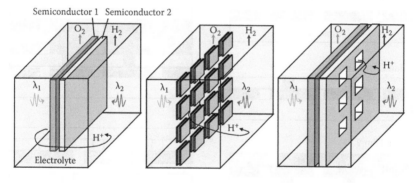

FIGURE 9.49 Positional configurations of thin films in thin-film photosynthesis devices.

collecting films described in Section 9.2. In the thin-film photo-voltaic device, a collected light beam is introduced onto a semiconductor film through an optical waveguide with a vertical mirror. This structure enables us to replace an expensive and electrically resistive transparent electrode with a low-cost metal electrode with low resistivity.

In the thin-film photosynthesis device, using 3-D optical circuits with embedded wavelength filters, collected light beams branch into two optical paths for wavelengths of λ_1 and λ_2, which are respectively introduced onto two semiconductor thin films in an electrolyte, generating H_2 and O_2.

The photosynthesis process using two semiconductors is developed by Sayama et. al [28] using a solution in which semiconductor powders are dispersed. Since control of the relative positions of the two kinds of semiconductors is important for improving the photosynthesis performance, a thin-film structure approach shown in Figure 9.48, where the thin film locations are presicely controlled, seems promising. Figure 9.49 shows several examples for positional configurations of

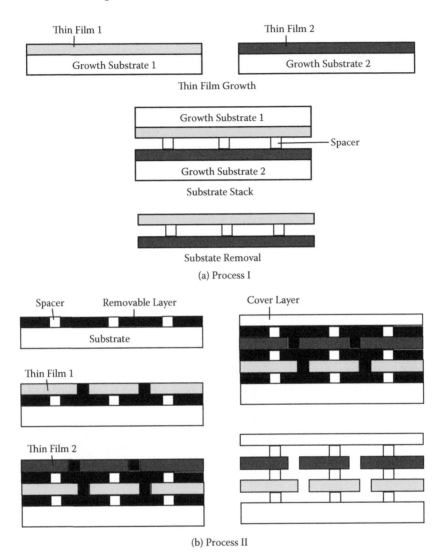

FIGURE 9.50   Possible fabrication processes for thin-film photosynthesis devices.

thin films in thin-film photosynthesis devices. For optimization of H^+ diffusion in the solution, dividing the thin films or opening windows in the thin films might be effective.

Possible fabrication processes for thin-film photosynthesis devices are shown in Figure 9.50. In process I, thin film 1 and thin film 2 are respectively grown on growth substrate 1 and growth substrate 2. The films are attached to each other with spacers between them. Then the growth substrates are removed to obtain stand-alone coupled thin films. In process II, thin film 1 is deposited on a removable layer of a substrate with spacers. Then, another removable layer and spacers are formed on thin

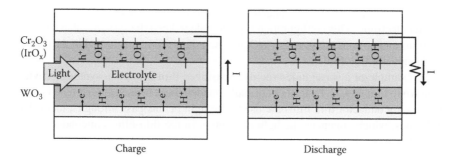

FIGURE 9.51 Waveguide-type photovoltaic device with a charge storage function.

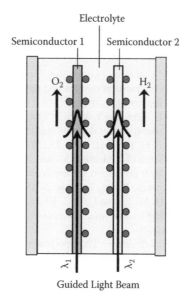

FIGURE 9.52 Waveguide-type dye-sensitized thin-film photosynthesis device.

film 1, and thin film 2 is deposited on them. After a cover layer is formed, all the removable layers are removed by etching to obtain a structure with thin films located in designated distributions three dimensionally.

9.4 WAVEGUIDE-TYPE PHOTOVOLTAIC DEVICES WITH A CHARGE STORAGE/PHOTOSYNTHESIS FUNCTION

The waveguide-type photovoltaic device shown in Figure 9.51 has a charge storage function. Electrochomic WO_3 thin film is an n-type semiconductor and exhibits a charge storage function as described in Section 2.6.1 as well as a coloration function. So, when a light beam is guided in a stack of WO_3/electrolyte, it is expected that a charge is generated and stored in the WO_3 thin film. In the model illustrated in Figure 9.51, a p-type semiconductor thin film (Cr_2O_3, Ir_xO, etc.) is placed on the counter electrode. This configuration is preferable to keep the system stable because electrons and protons are stored in WO_3 and holes and anions are stored in the p-type semiconductor thin film.

An example of a waveguide-type dye-sensitized thin-film photosynthesis device is shown in Figure 9.52. Instead of normally incident light beams, guided light beams in the semiconductor thin films are used. The light beams pass through a lot of dye molecules, which enhances the light absorption in the thin-film semiconductors to generate gases efficiently.

REFERENCES

1. T. Yoshimura, *Molecular nano systems: Applications to optoelectronic computers and solar energy conversion*, Corona Publishing Co., Ltd., Tokyo (2007) [in Japanese].
2. T. Yoshimura, Integrated photonic/electronic/chemical systems, solar energy conversion systems, light collectors, and optical waveguides" Japanese Patent Tokukai 2009-4717 (2009) [in Japanese].
3. T. Yoshimura and D. Kim, "A molecular-controlling deposition method and its applications to solar cells, electro-luminescent devices and optical switches" Japanese Patent Tokugan 2009-102408 [in Japanese].
4. T. Yoshimura, A. Oshima, D. Kim, and Y. Morita, "Quantum dot formation in polymer wires by three-molecule molecular layer deposition (MLD) and applications to electro-optic/photovoltaic devices," *ECS Transactions* **25**, No. 4 "Atomic Layer Deposition Applications 5," from the 216th ECS Meeting, Vienna, Austria, 15–25 (2009).
5. T. Yoshimura, "Molecular layer deposition and applications to solar cells and the cancer photodynamic therapy," Japanese Patent, Tokugan 2010-206098 (2010) [in Japanese].
6. C. Brauchle, U. P. Wild, D. M. Burland, G. C. Bjorkund, and D. C. Alvares, "Two-photon holographic recording with continuous-wave lasers in the 750-1100-nm range," *Opt. Lett.* **7**, 177–179 (1982).
7. V. D. McGinniss and R. E. Schwerzel, "Photopolymerizable composition containing a photosensitive donor and photoinitiating acceptor," US Patent 4,571,377 (1986).
8. B. O'Regan and M. Gratzel, "A low-cost high-efficiency solar cell based on dye-sensitized colloidal TiO_2 films," *Nature* **353**, 737–740 (1991).
9. H. Shin and C. Bae, "Nanoscale tubular structures and ampoules of oxides," AVS, 8th International Conference on Atomic Layer Deposition, TueM2-1, Bruges, Belgium (2008).
10. T. Yoshimura, S. Tatsuura, and W. Sotoyama, "Polymer films formed with monolayer growth steps by molecular layer deposition," *Appl. Phys. Lett.* **59**, 482–484 (1991).
11. T. Yoshimura, E. Yano, S. Tatsuura, and W. Sotoyama, Organic functional optical thin film, fabrication and use thereof, U.S. Patent 5,444,811 (1995).
12. T. Yoshimura, Liquid phase deposition, Japanese Patent, Tokukai Hei 3-60487 (1991) [in Japanese].
13. T. Yoshimura, Growth methods of polymer wires and thin films, Japanese Patent Tokukai 2008-216947 (2008) [in Japanese].
14. T. Yoshimura, K. Kiyota, H. Ueda, and M. Tanaka, "Contact potential difference of ZnO layer adsorbing p-type dye and n-type dye," *Jpn. J. Appl. Phys.* **18**, 2315–2316 (1979).
15. K. Kiyota, T. Yoshimura, and M. Tanaka, "Electrophotographic behavior of ZnO sensitized by two dyes," *Photogr. Sci. Eng.* **25**, 76–79 (1981).
16. T. Yoshimura, K. Kiyota, H. Ueda, and M. Tanaka, "Influence of illumination on the surface potential of ZnO adsorbing p-type dye and n-type dye," *Jpn. J. Appl. Phys.* **19**, 1007–1008 (1980).

17. T. Yoshimura, K. Kiyota, H. Ueda, and M. Tanaka, "Mechanism of spectral sensitization of ZnO coadsorbing p-type and n-type dyes," *Jpn. J. Appl. Phys.* **20**, 1671–1674 (1981).

18. H. J. Meier, "Sensitization of electrical effects in solids," *Phys. Chem.* **69**, 719–729 (1965).

19. H. Kokado, T. Nakayama, and E. Inoue, "A new model for spectral sensitization of photoconduction in zinc oxide powder," *J. Phys., & Chem. Solids* **34**, 1–8 (1973).

20. K. H. Hauffe, On spectral sensitization of Zinc oxide: *Photogr. Sci. & Eng.* **20**, 124 (1976).

21. T. Yoshimura, S. Ito, T. Nakayama, and K. Matsumoto, "Orientation-controlled molecule-by-molecule polymer wire growth by the carrier-gas-type organic chemical vapor deposition and the molecular layer deposition," *Appl. Phys. Lett.* **91**, 033103-1-3 (2007).

22. T. Yoshimura and Y. Kudo, "Monomolecular-step polymer wire growth from seed core molecules by the carrier-gas type molecular layer deposition (MLD)," *Appl. Phys. Express* **2**, 015502-1-3 (2009).

23. T. Yoshimura, K. Motoyoshi, S. Tatsuura, W. Sotoyama, A. Matsuura, and T. Hayano, "Selectively aligned polymer film growth on obliquely evaporated SiO$_2$ pattern by chemical vapor deposition," *Jpn. J. Appl. Phys.* **31**, L980–L982 (1992).

24. T. Yoshimura, K. Ogushi, Y. Kitabayashi, K Naito, Y. Miyamoto, and M. Miyazaki, "Optical waveguide films with two-layer skirt-type core end facets for beam leakage reduction at 45° mirrors," *Proc. SPIE* **6899**, from Photonics West 2008, San Jose, California, 689913-1-11 (2008).

25. T. Yoshimura, "Proposed applications of 3-D optical interconnect technologies to integrated chemical systems," *Digest of the LEOS Summer Topical Meetings* (Bio-Inspired Sensors and Application/Imprinting on Photonic Integrated Circuits), Portland, Oregon, 129–130 (2007).

26. T. Yoshimura and K. Asama, Integrated chemical systems and integrated light energy conversion systems, Japanese Patent Tokukai 2007-107085 (2007) [in Japanese].

27. R. Shioya and T. Yoshimura, "Design of solar beam collectors consisting of multi-layer optical waveguide films for integrated solar energy conversion systems" *Journal of Renewable and Sustainable Energy* **1**, 033106-1-15 (2009).

28. K. Sayama and H. Arakawa, "Dream of Artificial Synthesis," *Newton.* Newton Press, Tokyo, 102–109, September 2002 [in Japanese].

10 Proposed Applications to Biomedical Photonics

In the present chapter, expected applications of molecular layer deposition (MLD) to biomedical photonics are proposed. Although most of the proposals have little experimental confirmation, they will hopely provide some insights in the fields of biomedical research.

10.1 THERAPY FOR CANCER UTILIZING LIQUID-PHASE MLD

Liquid-phase MLD [1–4] is expected to be a powerful process in cancer therapy. Figure 10.1 shows a liquid-phase MLD process for stacking different kinds of functional molecules with designated arrangements on cancer cells [1]. The human body is the chamber and the cancer cells are the substrates for MLD in this case. In step 1, molecule A is introduced into a human body. Molecule A selectively adsorbs on cancer cells. In step 2, molecule B is introduced into the human body and is stacked on molecule A. In a similar manner, molecules C and D are introduced successively, and molecular wires having an A/B/C/D structure can be built on cancer cells. Such a tailored structure will be applied in cancer therapy.

10.1.1 PHOTODYNAMIC THERAPY USING TWO-PHOTON ABSORPTION WITH DIFFERENT WAVELENGTHS

An example of expected applications of liquid-phase MLD is photodynamic therapy (PDT) for cancer shown in Figure 10.2 [1]. In PDT, photosensitizers like Talaporfin sodium [5] are selectively adsorbed on cancer cells. The photosensitizers are then exposed to a light beam. When the photosensitizers are excited, by absorbing the light, they generate singlet oxygen, which damages the cancer cells.

In conventional PDT, a molecule is used as the photosensitizer. By using the tailored molecular wire fabricated by MLD for the photosensitizer, variations of the therapy will expand. As an example, we propose PDT utilizing the four-level two-photon absorption. Figure 10.3 shows this concept [1]. In the four-level two-photon absorption [6], as shown in Figure 10.3(a), a photon with a wavelength of λ_1 excites an electron from the S_0 state to the S_n state in a molecule. The excited electron in the S_n state transfers to the T_0 state. Then, a photon with another wavelength of λ_2 further excites the electron to the T_n state to induce a chemical reaction. This mechanism enables us to widen the region where PDT is effective as mentioned below.

In conventional PDT, the excitation light cannot reach the deep part of the region containing cancer cells. By using two-photon absorption with different wavelengths,

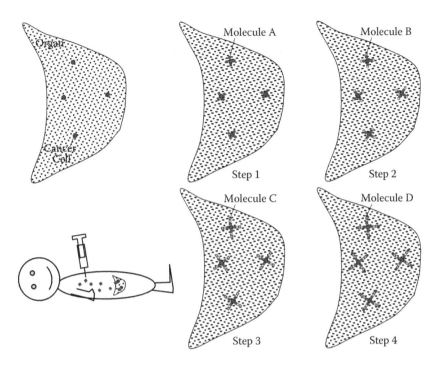

FIGURE 10.1 A liquid-phase MLD process for stacking functional molecules with designated arrangements on cancer cells.

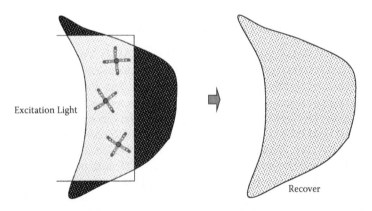

FIGURE 10.2 Schematic illustration of photodynamic therapy.

a three-dimensional attack on cancer cells will be possible as shown in Figure 10.3(b). In the four-level two-photon absorption, the chemical reaction occurs only in regions, where photons with λ_1 and photons with λ_2 coexist; therefore, the chemical reaction can be selectively induced in specific regions by controlling the direction of the light beam.

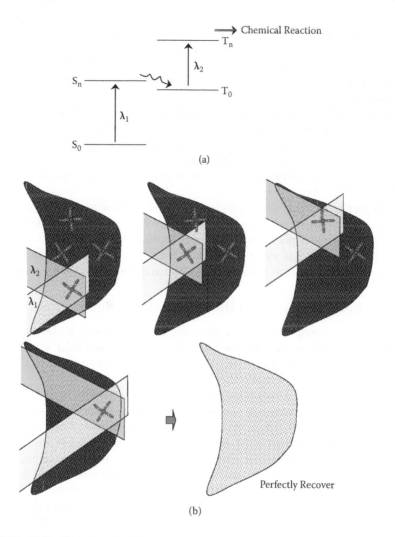

FIGURE 10.3 Concept of photodynamic therapy utilizing two-photon absorption. (a) Mechanism of two-photon absorption, and (b) three-dimensional attack on cancer cells using two light beams with different wavelengths.

Liquid-phase MLD is suitable for constructing the photosensitizers for two-photon absorption. For example, in step 1, molecule A, selectively adsorbs on the cancer cells, is introduced into a body by injection. In step 2, molecule B, with two-photon absorption, is introduced. Molecule B is then stacked on molecule A.

When Molecule A is porphyrin, which selectively adsorbs on cancer cells, and Molecule B is β-carotene, which is connected to the porphyrin on the cancer cells, by exposing the molecules to light of λ_1~500 nm and light of λ_2, an electron in a β-carotene molecule will be excited to S_1 state, then transferred to T_1 state by the intersystem crossing, and finally be excited from the T_1 state to T_n state, generating a singlet oxygen molecule by the energy transfer from the β-carotene molecule.

The wavelength of ~500 nm for λ_1 is too short for actual operation because hemoglobin in blood has strong absorption in the wavelength region. To resolve the problem, donor/acceptor group substitution into the β-carotene molecule, which moves the absorption spectra toward the longer wavelength side as mentioned in Section 5.7, might be effective.

For the selective connection of β-carotene to porphyrin, π-π interaction might be used since both β-carotene and porphyrin have widely-spread π electrons. Another way for the selective connection would be performed by chemical reactions described in Subsection 3.1.1. When –CHO group and –NH$_2$ group are respectively substituted into β-carotene and porphyrin, MLD can be performed by the reaction between –CHO and –NH$_2$ shown in Figure 3.12.

Protoporphyrin, tetraphenylporfin, biacetyl, comphorquinone, and benzyl, etc., for example, are known as molecules with the four-level two-photon-absorption characteristics. In order to apply to the human body applications, the molecules should be selected that are solved in water and exhibit the two-photon absorption with λ_1 and λ_2 in a range of 600–1000 nm, where light absorption due to hemoglobin is weak.

10.1.2 IN-SITU SYNTHESIS OF A DRUG WITHIN HUMAN BODIES

Finally, another example of expected applications of liquid-phase MLD is proposed. As Figure 10.1 shows, liquid-phase MLD can construct molecular structures by stacking molecules with designated sequencies. This can be regarded as a kind of in-situ syntheses within human bodies [1].

When a molecule of a drug is too large to deliver to target sites in a human body, the drug delivery might become possible by synthesizing the drug from small molecules at the target sites, where the drug is required, by liquid-phase MLD within the human body as follows. First, small molecules A, B, C, and D are prepared. Molecule A is used for selective adsorption onto a target site, for example, cancer cells. Namely, molecule A plays a roll of an anchor for initiation of the drug synthesis at the target site. Molecules B, C, and D are parts of the large molecule of the drug and are used for the synthesis of the large molecule. By injecting molecule A into a human body, it is selectively adsorbed at the target site. Next, molecule B is injected to be attached to molecule A selectivey. By an injection of molecule C, it reacts with molecule A and/or B. Finally, by an injection of molecules D, it reacts with molecule A, B and/or C to complete the construction of the large molecule of the drug. This may widen the range of drug choices.

When a molecule of a drug is poison for some parts in human bodies, a drug delivery with less poison conditions might become possible by synthesizing the drug from non-poison small molecules at target sites by liquid-phase MLD within human bodies in similar ways as mentioned above.

10.2 INDICATOR FOR REFLECTIVE OR EMISSIVE TARGETS UTILIZING R-SOLNET

Figure 10.4 shows the indicator concept for reflective or emissive materials using R-SOLNET [8]. The PRI materials are placed between an optical waveguide and

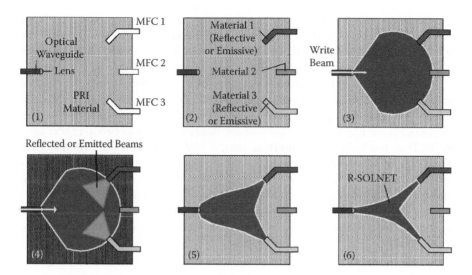

FIGURE 10.4   Indicator concept for reflective or emissive materials using R-SOLNET. From T. Yoshimura, "Proposed applications of 3-D optical interconnect technologies to integrated chemical systems," *Digest of the LEOS Summer Topical Meetings* (Bio-Inspired Sensors and Application/Imprinting on Photonic Integrated Circuits), Portland, Oregon, 129–130 (2007).

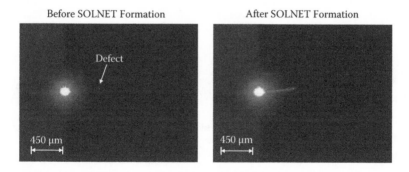

FIGURE 10.5   Indicator for defects and dusts utilizing R-SOLNET.

microfluidic circuits (MFCs). From the optical waveguide, a write beam is introduced into the PRI material. If materials 1 and 3 are reflective or emissive, the write beam and the reflected or emitted beams are superposed to increase the refractive index in the beam-overlapped regions because the refractive index of the PRI material increases with beam intensity. This effect induces self-focusing to create a SOLNET, that is, a self-organized path between the optical waveguide and MFCs 1 and 3, enabling us to indicate the locations of the reflective or emissive materials as well as to guide the write beam, the reflected beams, and the emitted beams.

Figure 10.5 shows an example of the indicator for defects and dusts [9]. In this case, a defect is the reflective material in a PRI sol-gel thin film described in Section 7.3.

A write beam with 0.5 mW of power and a wavelength of 405 nm was introduced into the PRI sol-gel thin film for 30 sec under heating at 200°C. It was found that SOLNET is drawn to a defect in the PRI sol-gel thin film, indicating the locations of the defect and, at the same time, guiding the write beam to the defect.

It should be noted that indicator mechanisms may be able to provide methods to find particular targets like cancer cells with emissive cap molecules within the human body and thus guide laser beams to the target. The proposed process is shown in Figure 10.6 [1]. First, emissive cap molecules are attached to cancer cells by MLD in the body. After inserting optical fibers and injecting PRI materials into a region close to the cancer cells, a write beam is introduced from the optical fiber. An R-SOLNET is then constructed to connect the optical fiber and the cancer cells. By introducing surgical laser beams into the SOLNET, cancer cells are destroyed. It may be possible to monitor, in situ, the degree of cancer cell destruction by detecting the backward luminescence emitted from the emissive cap molecules.

10.3 INTEGRATED PHOTOLUMINESCENCE ANALYSIS CHIPS

In integrated photoluminescence (PL) analysis chips [8], MFCs that contain emissive materials are inserted between two optical waveguide films, as shown in Figure 10.7.

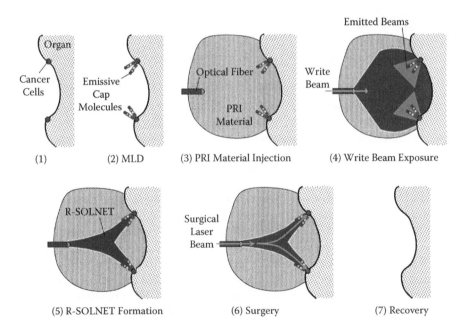

FIGURE 10.6 Possible example of cancer surgery to find cancer cells and guide laser beams to them by MLD and R-SOLNET.

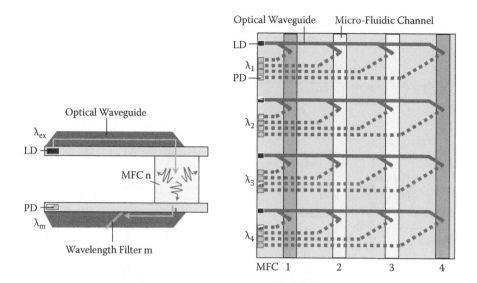

FIGURE 10.7   Integrated photoluminescence analysis chip with 3-D optical circuits. From T. Yoshimura, "Proposed applications of 3-D optical interconnect technologies to integrated chemical systems," *Digest of the LEOS Summer Topical Meetings* (Bio-Inspired Sensors and Application/Imprinting on Photonic Integrated Circuits), 129–130 (2007).

When excitation light beams with λ_{ex} in wavelength of a laser diode (LD) are introduced into MFC *n* from the upper optical waveguide, PL is generated. The PL is collected by the lower optical waveguide, and is guided to photodetector (PD) passing through wavelength filter *m*, which transmits light beams of λ_m. By using arrayed MFCs where different materials are injected, PL intensity measurements at λ_m can be done for the different materials at one time. The same operation can be performed by placing the wavelength filters in series to demultiplex the PL efficiently.

By setting an array of optical circuits with embedded wavelength filters for different wavelengths, spectral analysis of PL can be performed for many kinds of materials at one time.

10.4   MOLECULAR RECOGNITION CHIP

Molecular recognition chips consist of molecule-sensitive waveguides (MSWs), as shown in Figure 10.8 [8]. The MSW is, for example, a waveguide ring resonator with receptors for particular molecules. When wavelengths of light beams in an input waveguide are scanned, optical power transmitted to the PD at the edge of the output waveguide exhibits a peak at the resonant wavelength of the MSW. When molecules are adsorbed on the MSW, the refractive index of the MSW changes, which induces shifts of resonant wavelengths. For example, when molecule 1, molecule 3, and molecule 4 are contained in the atmosphere, resonant frequencies of MSW 1, MSW 3, and MSW 4 shift. Thus, constituent molecules can be recognized at one time.

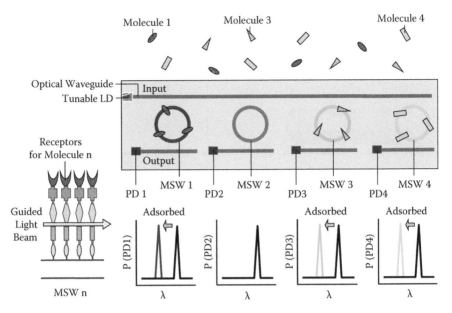

FIGURE 10.8 Molecular recognition chip with the molecule-sensitive waveguide (MSW). From T. Yoshimura, "Proposed applications of 3-D optical interconnect technologies to integrated chemical systems," *Digest of the LEOS Summer Topical Meetings* (Bio-Inspired Sensors and Application/Imprinting on Photonic Integrated Circuits), 129–130 (2007).

The MSW can also operate as a device to measure absorption spectra of the adsorbed molecules by introducing white light and analyzing the transmitted light by using a spectrometer.

REFERENCES

1. T. Yoshimura, "Molecular layer deposition and applications to solar cells and the cancer photodynamic therapy," Japanese Patent, Tokugan 2010-206098 (2010) [in Japanese].
2. T. Yoshimura, "Liquid phase deposition," Japanese Patent, Tokukai Hei 3-60487 (1991) [in Japanese].
3. T. Yoshimura, Growth methods of polymer wires and thin films, Japanese Patent Tokukai 2008-216947 (2008) [in Japanese].
4. T. Yoshimura, Integrated photonic/electronic/chemical systems, solar energy conversion systems, light collectors, and optical waveguides, Japanese Patent Tokukai 2009-4717 (2009) [in Japanese].
5. J. Akimoto and K. Aizawa, "Intraoperative photodynamic diagnosis and therapy for malignant glioma using novel photosensitizer Talaporfin," in *Meeting Digest of IPDA* 10–18, May 28, 2010, IEICE.
6. C. Brauchle, U. P. Wild, D. M. Burland, G. C. Bjorkund, and D. C. Alvares, "Two-photon holographic recording with continuous-wave lasers in the 750–1100-nm range," *Opt. Lett.* **7**, 177–179 (1982).
7. V. D. McGinniss and R. E. Schwerzel, Photopolymerizable composition containing a photosensitive donor and photoinitiating acceptor, U.S. Patent 4,571,377 (1986).

8. T. Yoshimura, "Proposed applications of 3-D optical interconnect technologies to integrated chemical systems," *Digest of the LEOS Summer Topical Meetings* (Bio-Inspired Sensors and Application/Imprinting on Photonic Integrated Circuits), Portland, Oregon, 129–130 (2007).

9. S. Ono, T. Yoshimura, T. Sato, and J. Oshima, "Fabrication of self-organized optical waveguides in photo-induced refractive index variation sol-gel materials with large index contrast," *J. Lightwave Technol.* **27**, 5308–5313 (2009).

Epilogue

October 14, 2007. From the left: Naoko, Tetsuzo, Yoriko, and Chikako.

In January 2009, I had two offers from different publishers to give me an opportunity to write books on different subjects. One of them was from CRC/Taylor & Francis for the present work. I was afraid that it might be very hard to complete two books at the same time, so I told my wife, Yoriko, "How do you think about writing two books at the same time in English?" Then, she said to me "You have accomplished anything you once told me." Her answer made me decide to accept the two offers. Therefore, this book would not be finished without the encouragement from Yoriko.

During the preparation of this book, Yoriko became ill and passed away on September 27, 2009, having received much encouragement and support from her friends of JOSHIGAKUIN, Musashino Academia Musicae and Morimura-Gokuan, parents, students of her private piano school, and her relatives and family. I thank Yoriko for her sincere support and encouragement for over 32 years since 1977, and would like to put her photograph with her family in the epilogue as a memorial to Yoriko.

Tetsuzo Yoshimura
June 12, 2010

Index

Printed and bound by CPI Group (UK) Ltd, Croydon, CR0 4YY

18/10/2024

01776269-0006